276
Current Topics in Microbiology and Immunology

Springer
Berlin
Heidelberg
New York
Hong Kong
London
Milan
Paris
Tokyo

A. Steinkasserer (Ed.)

Dendritic Cells and Virus Infection

With 24 Figures and 2 Tables

Springer

Prof. Dr. Alexander Steinkasserer

Dermatologische Klinik
Universität Erlangen
Hartmannstr. 14
91052 Erlangen
Germany
e-mail: steinkasserer@derma.imed.uni-erlangen.de

Cover Illustration by S. Schneider Schaulies (this volume):
Schematic Representation of the MV Particle

ISSN 0070-217X
ISBN 3-540-44290-1
Springer-Verlag Berlin Heidelberg New York

Springer-Verlag Berlin Heidelberg New York
a member of BertelsmannSpringer Science+Business Media GmbH

http://www.springer.de

Library of Congress Catalog Card Number 15-12910

Cover Design: *design & production GmbH*, Heidelberg
Typesetting: Stürtz AG, Würzburg
Production Editor: Angélique Gcouta
Printed on acid-free paper SPIN: 10892514 27/3020 5 4 3 2 1 0

Preface

In the first part of this volume the interactions between Dendritic cells (DCs) and several viruses will be presented and discussed.

DCs are the best antigen presenting cells (APCs) know to date, and since DCs are the only APCs able to induce naïve T cells they are also known as "nature's adjuvant". Thus, DC play a crucial role in the induction of anti-viral immune responses. On the other hand, many viruses are capable of developing a vast armentarium in order to interfere with DC-biology, thereby blocking anti-viral immune responses. DCs often play a dual role in viral infections, influencing both pathogenesis as well as protective mechanisms. This has been shown, for instance, in the case of HIV-1 and SIV, where HIV-1 infects selected DC-populations. In addition HIV-1 uses DCs as a vector to reach the lymphoid tissues where they infect T cells. The cell surface molecule DC-SIGN has been identified as such a DC-specific HIV-1 receptor, which plays a key-role in the dissemination of HIV-1 by DCs. On the other hand, DCs are efficient APCs for HIV-1 and SIV antigens to both $CD4^+$ and $CD8^+$ T cells. In addition DCs can cross-present non-replicating viral antigens on class I MHC molecules, thereby allowing the stimulation of $CD8^+$ T cells after the uptake of anti-body-coated HIV-1 and dying infected T cells.

DCs also play a crucial role in Epstein-Barr virus (EBV) infection. There is growing evidence that DCs, and not EBV-transformed B cells, are responsible for protective antiviral immunity and that EBV nuclear antigen 1 (EBNA1) is the crucial antigen regarding the resistance against all types of EBV-associated malignancies.

Measles virus (MV) infection is the major cause of childhood mortality in developing countries and is accompanied by a severe immunosuppression of the infected host. However, virus-specific immunity is efficiently induced, leading to viral clearance and long-term induction of protective immunity. On the one hand, DCs play a crucial role in the induction of antiviral immune responses, while on the other hand MV infects DCs, thereby interfering with DC biology. Inhibition of stimulated IL-12 production from MV-infected DCs might affect T cell responses and favour Th2 and suppressing Th1 responses. In addition, viral protein expressed on the cell surface of infected DCs most likely also inhibit T cell responses.

DCs also express the receptor (α-dystroglycan) for lymhocytic choriomeningitis virus (LCMV), Lassa fever virus (LFV) and several other arenaviruses. It has been shown that LCMV strains which bind to the receptor with high affinity replicate in the majority of DCs, causing a generalized immunosuppression, and establish a persistent infection. This immune suppression is caused by the loss of MHC class II molecules and costimulatory molecules on the cell surface of the DCs. Furthermore, the migratory capacity of DCs is hampered by LCMV infections. It is noted, however, that LCMV strains which bind with low affinity rarely replicate in DCs, and generate a potent anti-LCMV response, which is able to clear the infection.

HSV-1 is a further example of a virus which down-modulates DC functions by interfering with the expression of DC specific molecules such as CD83. In fact, CD83 is completely degraded in HSV-1 infected DCs, which coincides with a clearly reduced T cell stimulation, representing yet another new escape strategy.

The second part of this volume concentrates on DC-specific strategies in order to strengthen antiviral immune responses.

DC-based vaccination strategies represent one of the most promising approaches for the immunotherapy of cancer and infectious diseases. Several clinical trials have already been performed showing only minimal toxicity in tumor patients. Both induction of antigen specific T cells as well as some clinical responses have been reported using this strategy, even in far advanced tumor patients. Nevertheless, DC-based immunotherapy is still at an early stage and many variables have still to be addressed. However, the increasing knowledge of DC biology will help to improve and further develop this new strategy.

Data derived from animal models, regarding the induction of LCMV-specific immunity with various DC-vaccination approaches will be discussed. In addition, the interplay between human papillomaviruses (HPV) and DCs will be presented. Here, DCs (including Langerhans cells) are pivotal for the induction of T cell dependent immunity. This, however, depends exclusively on cross-presentation of viral antigens by DCs. This exogenous pathway of MHC class I-restricted antigen presentation plays an important role in the generation of antiviral immunity against several viruses. Finally, viral vectors which could be used for DC-based immunotherapy will be presented and their advantages and disadvantages will be discussed respectively.

April 2003 Alexander Steinkasserer

List of Contents

List of Contributors

(Their addresses can be found at the beginning of their respective chapters)

Chapter 1

The Interaction of Immunodeficiency Viruses with Dendritic Cells

R. M. Steinman, A. Granelli-Piperno, M. Pope, C. Trumpfheller, R. Ignatius, G. Arrode, P. Racz, K. Tenner-Racz

R.M. Steinman (✉), A. Granelli-Piperno, C. Trumpfheller, G. Arrode
Laboratory of Cellular Physiology and Immunology, The Rockefeller University, 1230 York Avenue, New York, NY 10021-6399, USA

M. Pope
Center for Biomedical Research, Population Council, New York, NY 10021-6399, USA

R. Ignatius
Department of Medical Microbiology and Immunology of Infection, Benjamin Franklin Medical Center, Free University of Berlin, Hindenburgdamm 27, 12203 Berlin, Germany

P. Racz, K. Tenner-Racz
Korber AIDS Research Unit, Bernhard Nocht Institute for Tropical Medicine, 20359 Hamburg, Germany

Abstract. Dendritic cells (DCs) can influence HIV-1 and SIV pathogenesis and protective mechanisms at several levels. First, HIV-1 productively infects select populations of DCs in culture, particularly immature DCs derived from blood monocytes and skin (Langerhans cells). However, there exist only a few instances in which HIV-1- or SIV-infected DCs have been identified in vivo in tissue sections. Second, different types of DCs reliably sequester and transmit infectious HIV-1 and SIV in culture, setting up a productive infection in T cells interacting with the DCs. This stimulation of infection in T cells may explain the observation that $CD4^+$ T lymphocytes are the principal cell type observed to be infected with HIV-1 in lymphoid tissues in vivo. DCs express a C-type lectin, DC-SIGN/CD209, that functions to bind HIV-1 (and other infectious agents) and transmit virus to T cells. When transfected into the THP-1 cell line, the cytosolic domain of DC-SIGN is needed for HIV-1 sequestration and transmission. However, DCs lacking DC-SIGN (Langerhans cells) or expressing very low levels of DC-SIGN (rhesus macaque monocyte-derived DCs) may use additional molecules to bind and transmit immunodeficiency viruses to T cells. Third, DCs are efficient antigen-presenting cells for HIV-1 and SIV antigens. Infection with several recombinant viral vectors as well as attenuated virus is followed by antigen presentation to $CD4^+$ and $CD8^+$ T cells. An intriguing pathway that is well developed in DCs is the exogenous pathway for nonreplicating viral antigens to be presented on class I MHC products. This should allow DCs to stimulate $CD8^+$ T cells after uptake of antibody-coated HIV-1 and dying infected T cells. It has been proposed that DCs, in addition to expanding effector helper and killer T cells, induce tolerance through T cell deletion and suppressor T cell formation, but this must be evaluated directly. Fourth, DCs are likely to be valuable in improving vaccine design. Increasing

DC uptake of a vaccine, as well as increasing their numbers and maturation, should enhance efficacy. However, DCs can also capture antigens from other cells that are initially transduced with a DNA vaccine or a recombinant viral vector. The interaction of HIV-1 and SIV with DCs is therefore intricate but pertinent to understanding how these viruses disrupt immune function and elicit immune responses.

1 Introduction

Dendritic cells (DCs) interface with several areas of research on the immunodeficiency viruses HIV-1 and SIV, as reviewed recently (Frank and Pope 2002). Four areas are discussed here: (1) the ability of immunodeficiency viruses to productively infect DCs, (2) the capacity of DCs to sequester and transmit infectious viruses to T cells, (3) the role of DCs in presenting viral antigens and influencing overall immune status, and (4) the ways in which DCs might be exploited in the field of vaccine design. Much of the research in this field has taken place with DCs and T cells in tissue culture. The challenge remains to evaluate and exploit these findings in vivo.

2 Productive Infection of DCs with HIV-1 and SIV

2.1 Monocyte-Derived DCs

Research on infection of DCs has often used cells derived from blood monocytes, the most accessible current source of DCs (Bender et al. 1996; Romani et al. 1996). In this widely used method, monocytes develop into DCs in two stages. First the monocytes are cultured for 5–6 days in medium supplemented with two recombinant cytokines, GM-CSF and IL-4. This leads to differentiation into cells termed immature DCs. The monocyte marker CD14 is lost (from human but less so in monkey monocytes), whereas others such as CD32 FcγR and CD68 are retained. The cells express several T cell interaction molecules such as CD40, CD86, MHC class II, and a new C-type lectin, DC-SIGN or CD209 (Geijtenbeek et al. 2000b). Endocytic uptake of solutes and particulates is evident, including several defined receptor-mediated pathways involving asialoglycoprotein R (ASGP-R; Valladeau et al. 2001), the macrophage mannose receptor (MMR, CD206; Sallusto et al. 1995), and DEC-205/CD205 (Guo et al. 2000). These DCs are termed immature because they are relatively weak stimulators of several T cell responses in culture (Bender et al. 1996; Romani et al. 1996) and in vivo (Dhodapkar et al. 2001). Nonetheless, immature DCs are the ideal cells in which to study the uptake and processing of complex antigens and the subsequent critical maturation events that control the quantity and quality of the immune response.

When immature, poorly immunogenic DCs are exposed to different kinds of stimuli (such as lipopolysaccharide or a combination of inflammatory products that includes IL-1β, TNFα, and PGE_2), there are marked changes over a period of 1–2 days, leading to the formation of terminally differentiated or mature DCs. The mature DCs have lost the capacity to stick to plastic surfaces, becoming motile and "veiled," i.e., extending sheetlike processes in many directions from the cell body. The cells reduce their endocytic activity and several of the uptake receptors mentioned above. However, coated pits (Garrett et al. 2000) and some adsorptive endocytosis continue, notably uptake of SIV itself (Frank et al. 2002). Maturing DCs express higher amounts of CD86 and CD40, as well as several new markers such as IL-2 Rα chain (CD25), DC-LAMP (CD208), and surface CD83. The repertoire of chemokine receptors also changes, because CCR5 is extinguished and CCR7 and CXCR4 appear (Delgado et al. 1998; Sallusto et al. 1998). Within the first 12 h of receipt of the maturation stimulus, large amounts of chemokines (e.g., CCL 3,4,5) and cytokines (e.g., IL-12) are produced (Langenkamp et al. 2000; Sallusto et al. 1999). Importantly, the mature DCs become potent stimulators of Th1 and CTL-type immune responses, as in patients with advanced melanoma and healthy human volunteers; these in vivo experiments involve the cutaneous injection of antigen-pulsed, autologous mature DCs (Dhodapkar et al. 2000, 2001; Schuler-Thurner et al. 2002). At this time, mature DCs are the only adjuvants known to polarize in a single dose the human immune response to the Th1 type of strong cell-mediated immunity.

Immature monocyte-derived DCs are susceptible to productive infection with M-tropic but not T-tropic forms of HIV-1 (Granelli-Piperno et al. 1998). However, when the same experiments have been done with cells from the rhesus macaque, productive infection with SIV has yet to be documented (Messmer et al. 2000). With human monocyte-derived

DCs, productive HIV-1 infection is detected by staining for the p24 gag protein at the single cell level and by the release of large amounts of p24 antigen, reverse transcriptase activity, and infectious virus into the culture medium (Granelli-Piperno et al. 1998). Interestingly, only a small fraction of the DCs are productively infected, as assessed by FACS labeling with anti-p24, generally less than 5% of the cells in the culture. The reason for this relatively low level of infection is not clear because most immature monocyte-derived DCs express CD4 and CCR5, although at much lower absolute levels than T cells.

Mature DCs, in contrast, do not support a productive infection with HIV-1 (Granelli-Piperno et al. 1998; Ignatius et al. 1998). In part, this is due to a reduction at the level of viral entry, which for M-tropic viruses relates to a decrease in CCR5 expression (Granelli-Piperno et al. 1996). In addition, postintegration blocks have been reported in mature DCs, i.e., that HIV-1 is able to reverse transcribe, enter the nucleus, and integrate, but then negative regulatory events take place at the transcriptional level (Bakri et al. 2001).

2.2
Infection of DCs from Skin

A key type of DC that demonstrates productive infection in culture is the Langerhans cell (LC). LCs are the prototype immature DCs (Inaba et al. 1986; Schuler and Steinman 1985) derived from the epidermis. Although skin is not a site for HIV-1 transmission, LCs are considered to be relevant subjects for research for two reasons. First, LCs in the skin are similar and possibly identical to LCs at surfaces for HIV-1 transmission, such as the vaginal and anal epithelium, e.g., in terms of distinct markers like Birbeck granules, CD1a, Langerin, or CD207. Second, LCs are permissive to M-tropic but not T-tropic forms of HIV-1, as is the case with transmission in patients. This at least in part is because LCs selectively express CCR5 receptors and not CXCR4 early during the tissue culture process (Patterson et al. 1998; Reece et al. 1998; Zaitseva et al. 1997).

Infection of LCs, isolated from noninfected individuals, was reported by Berger et al. (Berger et al. 1992) and then observed in skin explants when DCs emigrate from the skin into the culture medium. The skin can be cultured as epidermal sheets, obtained from what are called suction blisters (Kawamura et al. 2000), or from thicker combined epidermal-dermal explants (Reece et al. 1998). The virus is applied to these organ cultures, and then the explants are cultured for more than 1 day. A fraction (<10%) of the DCs that crawl out of the explants become strongly p24 positive. Perhaps these observations in tissue culture mimic what might take place at an inflamed genital surface where LCs would be expected to emigrate and mature.

Soto-Ramirez et al. (Soto-Ramirez et al. 1996) inoculated long-term primary LC cultures with primary HIV-1 isolates from U.S. homosexuals and Thai heterosexuals and found that the former, which are subtype B, replicated less vigorously than the latter subtype E. In contrast, this replication advantage was not observed when skin DC-T cell mixtures, or blood-derived DC-T cell mixtures, were infected with a panel of primary B and E subtypes (Pope et al. 1997b). Additional analysis was also performed on LCs that migrated from epidermal sheets, with primary HIV-1 isolates from subtypes A through F obtained from different risk groups; it was again not feasible to document preferential tropism by subtype E (Dittmar et al. 1997).

2.3
Infection of DCs Freshly Isolated from Blood and Lymphoid Tissues

The immature DCs that are normally resident in blood have not been extensively studied as targets for HIV-1 infection in recent years. The topic is now ripe for re-examination because the identification and isolation of DCs in fresh blood are now more easily accomplished. There are at least two major DC subsets, both immature and termed myeloid and plasmacytoid, each accounting for less than 1% of the peripheral blood mononuclear cells. The former express high levels of CD11c but are low in CD14, whereas the latter lack both CD11c and CD14. There are many differences between these two types of DCs, too numerous to be detailed here. To illustrate, the plasmacytoid DC produces exceptionally high levels of type I interferons in response to viral infection and to ligation of TLR9 receptors with select CpG deoxyoligonucleotide sequences, whereas myeloid DCs in blood express TLR3 and TLR4 and produce large amounts of IL-12 in response to double-stranded RNA and lipopolysaccharide, respectively (Kadowaki et al. 2001; Krug et al. 2001).

Unfortunately, it has been difficult to isolate DCs from such critical tissues as lymph nodes and mucosa-associated lymphoid tissue (MALT), the sites where HIV-1 and SIV replication is most active in vivo (Stahl-Hennig et al. 1999; Veazey et al. 1998).

2.4
HIV-1 Infection of LCs in Skin Samples from Infected Individuals

Conflicting conclusions have been reached on the presence and extent of LC infection in vivo. LCs as targets of infection with HIV-1 were reported early on, including the budding of HIV-1 virions from an LC by electron microscopy (Tschachler et al. 1987) and the recovery of infectious virus (Rappersberger et al. 1988). In purified LCs obtained from chronically infected patients, proviral DNA and HIV RNA could be detected by PCR (Cimarelli et al. 1994; Giannetti et al. 1993). Also, in the skin rash that occurs during primary HIV-1 infection, epidermal LCs were reported to be a major site of infection (Simonitsch et al. 2000). Henry et al. (Henry et al. 1994), by detecting mRNA for the multiply spliced regulatory genes of HIV-1 in highly enriched LC but not LC-depleted epidermal cells, concluded that LCs are the main if not the only cell type infected within the epidermis of AIDS patients.

Other reports have found it difficult to document infection of LCs. Structural HIV-1 proteins were present in the oral mucosa of only 2 of 26 biopsies obtained from infected patients, but even in these 2 cases, positive cells could not be detected in the epithelium (Becker et al. 1988). Similarly, no epidermal LCs could be stained for gag and env proteins in 44 lesional and nonlesional skin biopsies (Kanitakis et al. 1989). When PCR was carried out separately on the dermis and epidermis in 21 specimens, HIV-1 DNA and RNA were more frequently found in the dermis (Kanitakis et al. 1991). Kalter and her coworkers examined 28 skin samples by immunohistochemistry, transmission electron microscopy, in situ hybridization, PCR, and coculture assays. Very little evidence for latent or productive infection could be demonstrated in the epidermis by any of these assays (Kalter et al. 1991).

Some of the present authors have also addressed this issue by using monoclonal p24 antibody to screen 137 skin samples from 49 HIV-1 infected patients, comprising both biopsy and autopsy material (Ramsauer et al. 1991). Only three positive LCs were found. After in situ hybridization for HIV-1 RNA in 52 skin biopsies, only five instances with one or two RNA-positive cells in the dermis were noted, but no positive cells were present in the epidermis of 252 sections analyzed (Stemm v et al.

1993). The latter study included 19 patients who had lived in Zaire, so that the study did include material infected with subtypes of HIV-1 from Africa. Together, these findings indicate that, independent of the stage of disease, infection of epidermal LCs is rare.

2.5 HIV-1 Infection of LCs and DCs from Other Tissues In Vivo

We have reported DC infection with HIV-1 at mucosal surfaces in select instances. LCs positive for S100 protein and HIV-1 RNA were noted in the epithelial lining of benign lymphoepithelial cysts of the parotid gland, a condition in which ectopic salivary gland tissue is incorporated during embryonic life into the parotid (Tenner-Racz et al. 1994). Syncytia expressing the S100 and p55 markers of LCs and p24 protein of HIV-1 have been found in the adenoid (Frankel et al. 1996) and tonsils (Frankel et al. 1997) of surgical specimens removed from patients because of enlargement, although the patients were in many cases not otherwise immunocompromised and often unaware of their HIV-1 infection. Therefore, productive infection can occur, and at high levels in select circumstances.

However, the detection of infection of DCs in lymphoid tissues has been problematic. We only found infection in T cells, and not DCs in the T-dependent zones, by using in situ hybridization for HIV-1 RNA in lymph nodes and benign lymphoepithelial cysts of untreated patients (Stemm v et al. 1993; Tenner-Racz et al. 1994, 1998). Similar observations were made in spleen (McIlroy et al. 1995), where only 1 in 3,000 of the cells in an enriched DC fraction contained HIV-1 DNA. Therefore, it is evident that for both LCs and DCs productive infection with HIV-1 can take place in vivo, but it appears to be rare.

2.6 SIV Infection of LCs and DCs in Rhesus Macaques In Vivo

It likewise has not been straightforward to detect SIV infection of DCs in sections from many different tissues. A detailed analysis of SIV transmission at a portal of viral entry was carried out in monkeys infected by the atraumatic application of SIVmac251 (Stahl-Hennig et al. 1999) and the SIV239 Δnef vaccine (Stahl-Hennig et al. 2002) to the surface of the palatine and lingual tonsils. Within 4–7 days, robust infection was evident by in situ hybridization of T cells in the oral MALT and later in peripheral and mucosal lymphoid tissues to which the virus spread. However, infection was difficult to detect in either CD1a$^+$ or DC-LAMP$^+$ DCs within the crypt epithelium and the T-dependent zones of the MALT and within the overlying stratified squamous epithelium to which the virus had been applied. Similarly, scattered T cells but not DCs were identified as the main virus producers in the epithelium and lamina propria of the vagina of macaques infected intravaginally with SIVmac 251 (Zhang et al. 1999). Morphologic alterations, but not productive infection, were reported for DCs in the thymus of infected macaques (Muller et al. 1993). Nevertheless, infected LCs beneath the vaginal epithelium have been described within the first day of vaginal infection of rhesus macaques with SIV or SHIV (Hu et al. 2000; Miller and Hu 1999; Spira et al. 1996). It was reported that cell-free SIV enters the mucosa within 60 min of exposure and infected intraepithelial DC are found in the first 18 h (Hu et al. 2000). These experiments would represent a counterpart to the tissue culture studies (above) in which DCs migrating from skin become infected with immunodeficiency viruses. In chronically infected animals, rare SIV-positive DCs have been identified in mucosal surfaces and lymphoid tissues, but CD4$^+$ lymphocytes were by far the principal infected cells that were detected (Hu et al. 1998, 1999).

2.7
Comments

The difficulty in identifying productive infection of DCs in situ brings up several points for discussion. If infection fails to take place, what is the mechanism? Is there a complete block to infection, e.g., due to the production of antiviral chemokines (CCL3, CCL4, CCL5) (McDyer et al. 1999)? Alternatively, infection may be more prevalent than currently appreciated because of several possibilities. First, the physiologic counterpart of cultured monocyte-derived DCs is not well established, and this must be pinpointed to identify infected DCs. Monocyte-derived DCs may correspond to monocytes that are differentiating into DCs during entry into afferent lymphatics (Randolph et al. 1998, 1999). Second, there is a lack of data on the viral life cycle in many pertinent compartments of DCs, such as those in blood, lymph, and lymphoid tissues. Third, it remains possible that DCs are more frequently infected in vivo than is currently apparent but then the cells either die quickly or accumulate low levels of viral RNA relative to T cells. Fourth, DCs might be latently infected in vivo, with only low levels for HIV-1 mRNA being transcribed, but this infection can be activated during interaction with stimulated T cells, which express substantial levels of active CD40L (Granelli-Piperno et al. 1999; Tsunetsugu-Yokota et al. 1995). Fifth, DCs also may be such active antigen-presenting cells that as soon as they start to express viral RNA, they become targets for virus-specific CTL and/or the infected DCs produce high levels of antiviral cytokines (interferons) and chemokines (CCL3, 4, 5). Nevertheless, even without extensive infection, DCs may be critical for fueling the observed robust infection of T cells in situ, our next topic. In other words, even in the absence of extensive infection, DCs may drive virus replication in T lymphocytes by transporting infectious virus to these cells and making T cells permissive for productive viral infection.

3
Transmission of Infectious Virus via DCs to T Cells

3.1
Transmission of HIV from Human Blood DCs to T Cells

In the early period of research on the DC-HIV-1 interaction, when there was considerable debate on whether DCs from blood were frequent and explosive sites of HIV-1 infection, experiments were carried out in which DCs from blood (not monocyte-derived DCs, but cells that are already showing signs of DC differentiation in fresh blood) were exposed to HIV-1 for 1–2 h, washed, and then added to quiescent T cells from blood. There was relatively little infection of DC-T cells under the conditions used, which involved culture in human serum and low ratios of DCs to T cells. However, when the virus-pulsed DCs were presenting alloantigens or superantigens to T cells, there was a striking productive infection including syncytium formation (Cameron et al. 1992). The infected cells and syncytia were largely found in clusters of DCs and T cells (Cameron et al. 1992), which represent the microenvironment for the generation of many T cell responses in culture. The T cells and not the DCs were the major sites for productive infection, and many of the T cells underwent apoptotic cell death (Cameron et al. 1994). The DCs from blood did not have to undergo productive infection to transmit virus to T cells. For example, AZT treatment of the DCs did not block the capacity to initiate reverse transcription in permissive T cells 24 h later (Cameron et al. 1994), and mouse DCs could transmit HIV-1 to human T cells (Cameron et al. 1992).

These findings were noteworthy in the sense that virus production was rapid and reached high levels. Virus production was comparable to that seen in standard experiments in which HIV-1 was added to peripheral blood mononuclear cells stimulated with a mitogen like PHA in the further presence of IL-2 and fetal calf serum. The design of the DC experiments also seemed pertinent to in vivo situations, i.e., that DCs might be exposed to virus in blood or at a mucosal surface and then carry the virus to T cells in the lymphoid organs, the known sites of vigorous replication (Stahl-Hennig et al. 1999; Tenner-Racz et al. 1988; van der Ende et al. 1999).

3.2
Transmission of HIV-1 and SIV from Monocyte-Derived DCs to T Cells

When methods were developed to obtain DCs in larger numbers from blood monocytes (above), many additional parameters became amenable to study. To begin with, monocyte-derived DCs would also transmit HIV to T cells, even if the virus-pulsed DCs were cultured for a day before addition of the T cells (Granelli-Piperno et al. 1998). With HIV-1, mature DCs were much more active in transmitting virus to T cells (Granelli-Piperno et al. 1998), whereas for SIV, both immature and mature DCs were active (Ignatius et al. 1998; Messmer et al. 2000). In contrast to the initial studies of Cameron et al., in which the number of added DCs was low, these studies with syngeneic DC-T cell mixtures used high DC-to-T cell ratios, 1 to10 or higher, but again did not require any exogenous activating stimuli. Furthermore, either the DC or the T cell could introduce infectious virus to the permissive cocultures of the two cell types (Messmer et al. 2000). In both human and monkey DCs, the HIV-1 and SIV *nef* gene product contributed to efficient virus replication (Petit et al. 2001), although it seems that *nef* is mainly required when immature DCs are driving viral replication (Messmer et al. 2000). Nef also was able to stimulate other key functions of DCs without overt maturation of the cell; these include an increase in IL-12 and β-chemokine production (Messmer et al. 2002) and increased binding to lymphocytes (Sol-Foulon et al. 2002). In all these studies with monocyte-derived human and monkey DCs, the bulk of the infected cells in DC-T cell cocultures were T cells. This indicated efficient HIV-1 and SIV transmission to T cells from the DCs, which had been exposed initially to virus and were then cultured in virus-free medium before addition to the permissive T cells.

The HIV-1 transmission studies were also carried out with cells from two individuals missing functional CCR5 coreceptors, as a result of the Δ32 base pair deletion. Both DCs and T cells had to express CCR5 chemokine receptors, implying a role for viral entry into both cell types for transmission and productive infection to take place (Granelli-Piperno et al. 1998). Subsequently, with the use of DCs that had been infected with a single cycle, VSV-pseudotyped HIV-1, it could be shown that activated T cells also stimulate productive infection of the DCs, which made readily detectable HIV-1 p24 antigen (Granelli-Piperno et al. 1999). The effect of T cells could be reproduced by CD40L-transfected fibroblasts (CD40L is expressed by activated T cells). In sum, DCs transmit immunodeficiency viruses to T cells in culture, and the resulting infection is rapid and productive. Nevertheless, infection also can take place in some of the DCs in the coculture, and according to the data with CCR5-deficient DCs, entry into DCs plays an important role in enhancing infection in T cells in this particular experimental model.

Recent experiments have applied SIV to monocyte-derived macaque DCs and then reinjected the autologous DCs subcutaneously. A robust infection could be established in the lymphoid tissues of the recipients. Interestingly, the lymph nodes of these animals were found to contain large numbers of infected macrophages (Ignatius et al. 2002). Because it is difficult to identify infected macrophages in lymph nodes before frank immunodeficiency (Igarashi et al. 2001), it was hypothesized that the injected DCs had died on reaching the lymph node or that factors may be released from virus-carrying cells to encourage recruitment of permissive macrophages.

3.3
Transmission of HIV-1 and SIV by DCs from Skin and Lymph Nodes

With the above evidence for transmission of HIV-1 via blood DCs, studies were undertaken with skin-derived DCs (mixtures of LCs and dermal DCs) to model events at a mucosal surface for virus transmission, e.g., the vagina, anus, or oral pharynx (Pope et al. 1994, 1995). Mixtures of DCs and T cells that emigrated from skin explants were exposed to virus (whereas above, we described experiments in which the skin explant itself was exposed to virus). Again, robust infection took place even though the system was totally syngeneic and no other stimuli were applied (fetal calf serum, IL-2, mitogens). In fact, the T cells proliferated weakly if at all. When analyzed, it became evident that the infection was being initiated by tight conjugates of DCs and T cells that were present in the cell population that emigrated from the skin. Free DCs were not infected, but the heterologous syncytia formed from DCs and T cells were strongly stained for viral p24 antigen, and virions were observed by electron microscopy (Pope et al. 1994). As in the case of DCs from blood, if the cells were exposed to HIV-1, cultured for 1.5 days, and then added to syngeneic T cells, infection was transmitted. In contrast to the transmission of virus from blood DCs mentioned above (Cameron et al. 1994), the transmission was blocked if the skin DCs were treated with AZT during the 1.5 days of culture before addition to T cells (Pope et al. 1995). These findings indicate that in the case of skin-derived DCs, some infection may take place before transmission to T cells, consistent with the above observations on monocyte-derived DCs obtained from CCR5Δ32 donors.

With the SIV system, it became possible to isolate DCs from different tissues of the rhesus macaque and to mix these with T cells from blood and other sources. DC-T cell mixtures isolated from explants of vaginal and cervical mucosa, as well as adenoid and tonsil, were able to support extensive replication of SIV (Pope et al. 1997a). In addition, skin- or monocyte-derived DCs loaded with SIV amplified SIV replication when added to syngeneic $CD4^+$ T cells from skin, blood, lymph node, or spleen (Ignatius et al. 1998). Populations of memory T cells were more supportive of viral replication than populations enriched in naïve T cells (Frank and Pope 2002), consistent with the in vivo observations of memory T cell infection within the MALT (Veazey et al. 1998; Veazey et al. 2000). Therefore, the capacity of DCs to bind SIV and then set up an infection in T cells is evident with DCs and T cells from many different tissues.

The underlying mechanism that brings DCs and T cells together in the absence of added antigens or foreign sera is still not worked out. A clue is suggested by recent studies of DCs taken from the lungs of immunized mice challenged with an allergen via the airway. The DCs in the lung retained antigen for presentation to antigen-specific T cells for very long periods, at least a month; likewise, antigen-specific memory T cells also persisted in the lung (Julia et al. 2002). Possibly the DCs that are active in supporting HIV-1 replication in T cell cultures are actually DCs presenting very small amounts of residual antigen to their cognate memory T cells.

3.4
DC-SIGN Function During DC-Based Transmission of HIV-1 to T cells

A fascinating set of findings emerged during a search for ligands for ICAM-3, an Ig superfamily member expressed by resting T cells. The work began by investigating the possibility that the LFA-1 integrin was not the only ligand for ICAM-3. With the use of an ICAM-3 fusion protein to probe for ligands, a new binding partner was found. It was not an integrin but rather a C-type lectin. The lectin was expressed primarily by monocyte-derived DCs and by select DCs in sections of human tissue. It was termed DC-SIGN (CD209) for DC-specific ICAM-3 grabbing nonintegrin (Geijtenbeek et al. 2000b). When cloned, the same cDNA had been identified almost 10 years earlier as a placental molecule that could bind the gp120 envelope protein from HIV-1. In fact, antibodies to DC-SIGN could interfere with the capacity of DC-SIGN transfectants to transmit virus to permissive T cells in culture (Geijtenbeek et al. 2000a). Therefore monocyte-derived DCs selectively expressed a C-type lectin that is able to mediate transmission of HIV-1 to T cells.

Subsequent studies of DC-SIGN emphasized the relevance of its cytosolic domain, which has three motifs known to be required in endocytic uptake and trafficking. When DC-SIGN was transfected into the THP-1 monocytic cell line, the THP-1 cells retained HIV-1 in an infectious state for days and then could transmit virus to activated T cells. However, when the transfected DC-SIGN had been deleted of its cytosolic domain, the THP-1 cells no longer could retain infectious HIV-1 for transmission to T cells (Kwon et al. 2002). It may be important to use THP-1 cells to observe this transmission function of DC-SIGN (Trumpfheller et al. 2002). When the cDNA for DC-SIGN was transfected into 293 and HOS cells (which are derived from kidney epithelium and osteosarcoma, respectively), the DC-SIGN bound HIV-1 but was unable to retain the virus for days in an infectious form for T cells. It therefore appears that DC-SIGN must maintain its endocytic function and be in the correct cellular environment to be able to sequester HIV-1 for days in an infectious state. Sequestration in both THP-1/DC-SIGN transfectants and DCs themselves occurs within peripheral compartments that are distinct from typical caveolae, endosomes, and lysosomes, because the virus-positive vacuoles do not double label (two-color confocal immunofluorescence microscopy) for caveolin, transferrin, or LAMP/CD107a (Frank et al. 2002; Kwon et al. 2002; Trumpfheller et al. 2002).

3.5 Is DC-SIGN Essential for Sequestration and Transmission of HIV-1 to T Cells?

The work on DC-SIGN provides a major new molecular basis for the capacity of DCs to sequester, protect, and transmit infectious immunodeficiency viruses to T cells. Nevertheless, some experiments suggest a more intricate situation. When DCs were prepared from monocytes with the Δ32 bp deletion in CCR5, the cells were found to be deficient in their ability to transmit HIV-1 to T cells (Granelli-Piperno et al. 1998). In the case of Ba-L virus, which requires CCR5 for entry, infection of the T cells 2 days later was dramatically reduced, whereas T-tropic virus was transmitted well by the DCs from Δ32 CCR5 donors. These results imply that CCR5-dependent entry into DCs is important for transmission of virus from monocyte-derived DCs to T cells. Nevertheless, blocking antibodies to DC-SIGN also can have a significant inhibitory effect on the capacity of DCs to transmit HIV-1 to T cells (Trumpfheller et al. 2002). Possibly at lower doses of virus, DC-SIGN enhances direct infection of monocyte-derived DCs in "cis" (Lee et al. 2001) as well as infection of T cells in "trans" (Geijtenbeek et al. 2000a; Kwon et al. 2002).

DCs conceivably express other C-type lectins that have a function analogous to that of DC-SIGN. For example, LCs transmit HIV-1 to T cells, but they lack DC-SIGN. Likewise, DCs derived from monkey blood monocytes have very low levels of DC-SIGN but actively transfer SIV to T cells (Wu et al. 2002). Interestingly, many of the markers that are used to identify DCs or their subsets are C-type lectins with some homology to DC-SIGN, including the capacity to mediate endocytosis, e.g., Langerin on LCs, BDCA-2 on plasmacytoid DCs, and ASGP-R and CD23 on monocyte-derived DCs. Therefore, DCs may have other molecules, in addition to DC-SIGN, that can explain their impressive capacity to set up a productive infection in T cells in culture and to sequester the virus in an infectious state for days before T cell encounter. Reciprocally, recent reports suggest that other infectious agents, not just HIV-1, may be able to exploit DCs through DC-SIGN (Alvarez et al. 2002; Colmenares et al. 2002).

3.6 Comments

HIV-1 productively infects a large number of T cells within the lymphoid tissues of patients (Tenner-Racz et al. 1988; van der Ende et al. 1999); likewise, SIV replicates vigorously in the T cells of lymphoid organs from rhesus macaques (Stahl-Hennig et al. 1999). Although $CD4^+$ T cells are the major site for viral RNA production, the many tissue culture studies discussed above make it possible that DCs play a crucial role in sequestering and carrying virus to T cells and in providing stimuli that drive viral replication there. HIV-1 loses infectivity rapidly in standard culture medium, but within DCs, the virus is retained in an infectious state in distinct intracellular compartments. DC-SIGN is one receptor that can explain these uptake and protective functions of DCs. Nevertheless, there may be other homologous lectins that enable DCs or different subsets of DCs to play a major role in transmitting and fueling HIV-1 replication in situ.

4 Presentation of HIV-1 and SIV Antigens by DCs

4.1 The Uptake and Fate of SIV Virions in DCs

In Sect. 3.4, we mentioned studies on the uptake of chemically inactivated (with AT-2 or aldrithiol-2) HIV-1/SIV by monocyte-derived DCs and on the persistence of virus in nonlysosomal compartments with two-color immunofluorescence confocal microscopy. Extensive studies with SIV AT-2 and infectious SIV also have been carried out in mature and immature monocyte-derived DCs (Frank et al. 2002). Both types of DC took up virions via coated pits. At a dose of roughly 1,000 virions/DC, both immature and mature cells took up several hundred RNA copies within 2 h. However, in mature DCs, the virions persisted in large endocytic vacuoles, implying a nonlysosomal destination deep within the cell. In contrast, immature DCs held fewer virus particles in smaller compartments close to the cell periphery. Perhaps the differential localization of virus particles within immature and mature DCs influences antigen presentation versus retention of virus for transmission to T cells.

4.2
CD4 and CD8 T Cell Responses to DCs Expressing HIV-1/SIV Antigens

HIV-1 reactive $CD4^+$ and $CD8^+$ T cells can be detected in infected individuals (Pitcher et al. 1999), and these T cells can respond to DCs presenting both live and defective forms of HIV-1 (Buseyne et al. 2001). Likewise, chemically inactivated (with aldrithiol) SIV is presented by DCs to immune $CD4^+$ and $CD8^+$ T cells (Mehlhop et al. 2002). Because of the interest in developing ways to deliver HIV antigens in safe vaccines, a number of recombinant forms of HIV-1 have been developed (Jenne et al. 2001) and also tested for their capacity to be presented by DCs to virus-specific T cells in culture. All of these successfully deliver antigens to DCs for presentation on MHC class I and II molecules. These include single-cycle vectors pseudotyped with the VSV envelope (Granelli-Piperno et al. 2000), recombinant adenovirus (Zhong et al. 2000), liposomes entrapping HIV-1 protein (Zheng et al. 1999), HIV-1 peptides linked to a lipid tail (Andrieu et al. 2000), and recombinant canary pox (Engelmayer et al. 2001). Such vectors might be used to charge DCs with HIV-1 antigens for use in therapeutic cellular vaccines, i.e., to boost the immune system of infected individuals who wish to stop taking antiretroviral therapy. In this context, RNA transfection could be an attractive way to deliver HIV-1 antigens, because RNA can efficiently transduce DCs and provide a broad spectrum of HIV-1 epitopes for presentation to both CD8 and CD4 T cells (Weissman et al. 2000).

4.3
The Exogenous Pathway of HIV-1 Presentation on Class I MHC Products

In the studies sited above, the antigen delivery strategy (e.g., adenoviral vector or RNA transduction) involves synthesis of viral proteins in the cytoplasm of the DCs and then presentation by the classic or "endogenous" pathway of MHC class I presentation (Pamer and Cresswell 1998). DCs also are particularly efficient in what is called the "exogenous" pathway. Here nonreplicating viral antigens are first endocytosed by DCs, and then antigen presentation takes place in the absence of further protein synthesis, although typically involving the classic transporters of antigenic peptides (TAPs). The well-developed exogenous pathway in DCs means that these cells could present several noninfectious forms of HIV-1. These include (1) defective (Buseyne et al. 2001) and inactive (Larsson et al. 2002) HIV-1 virions; (2) HIV-1 coated with antibody, because the Fcγ receptor can lead to presentation on MHC class I of DCs (Dhodapkar et al. 2002; Regnault et al. 1999); and (3) dying infected cells, as has been documented for HIV-1 (Larsson et al. 2002; Zhao et al. 2002) and other viruses like influenza (Albert et al. 1998) and CMV (Arrode et al. 2000). DCs may also die as a result of HIV-1 infection (Rasola et al. 2001) and then be cannibalized by other DCs. The exogenous pathway may be a major one whereby DCs present antigens in patients, because the cells are continually exposed to all three sources of HIV-1 antigens, i.e., defective virus, antibody-complexed virions and dying HIV-1-infected cells.

4.4 Comments

DCs efficiently present HIV-1 and SIV antigens to $CD4^+$ and $CD8^+$ T cells. Many important topics remain to be explored. The number of DCs taken from the blood of infected individuals is now under study in many laboratories, with the use of new markers for DCs and their subsets; the antigen presenting and other functions of these DCs from patients require evaluation. In vivo, DCs may be presenting HIV-1 antigens from many sources (the virus itself, antibody-coated virions, and dying infected cells), but the consequences may extend beyond the stimulation of effector $CD4^+$ helper and $CD8^+$ killer T cells. During chronic infection, patients typically show few signs of inflammation, and possibly HIV-1 is being regarded as a "self" antigen. This is because a normal function of DCs in the steady state, i.e., in the absence of infection and inflammation, is to induce peripheral tolerance (Hawiger et al. 2001; Liu et al. 2002; Mahnke et al. 2002). Therefore, it is possible that DCs in patients are presenting antigens to delete HIV-1-specific T cells and/or to induce HIV-1 reactive suppressors (Steinman and Nussenzweig 2002). Such deletion and suppression would then counteract the capacity of antigen presentation to generate functional effector CD4 and CD8 T cells. Another underexplored area is the role of DCs in eliciting antibody responses. DCs have important capacities to stimulate B cells, including mucosal type IgA responses (Fayette et al. 1997) as well as IgG responses to protein and polysaccharide antigens (Colino et al. 2001; Le Bon et al. 2001). Antigen presentation mechanisms are not yet characterized, but it appears that IFNα must stimulate the DCs (Le Bon et al. 2001), which also may produce IL-6 and ligate receptors for BAFF/BlyS to stimulate the antibody-producing B cells. Finally, the work on antigen presentation has not yet developed the means to target viral antigens to DCs in vivo. Adenoviruses, for example, can exploit the $\alpha_v\beta_5$ integrins as receptors on DCs and thereby infect with high efficiency (Zhong et al. 2000). The capacity to direct HIV-1 antigens to DCs in vivo could prove valuable in vaccine design (next section).

5 Role of DCs as Mediators of Effective Vaccination

5.1 DCs and the Design of Improved HIV-1 and SIV Vaccines

HIV-1 vaccines should express a large number of viral proteins, so that the immune response is broad and also difficult for these rapidly replicating viruses to escape. On the other hand, standard attenuated whole-virus vaccines are considered too dangerous. Therefore, vaccine design is turning to safe but multigenic forms of the immunodeficiency viruses, both DNA vaccines and vectors recombinant for several viral genes (Amara et al. 2001; Barouch et al. 2000; Shiver et al. 2002).

Vaccine design might benefit from more direct exploitation of DC biology, by ensuring that sufficient numbers of mature DCs present viral antigens, as discussed elsewhere (Steinman and Pope 2002). A cytokine like GM-CSF might help to recruit more DCs to a vaccine site. The vaccine also might be designed to be taken up more efficiently by DCs by targeting DC receptors that lead to presentation on MHC class I and II products. The vaccine additionally must contain stimuli that mature the antigen-presenting DCs. For example, DNA vaccines can stimulate DC maturation through Toll-like receptors, but at the moment, it appears that the responsible TLR9 in humans is expressed primarily by plasmacytoid DCs and B cells, and not on the more abundant myeloid DCs (Kadowaki et al. 2001; Krug et al. 2001). More work is needed to see whether TLR9 is the correct path to DC maturation in vivo in the setting of vaccination against immunodeficiency viruses.

5.2
Do Vaccines Based on Plasmid DNA and Poorly Replicating Viral Vectors Exploit DCs?

Clearly, DNA vaccines, in sufficient amounts and with a sufficient number of injections, are able to lower the viral set point of monkeys challenged with some SHIV strains (Amara et al. 2001; Barouch et al. 2000; Shiver et al. 2002), although not others like SIVmac251 (Horton et al. 2002). How do DNA vaccines work, particularly considering that the DNA is expressed primarily in muscle or skin cells that are not strong antigen-presenting cells for initiating immunity?

In mice, some DCs are directly transduced during DNA vaccination, and these appear responsible for immune priming (Akbari et al. 1999; Bot et al. 2000; Casares et al. 1997; Porgador et al. 1998). Furthermore, DNA broadly matures the DCs via TLR9 signaling, again in mice (Akbari et al. 1999), where TLR9 receptors are expressed in many DCs. Once primed, the cell-mediated immune response may be significantly expanded if DCs pick up dying HIV-1-transfected nonhematopoietic cells.

In fact, the death of cells infected with a recombinant viral vaccine leads to reprocessing and presentation by DCs. It is known that direct DC infection with pox virus vectors can retard maturation (Engelmayer et al. 1999) and also be cytopathic (Ignatius et al. 2000; Subklewe et al. 1999). In contrast, dying infected non-DCs are efficiently captured and presented by DCs. Cell death, for example of cells infected with recombinant canary pox HIV-1 vaccines, also can be accompanied by TNF release that matures the phagocytic DCs (Ignatius et al. 2000; Motta et al. 2001). In other words, some measure of cytopathic effects in non-DCs may ultimately lead to the targeting of vaccine antigens to DCs and their maturation. In contrast, direct targeting of cytopathic vaccine viruses (or those that inhibit DC maturation) could be less effective, and even counterproductive. There are in vivo data (Chattergoon et al. 2000; Sasaki et al. 2001) that cytopathicity is valuable for vaccination.

5.3
Comments

The design of vaccines has yet to exploit DC biology directly. Nonetheless, current formulations of DNA vaccines and recombinant cytopathic viral vectors may be operating via DCs. In both cases, non-DCs predominantly express the vaccine antigens, but the cells can die and be picked up for presentation by DCs. In addition, these vaccines have the capacity to mature DCs, e.g., through Toll-like and TNF receptor signaling. The efficacy of other noncytopathic vaccines might be enhanced considerably if improvements could be made at the level of vaccine targeting to DCs, if DC numbers were increased, and if DC maturation were stimulated.

References

Akbari O, Panjwani N, Garcia S, Tascon R, Lowrie D, and Stockinger B. (1999). DNA vaccination: transfection and activation of dendritic cells as key events for immunity. J Exp Med 189 169–178

Albert M. L, Sauter B, and Bhardwaj N. (1998). Dendritic cells acquire antigen from apoptotic cells and induce class I-restricted CTLs. Nature 392 86–89

Alvarez C. P, Lasala F, Carrillo J, Muniz O, Corbi A. L, and Delgado R. (2002). C-type lectins DC-SIGN and L-SIGN mediate cellular entry by Ebola virus in cis and in trans. J Virol 76 6841–4

Amara R. R, Villinger F, Altman J. D, Lydy S. L, O'Neil S. P, Staprans S. I, Montefiori D. C, Xu Y, Herndon J. G, Wyatt L. S, et al. (2001). Control of musosal challenge and prevention of AIDS by a multiprotein DNA/MVA vaccine. Science 292 69–74

Andrieu M, Loing E, Desoutter J. F, Connan F, Choppin J, Gras-Masse H, Hanau D, Dautry-Varsat A, Guillet J. G, and Hosmalin A. (2000). Endocytosis of an HIV-derived lipopeptide into human dendritic cells followed by class I-restricted CD8(+) T lymphocyte activation. Eur J Immunol 30 3256–3265

Arrode G, Boccaccio C, Lule J, Allart S, Moinard N, Abastado J. P, Alam A, and Davrinche C. (2000). Incoming human cytomegalovirus pp65 (UL83) contained in apoptotic infected fibroblasts is cross-presented to $CD8^+$ T cells by dendritic cells. J Virol 74 10018–24

Bakri Y, Schiffer C, Zennou V, Charneau P, Kahn E, Benjouad A, Gluckman J. C, and Canque B. (2001). The maturation of dendritic cells results in postintegration inhibition of HIV-1 replication. J Immunol 166 3780–3788

Barouch D. H, Santra S, Schmitz J. E, Kuroda M. J, Fu T. M, Wagner W, Bilska M, Craiu A, Zheng X. X, Krivulka G. R, et al. (2000). Control of viremia and prevention of clinical AIDS in rhesus monkeys by cytokine-augmented DNA vaccination. Science 290 486–92

Becker J, Ulrich P, Kunze R, Gelderblom H, Langford A, and Reichart P. (1988). Immunohistochemical detection of HIV structural proteins and distribution of T-lymphocytes and Langerhans cells in the oral mucosa of HIV infected patients. Virchows Arch A Pathol Anat Histopathol 412 413–419

Bender A, Sapp M, Schuler G, Steinman R. M, and Bhardwaj N. (1996). Improved methods for the generation of dendritic cells from nonproliferating progenitors in human blood. J Immunol Methods 196 121–135

Berger R, Gartner S, Rappersberger K, Foster C. A, Wolff K, and Stingl G. (1992). Isolation of human immunodeficiency virus type 1 from human epidermis: virus replication and transmission studies. J Invest Dermatol 99 271–277

Bonifaz L, Bonnyay D, Mahnke K, Rivera M, Enk A. H, Nussenzweig M. C, and Steinman R. M. (2002). Efficient targeting of protein antigen to the dendritic cell receptor DEC-205 leads in the steady state to antigen presentation on MHC class 1 and peripheral $CD8^+$ T cell tolerance. J Exp Med 196 1627–1638

Bot A, Stan A. C, Inaba K, Steinman R. M, and Bona C. (2000). Dendritic cells at a DNA vaccination site express the encoded influenza nucleoprotein and prime MHC class I-restricted cytolytic lymphocytes upon adoptive transfer. Int Immunol 12 825–832

Buseyne F, Gall S. L, Boccaccio C, Abastado J. P, Lifson J. D, Arthur L. O, Riviere Y, Heard J. M, and Schwartz O. (2001). MHC-I-restricted presentation of HIV-1 virion antigens without viral replication. Nat Med 7 344–349

Cameron P. U, Freudenthal P. S, Barker J. M, Gezelter S, Inaba K, and Steinman R. M. (1992). Dendritic cells exposed to human immunodeficiency virus type-1 transmit a vigorous cytopathic infection to CD4+ T cells. Science 257 383–387

Cameron P. U, Pope M, Gezelter S, and Steinman R. M. (1994). Infection and apoptotic cell death of CD4+ T cells during an immune response to HIV-1 pulsed dendritic cells. AIDS Res Hum Retroviruses 10 61–71

Casares S, Inaba K, Brumeanu T.-D, Steinman R. M, and Bona C. A. (1997). Antigen presentation by dendritic cells following immunization with DNA encoding a class II-restricted viral epitope. J Exp Med 186 1481–1486

Chattergoon M. A, Kim J. J, Yang J. S, Robinson T. M, Lee D. J, Dentchev T, Wilson D. M, Ayyavoo V, and Weiner D. B. (2000). Targeted antigen delivery to antigen-presenting cells including dendritic cells by engineered fas-mediated apoptosis. Nat Biotechnol 18 974–9

Cimarelli A, Zambruno G, Marconi A, Girolomoni G, Bertazzoni U, and Giannetti A. (1994). Quantitation by competitive PCR of HIV-1 proviral DNA in epidermal Langerhans Cells of HIV-infected patients. J Acquir Immune Defic Syndr 7 230–235

Colino J, Shen Y, and Snapper C. M. (2001). Dendritic cells pulsed with intact *Streptococcus pneumoniae* elicit both protein- and polysaccharide-specific immunoglobulin isotype responses in vivo through distinct mechanisms. J Exp Med 195 1–14

Colmenares M, Puig-Kroger A, Muniz Pello O, Corbi A. L, and Rivas L. (2002). Dendritic-cell specific ICAM-3 grabbing nonintegrin (DC-SIGN CD209) a C-type surface lectin in human dendritic cells is a receptor for *Leishmania amastigotes.* J Biol Chem 16 16

Delgado E, Finkel V, Baggiolini M, Clark-Lewis I, Mackay C. R, Steinman R. M, and Granelli-Piperno A. (1998). Mature dendritic cells respond to SDF-1 but not to several β chemokines. Immunobiology 198 490–500

Dhodapkar K. M, Krasovsky J, Williamson B, and Dhodapkar M. V. (2002). Antitumor monoclonal antibodies enhance cross-presentation of cellular antigens and the generation of myeloma-specific killer T cells by dendritic cells. J Exp Med 195 125–133

Dhodapkar M. V, Krasovsky J, Steinman R. M, and Bhardwaj N. (2000). Mature dendritic cells boost functionally superior T cells in humans without foreign helper epitopes. J Clin Invest 105 R9-R14

Dhodapkar M. V, Steinman R. M, Krasovsky J, Munz C, and Bhardwaj N. (2001). Antigen-specific inhibition of effector T cell function in humans after injection of immature dendritic cells. J Exp Med 193 233–8

Dittmar M. T, Simmons G, Hibbitts S, O'Hare M, Louisirirotchanakul S, Beddows S, Weber J, Clapham P. R, and Weiss R. A. (1997). Langerhans cell tropism of human immunodeficiency virus type 1 subtype A through F isolates derived from different transmission groups. J Virol 71 8008–8013

Engelmayer J, Larsson M, Lee M, Cox W. I, Steinman R. M, and Bhardwaj N. (2001). Mature dendritic cells infected with canarypox virus elicit strong anti-human immunodeficiency virus CD8+ and CD4+ T cell responses from chronically infected individuals. J Virol 75 2142–2153

Engelmayer J, Larsson M, Subklewe M, Chahroudi A, Schmaljohn A, William C, Steinman R. M, and Bhardwaj N. (1999). Vaccinia virus inhibits the maturation of human dendritic cells: a novel mechanism of immune evasion. J Immunol 163 6762–6768

Fayette J, Dubois B, Vandenabelle S, Bridon J.-M, Vanbervliet B, Durand I, Banchereau J, Caux C, and Briere F. (1997). Human dendritic cells skew isotype switching of CD40-activated naive B cells towards IgA1 and IgA2. J Exp Med 185 1909–1918

Frank I, Piatak M, Jr, Stoessel H, Romani N, Bonnyay D, Lifson J. D, and Pope M. (2002). Infectious and whole inactivated simian immunodeficiency viruses interact similarly with primate dendritic cells (DCs): Differential intracellular fate of virions in mature and immature DCs. J Virol 76 2936–2951

Frank I, and Pope M. (2002). The enigma of dendritic cell-immunodeficiency virus interplay. Curr Molec Med 2 229–248

Frankel S. S, Tenner-Racz K, Racz P, Wenig B. M, Hansen C. H, Listinsky C, Heffner D, Nelson A. M, Pope M, and Steinman R. M. (1997). Active replication of HIV-1 at the lymphoepithelial surface of the tonsil. Am J Pathol 151 89–96

Frankel S. S, Wenig B. M, Burke A. P, Mannan P, Thompson L. D. R, Abbondanzo S. L, Nelson A. M, Pope M, and Steinman R. M. (1996). Replication of HIV-1 in dendritic cell-derived syncytia at the mucosal surface of the adenoid. Science 272 115–117

Garrett W. S, Chen L. M, Kroschewski R, Ebersold M, Turley S, Trombetta S, Galan J. E, and Mellman I. (2000). Developmental control of endocytosis in dendritic cells by Cdc42. Cell 102 325–34

Geijtenbeek T. B. H, Kwon D. S, Torensma R, van Vliet S. J, van Duijnhoven G. C. F, Middel J, Cornelissen I. L. M. H. A, Nottet H. S. L. M, KewalRamani V. N, Littman D. R, et al. (2000a). DC-SIGN a dendritic cell specific HIV-1 binding protein that enhances trans-infection of T cells. Cell 100 587–597

Geijtenbeek T. B. H, Torensma R, van Vliet S. J, van Duijnhoven G. C. F, Adema G. J, van Kooyk Y, and Figdor C. G. (2000b). Identification of DC-SIGN a novel dendritic cell-specific ICAM-3 receptor that supports primary immune responses. Cell 100 575–585

Giannetti A, Zambruno G, Cimarelli A, Marconi A, Negroni M, Girolomoni G, and Bertazzoni U. (1993). Direct detection of HIV-1 RNA in epidermal Langerhans cells of HIV-infected patients. J Acquir Immune Defic Syndr 6 329–33

Granelli-Piperno A, Delgado E, Finkel V, Paxton W, and Steinman R. M. (1998). Immature dendritic cells selectively replicate M-tropic HIV-1 while mature cells efficiently transmit both M- and T-tropic virus to T cells. J Virol 72 2733–2737
Granelli-Piperno A, Finkel V, Delgado E, and Steinman R. M. (1999). Virus replication begins in dendritic cells during the transmission of HIV-1 from mature dendritic cells to T cells. Current Biol 9 21–29
Granelli-Piperno A, Moser B, Pope M, Chen D, Wei Y, Isdell F, O'Doherty U, Paxton W, Koup R, Mojsov S, et al. (1996). Efficient interaction of HIV-1 with purified dendritic cells via multiple chemokine coreceptors. J Exp Med 184 2433–8
Granelli-Piperno A, Zhong L, Haslett P, Jacobson J, and Steinman R. M. (2000). Dendritic cells infected with VSV-pseudotyped HIV-1 present antigens to CD4+ and CD8+T cells from HIV-1 infected individuals. J Immunol 165 6620–6626
Guo M, Gong S, Maric S, Misulovin Z, Pack M, Mahnke K, Nussenzweig M, and Steinman R. M. (2000). A monoclonal antibody to the DEC-205 endocytosis receptor on human dendritic cells. Hum Immunol 61 729–738
Hawiger D, Inaba K, Dorsett Y, Guo K, Mahnke K, Rivera M, Ravetch J. V, Steinman R. M, and Nussenzweig M. C. (2001). Dendritic cells induce peripheral T cell unresponsiveness under steady state conditions in vivo. J Exp Med 194 769–780
Henry M, Uthman A, Ballaun C, Stingl G, and Tschachler E. (1994). Epidermal Langerhans cells of AIDS patients express HIV-1 regulatory and structural genes. J Invest Dermatol 103 593–596
Horton H, Vogel T. U, Carter D. K, Vielhuber K, Fuller D. H, Shipley T, Fuller J. T, Kunstman K. J, Sutter G, Montefiori D. C, et al. (2002). Immunization of rhesus macaques with a DNA prime/modified vaccinia virus ankara boost regimen induces broad simian immunodeficiency virus (SIV)-specific T-cell responses and reduces initial viral replication but does not prevent disease progression following challenge with pathogenic SIVmac239. J Virol 76 7187–7202
Hu J, Gardner M. B, and Miller C. J. (2000). Simian immunodeficiency virus rapidly penetrates the cervicovaginal mucosa after intravaginal inoculation and infects intraepithelial dendritic cells. J Virol 74 6087–6095
Hu J, Miller C. J, O'Doherty U, Marx P. A, and Pope M. (1999). The dendritic cell-T cell milieu of the lymphoid tissue of the tonsil provides a locale in which SIV can reside and propagate at chronic stages of infection. AIDS Res Hum Retroviruses 15 1305–14
Hu J, Pope M, O'Doherty U, Brown C, and Miller C. J. (1998). Immunophenotypic characterization of SIV-infected cells in cervix vagina and draining lymph nodes of chronically infected rhesus macaques. Lab Invest 78 435–451
Igarashi T, Brown C. R, Endo Y, Buckler-White A, Plishka R, Bischofberger N, Hirsch V, and Martin M. A. (2001). Macrophages are the principal reservoir and sustain high virus loads in rhesus macaques after the depletion of $CD4^+$ T cells by a highly pathogenic simian immunodeficiency virus/HIV type 1 chimera (SHIV): Implications for HIV-1 infections of humans. Proc Natl Acad Sci USA 98 658–663
Ignatius R, Isdell F, O'Doherty U, and Pope M. (1998). Dendritic cells from skin and blood of macaques both promote SIV replication with T cells from different anatomical sites. J Med Primatol 27 121–128
Ignatius R, Marovich M, Mehlhop E, Villamide L, Mahnke K, Cox W. I, Isdell F, Frankel S, Mascola J. R, Steinman R. M, and Pope M. (2000). Canarypox-induced

maturation of dendritic cells is mediated by apoptotic cell death and tumor necrosis factor-α secretion. J Virol 74 11329–11338

Ignatius R, Tenner-Racz K, Messmer D, Gettie A, Blanchard J, Luckay A, Russo C, Smith S, Marx P. A, Steinman R. M, Racz P, and Pope M. (2002). Increased macrophage infection upon subcutaneous inoculation of rhesus macaques with simian immunodeficiency virus-loaded dendritic cells or T cells but not with cell-free virus. J Virol 76 9787–9797

Inaba K, Schuler G, Witmer M. D, Valinsky J, Atassi B, and Steinman R. M. (1986). The immunologic properties of purified Langerhans cells: distinct requirements for the stimulation of unprimed and sensitized T lymphocytes. J Exp Med 164 605–613

Jenne L, Schuler G, and Steinkasserer A. (2001). Viral vectors for dendritic cell-based immunotherapy. Trends Immunol 22 102–107

Julia V, Hessel E. M, Malherbe L, Glaichenhaus N, O'Garra A, and Coffman R. L. (2002). A restricted subset of dendritic cells captures airborne antigens and remains able to activate specific T cells long after antigen exposure. Immunity 16 271–83

Kadowaki N, Ho S, Antonenko S, de Waal Malefyt R, Kastelein R. A, Bazan F, and Liu Y.-J. (2001). Subsets of human dendritic cell precursors express different toll-like receptors and respond to different microbial antigens. J Exp Med 194 863–70

Kalter D. C, Greenhouse J. J, Orenstein J. M, Schnittman S. M, Gendelman H. E, and Meltzer M. S. (1991). Epidermal Langerhans cells are not principal reservoirs of virus in HIV disease. J Immunol 146 3396–3404

Kanitakis J, Escaich S, Trepo C, and Thivolet J. (1991). Detection of human immunodeficiency virus-DNA and RNA in the skin of HIV-1 infected patients using the polymerase chain reaction. J Invest Dermatol 97 91–96

Kanitakis J, Marchand C, Su H, Trivolet J, Zambruno G, Schmitt D, and Gazzolo L. (1989). Immunohistochemical study of normal skin of HIV-1 infected patients shows no evidence of infection of epidermal Langerhans cells by HIV. Aids Res Hum Retroviruses 5 293–302

Kawamura T, Cohen S. S, Borris D. L, Aquilino E. A, Glushakova S, Margolis L. B, Orenstein J. M, Offord R. E, Neurath A. R, and Blauvelt A. (2000). Candidate microbicides block HIV-1 infection of human immature Langerhans cells within epithelial tissue explants. J Exp Med 192 1491–500

Krug A, Towarowski A, Britsch S, Rothenfusser S, Hornung V, Bals R, Giese T, Engelmann H, Endres S, Krieg A. M, and Hartmann G. (2001). Toll-like receptor expression reveals CpG DNA as a unique microbial stimulus for plasmacytoid dendritic cells which synergizes with CD40 ligand to induce high amounts of IL-12. Eur J Immunol 31 3026–37

Kwon D. S, Gregario G, Bitton N, Hendrickson W. A, and Littman D. R. (2002). DC-SIGN mediated internalization of HIV is required for *trans*-enhancement of T cell infection. Immunity 16 135–144

Langenkamp A, Messi M, Lanzavecchia A, and Sallusto F. (2000). Kinetics of dendritic cell activation: impact on priming of Th1 Th2 and nonpolarized T cells. Nat Immunol 1 311–316

Larsson M, Wilkens D. T, Fonteneau J. F, Beadle T. J, Merritt M. J, Kost R. G, Haslett P. A, Cu-Uvin S, Bhardwaj N, Nixon D. F, and Shacklett B. L. (2002). Am-

plification of low-frequency antiviral CD8 T cell responses using autologous dendritic cells. AIDS 16 171–80
Le Bon A, Schiavoni G, D'Agostinio G, Gresser I, Belardelli F, and Tough D. F. (2001). Type I interferons potently enhance humoral immunity and can promote isotype switching by stimulating dendritic cells in vivo. Immunity 14 461–470
Lee B, Leslie G, Soilleux E, O'Doherty U, Baik S, Levroney E, Flummerfelt K, Swiggard W, Coleman N, Malim M, and Doms R. W. (2001). Cis expression of DC-SIGN allows for more efficient entry of human and simian immunodeficiency viruses via CD4 and a coreceptor. J Virol 75 12028–38
Liu K, Iyoda T, Saternus M, Kimura Y, Inaba K, and Steinman R. M. (2002). Immune tolerance following delivery of dying cells to dendritic cells in situ. J Exp Med 196 1091–1097
McDyer J. F, Dybul M, Goletz T. J, Kinter A. L, Thomas E. K, Berzofsky J. A, Fauci A. S, and Seder R. A. (1999). Differential effects of CD40 ligand/trimer stimulation on the ability of dendritic cells to replicate and transmit HIV infection: evidence for CC-chemokine-dependent and -independent mechanisms. J Immunol 162 3711–7
McIlroy D, Autran B, Cheynier R, Wain-Hobson S, Clauvel J.-P, Oksenhendler E, Debre P, and Hosmalin A. (1995). Infection frequency of dendritic cells and CD4+ T lymphocytes in spleens of human immunodeficiency virus-positive patients. J Virol 69 4737–4745
Mehlhop E, Villamide L. A, Frank I, Gettie A, Santisteban C, Messmer D, Ignatius R, Lifson J. D, and Pope M. (2002). Enhanced in vitro stimulation of rhesus macaque dendritic cells for activation of SIV-specific T cell responses. J Immunol Meth 260 219–34
Messmer D, Ignatius R, Santisteban C, Steinman R. M, and Pope M. (2000). The decreased replicative capacity of SIV mac239 delta nef is manifest in cultures of immature dendritic cells and T cells. J Virol 74 2406–2413
Messmer D, Jacque J.-M, Santisteban C, Bristow C, Han S.-Y, Villamide-Herrera L, Mehlhop E, Marx P. A, Steinman R. M, Gettie A, and Pope M. (2002). Endogenously expressed *Nef* uncouples cytokine and chemokine production from membrane phenotypic maturation in dendritic cells. J Immunol 169 4172–41
Miller C. J, and Hu J. (1999). T cell-tropic simian immunodeficiency virus (SIV) and simian-human immunodeficiency viruses are readily transmitted by vaginal inoculation of rhesus macaques and Langerhans' cells of the female genital tract are infected with SIV. J Infect Dis 179 Suppl 3 S413–7
Motta I, Andre F, Lim A, Tartaglia J, Cox W. I, Zitvogel L, Angevin E, and Kourilsky P. (2001). Cross-presentation by dendritic cells of tumor antigen expressed in apoptotic recombinant canarypox virus-infected dendritic cells. J Immunol 167 1795–802
Muller J. G, Krenn V, Schindler C, Czub S, Stahl-Hennig C, Coulibaly C, Hunsmann G, Kneitz C, Kerkau T, Rethwilm A, and et al. (1993). Alterations of thymus cortical epithelium and interdigitating dendritic cells but no increase of thymocyte cell death in the early course of simian immunodeficiency virus infection. Am J Pathol 143 699–713
Pamer E, and Cresswell P. (1998). Mechanisms of MHC class I-restricted antigen processing. Annu Rev Immunol 16 323–58

Patterson B. K, Landay A, Andersson J, Brown C, Behbahani H, Jiyamapa D, Burki Z, Stanislawski D, Czerniewski M. A, and Garcia P. (1998). Repertoire of chemokine receptor expression in the female genital tract: implications for human immunodeficiency virus transmission. Am J Pathol 153 481–90

Petit C, Buseyne F, Boccaccio C, Abastado J. P, Heard J. M, and Schwartz O. (2001). Nef is required for efficient HIV-1 replication in cocultures of dendritic cells and lymphocytes. Virology 286 225–36

Pitcher C. J, Quittner C, Peterson D. M, Connors M, and Koup R. A. (1999). HIV-1-specific CD4+ T cells are detectable in most individuals with active HIV-1 infection but decline with prolonged viral suppression. Nat Med 5 518–525

Pope M, Betjes M. G. H, Romani N, Hirmand H, Cameron P. U, Hoffman L, Gezelter S, Schuler G, and Steinman R. M. (1994). Conjugates of dendritic cells and memory T lymphocytes from skin facilitate productive infection with HIV-1. Cell 78 389–398

Pope M, Elmore D, Ho D, and Marx P. (1997a). Dendritic cell-T cell mixtures isolated from the skin and mucosae of macaques support the replication of SIV. AIDS Res Hum Retroviruses 13 819–827

Pope M, Frankel S. S, Mascola J. R, Trkola A, Isdell F, Birx D. L, Burke D. S, Ho D. D, and Moore J. P. (1997b). HIV-1 strains from subtypes B and E replicate in cutaneous dendritic cell-T cell mixtures without displaying subtype-specific tropism. J Virol 71 8001–8007

Pope M, Gezelter S, Gallo N, Hoffman L, and Steinman R. M. (1995). Low levels of HIV-1 in cutaneous dendritic cells promote extensive viral replication upon binding to memory CD4+ T cells. J Exp Med 182 2045–2056

Porgador A, Irvine K. R, Iwasaki A, Barber B. H, Restifo N. P, and Germain R. N. (1998). Predominant role for directly transfected dendritic cells in antigen presentation to CD8(+) T cells after gene gun immunization. J Exp Med 188 1075–1082

Ramsauer J, Tenner-Racz K, Meigel W, and Racz P. (1991). Immunohistochemical study of Langerhans' cells in the skin of HIV-1 infected patients with and without Kaposi's sarcoma. In Accessory Cells in HIV and Other Retroviral Infections P. Racz C. D. Dijkstra and J. C. Gluckman eds. (Basel Karger) pp. 155–161

Randolph G. J, Beaulieu S, Steinman R. M, and Muller W. A. (1998). Differentiation of monocytes into dendritic cells in a model that mimics entry of cells into afferent lymph. Science 282 480–483

Randolph G. J, Inaba K, Robbiani D. F, Steinman R. M, and Muller W. A. (1999). Differentiation of phagocytic monocytes into lymph node dendritic cells in vivo. Immunity 11 753–761

Rappersberger K, Gartner S, Schenk P, Stingl G, Groh V, Tschachler E, Mann D. L, Wolff K, Konrad K, and Popovic M. (1988). Langerhans cells are an actual site of HIV-1 replication. Intervirol 29 185–194

Rasola A, Gramaglia D, Boccaccio C, and Comoglio P. M. (2001). Apoptosis enhancement by the HIV-1 Nef protein. J Immunol 166 81–8

Reece J. C, Handley A, Anstee J, Morrison W, Crowe S. M, and Cameron P. U. (1998). HIV-1 selection by epidermal dendritic cells during transmission across human skin. J Exp Med 187 1623–1631

Regnault A, Lankar D, Lacabanne V, Rodriguez A, Thery C, Rescigno M, Saito T, Verbeek S, Bonnerot C, Ricciardi-Castagnoli P, and Amigorena S. (1999). Fcγ receptor-mediated induction of dendritic cell maturation and major histocompatibility complex class I-restricted antigen presentation after immune complex internalization. J Exp Med 189 371–80

Romani N, Reider D, Heuer M, Ebner S, Eibl B, Niederwieser D, and Schuler G. (1996). Generation of mature dendritic cells from human blood: an improved method with special regard to clinical applicability. J Immunol Methods 196 137–151

Sallusto F, Cella M, Danieli C, and Lanzavecchia A. (1995). Dendritic cells use macropinocytosis and the mannose receptor to concentrate antigen in the major histocompatibility class II compartment. Downregulation by cytokines and bacterial products. J Exp Med 182 389–400

Sallusto F, Palermo B, Lenig D, Miettinen M, Matikainen S, Julkunen I, Forster R, Burgstahler R, Lipp M, and Lanzavecchia A. (1999). Distinct patterns and kinetics of chemokine production regulate dendritic cell function. Eur J Immunol 29 1617–25

Sallusto F, Schaerli P, Loetscher P, Schaniel C, Lenig D, Mackay C. R, Qin S, and Lanzavecchia A. (1998). Rapid and coordinated switch in chemokine receptor expression during dendritic cell maturation. Eur J Immunol 28 2760–2769

Sasaki S, Amara R. R, Oran A. E, Smith J. M, and Robinson H. L. (2001). Apoptosis-mediated enhancement of DNA-raised immune responses by mutant caspases. Nat Biotechnol 19 543–547

Schuler G, and Steinman R. M. (1985). Murine epidermal Langerhans cells mature into potent immunostimulatory dendritic cells in vitro. J Exp Med 161 526–546

Schuler-Thurner B, Schultz E. S, Berger T. G, Weinlich G, Ebner S, Woerl P, Bender A, Feuerstein B, Fritsch P. O, Romani N, and Schuler G. (2002). Rapid induction of tumor-specific type 1 T helper cells in metastatic melanoma patients by vaccination with mature cryopreserved peptide-loaded monocyte-derived dendritic cells. J Exp Med 195 1279–88

Shiver J. W, Fu T. M, Chen L, Casimiro D. R, Davies M. E, Evans R. K, Zhang Z. Q, Simon A. J, Trigona W. L, Dubey S. A, et al. (2002). Replication-incompetent adenoviral vaccine vector elicits effective anti-immunodeficiency-virus immunity. Nature 415 331–5

Simonitsch I, Geusau A, Chott A, and Jurecka W. (2000). Cutaneous dendritic cells are main targets in acute HIV-1-infection. Mod Pathol 13 1232–7

Sol-Foulon N, Moris A, Engering A, Abastado J.-P, Heard J.-M, van Kooyk Y, and Schwartz O. (2002). HIV-1 nef-induced upregulation of DC-SIGN in dendritic cells promotes lymphocyte clustering and viral spread. Immunity 16 145–155

Soto-Ramirez L. E, Renijifo B, McLane M. F, Marlink R, O'Hara C, Sutthent R, Wasi C, Vithayassai P, Vithayassai V, Apichartpiyakui C, et al. (1996). HIV-1 Langerhans cells tropism associated with heterosexual transmission of HIV. Science 271 1291–1293

Spira A. I, Marx P. A, Patterson B. K, Mahoney J, Koup R. A, Wolinsky S. M, and Ho D. D. (1996). Cellular targets of infection and route of viral dissemination following an intravaginal inoculation of SIV into rhesus macaques. J Exp Med 183 215–225

Stahl-Hennig C, Steinman R. M, Ten Haaft P, Uberla K, Stolte N, Saeland S, Tenner-Racz K, and Racz P. (2002). The SIVΔ*nef* vaccine after application to the tonsils of Rhesus macaques replicates primarily within $CD4^+$ T cells and elicits a local perforin positive $CD8^+$ T cell response. J Virol 76 688–696

Stahl-Hennig C, Steinman R. M, Tenner-Racz K, Pope M, Stolte N, Matz-Rensing K, Grobschupff G, Raschdorff B, Hunsmann G, and Racz P. (1999). Rapid infection of oral mucosal-associated lymphoid tissue with simian immunodeficiency virus. Science 285 1261–1265

Steinman R. M, and Nussenzweig M. C. (2002). Avoiding horror autotoxicus: the importance of dendritic cells in peripheral T cell tolerance. Proc Natl Acad Sci USA 99 351–358

Steinman R. M, and Pope M. (2002). Exploiting dendritic cells to improve vaccine efficacy. J Clin Invest 109 1519–26

Stemm v A. M, Ramsauer J, Tenner-Racz K, Schmidt H. F, Gigli I, and Racz P. (1993). Langerhans cells and interdigitating cells in HIV-infection. Adv Exp Med Biol 329 539–544

Subklewe M, Chahroudi A, Schmaljohn A, Kurilla M. G, Bhardwaj N, and Steinman R. M. (1999). Induction of Epstein-Barr Virus-specific cytotoxic T-lymphocyte responses using dendritic cells pulsed with EBNA-3A peptides or UV-inactivated recombinant EBNA-3A vaccinia virus. Blood 94 1372–1381

Tenner-Racz K, Racz P, Schmidt H, Dietrich M, Kern P, Louie A, Gartner S, and Popovic M. (1988). Immunohistochemical electron microscopic and in situ hybridization evidence for the involvement of lymphatics in the spread of HIV-1. AIDS 2 299–309

Tenner-Racz K, Stellbrink H.-J, van Lunzen J, Schneider C, Jacobs J.-P, Raschodorff B, Grobschupff G, Steinman R. M, and Racz P. (1998). The unenlarged lymph nodes of HIV-1 infected asymptomatic patients with high CD4 T cell counts are sites for virus replication and CD4 T cell proliferation. The impact of active antiretroviral therapy. J Exp Med 187 949–959

Tenner-Racz K, von Stemm A. M, Guhlk B, Schmitz J, and Racz P. (1994). Are follicular dendritic cells macrophages and interdigitating cells of the lymphoid tissue productively infected by HIV? Res Virol 145 177–82

Trumpfheller C, Park C.-G, Finke J, Steinman R. M, and Granelli-Piperno A. (2002). Cell type dependent interactions of DC-SIGN with HIV-1. Int Immunol (in press)

Tschachler E, Groh V, Popovic M, Mann D. L, Konrad K, Safai B, Eron L, diMarzo Veronese F, Wolff K, and Stingl G. (1987). Epidermal Langerhans cells – a target for HTLV-III/LAV infection. J Invest Dermatol 88 233–237

Tsunetsugu-Yokota Y, Akagawa K, Kimoto H, Suzuki K, Iwasaki M, Yasuda S, Hausser G, Hultgren C, Meyerhans A, and Takemori T. (1995). Monocyte-derived cultured dendritic cells are susceptible to human immunodeficiency virus infection and transmit virus to resting T cells in the process of nominal antigen presentation. J Virol 69 4544–4547

Valladeau J, Duvert-Frances V, Pin J.-J, Kleijmeer M. J, Ait-Yahia S, Ravel O, Vincent C, Vega F, Jr, Helms A, Gorman D, et al. (2001). Immature human dendritic cells express asialoglycoprotein receptor isoforms for efficient receptor-mediated endocytosis. J Immunol 167 5767–74

van der Ende M. E, Schutten M, Raschdorff B, Grossschupff G, Racz P, Osterhaus A. D, and Tenner-Racz K. (1999). CD4 T cells remain the major source of HIV-1 during end stage disease. AIDS 13 1015–9

Veazey R. S, DeMaria M, Chalifoux L. V, Shvetz D. E, Pauley D. R, Knight H. L, Rosenzweig M, Johnson R. P, Desrosiers R. C, and Lackner A. A. (1998). Gastrointestinal tract as a major site of CD4+ T cell depletion and viral replication in SIV infection. Science 280 427–31

Veazey R. S, Tham I. C, Mansfield K. G, DeMaria M, Forand A. E, Shvetz D. E, Chalifoux L. V, Sehgal P. K, and Lackner A. A. (2000). Identifying the target cell in primary simian immunodeficiency virus (SIV) infection: highly activated memory $CD4^+$ T cells are rapidly eliminated in early SIV infection in vivo. J Virol 74 57–64

Weissman D, Ni H, Scales D, Dude A, Capodici J, McGibney K, Abdool A, Isaacs S. N, Cannon G, and Kariko K. (2000). HIV gag mRNA transfection of dendritic cells (DC) delivers encoded antigen to MHC class I and II molecules causes DC maturation and induces a potent human in vitro primary immune response. J Immunol 165 4710–7

Wu L, Bashirova A. A, Martin T. D, Villamide L, Mehlhop E, Chertov A. O, Unutmaz D, Pope M, Carrington M, and KewalRamani V. N. (2002). Rhesus macaque dendritic cells efficiently transmit primate lentiviruses independently of DC-SIGN. Proc Natl Acad Sci USA 99 1568–1573

Zaitseva M, Blauvelt A, Lee S, Lapham C. K, Klaus-Kovtun V, Mostowski H, Manischewitz J, and Golding H. (1997). Expression and function of CCR5 and CXCR4 on human Langerhans cells and macrophages: implications for HIV primary infection. Nat Med 3 1369–1375

Zhang Z, Schuler T, Zupancic M, Wietgrefe S, Staskus K. A, Reimann K. A, Reinhart T. A, Rogan M, Cavert W, Miller C. J, et al. (1999). Sexual transmission and propagation of SIV and HIV in resting and activated $CD4^+$ T cells. Science 286 1353–1357

Zhao X. Q, Huang X. L, Gupta P, Borowski L, Fan Z, Watkins S. C, Thomas E. K, and Rinaldo C. R, Jr. (2002). Induction of anti-human immunodeficiency virus type 1 (HIV-1) $CD8^+$ and $CD4^+$T-cell reactivity by dendritic cells loaded with HIV-1 X4-infected apoptotic cells. J Virol 76 3007–14

Zheng L, Huang X. L, Fan Z, Borowski L, Wilson C. C, and Rinaldo C. R, Jr. (1999). Delivery of liposome-encapsulated HIV type 1 proteins to human dendritic cells for stimulation of HIV type 1-specific memory cytotoxic T lymphocyte responses. AIDS Res Hum Retroviruses 15 1011–20

Zhong L, Granelli-Piperno A, Pope M, Ignatius R, Lewis M, Frankel S. S, and Steinman R. M. (2000). Presentation of SIVgag to monkey T cells using dendritic cells transfected with a recombinant adenovirus. Eur J Immunol 30 3281–3290

Chapter 2

DC-SIGN: A Novel HIV Receptor on DCs That Mediates HIV-1 Transmission

T. B. H. Geijtenbeek, Y. van Kooyk

Abstract. The dendritic cell (DC)-specific HIV-1 receptor DC-SIGN plays a key-role in the dissemination of HIV-1 by DCs. DC-SIGN captures HIV-1 at sites of entry, enabling its transport to lymphoid tissues, where DC-SIGN efficiently transmits low amounts of HIV-1 to T cells. The expression pattern of DC-SIGN in mucosal tissue, lymph nodes, placenta and blood suggests a function for DC-SIGN in both horizontal and vertical transmission of HIV-1. Moreover, the efficiency of DC-SIGN^{+} blood DC to transmit HIV-1 to T cells supports a role in HIV-1 transmission via blood. To date, DC-SIGN represents a novel class of HIV-1 re-

T.B.H. Geijtenbeek, Y. van Kooyk (✉)
Department of Molecular Cell Biology, Vrije Universiteit Medical Center Amsterdam, van de Boechorststraat 7, 1081 BT Amsterdam, The Netherlands

ceptor, because it does not allow viral infection but binds HIV-1 and enhances its infection of T cells *in trans*. Its unique function is further underscored by its restricted expression on DCs. Although DC-SIGN is a C-type lectin with an affinity for carbohydrates exemplified by its interaction with its immunological ligand ICAM-3, recent evidence demonstrates that glycosylation of gp120 is not necessary for its interaction with DC-SIGN. Moreover, mutational analysis demonstrates that the HIV-1 gp120 binding site in DC-SIGN is different from that of ICAM-3.

Besides its role in DC-mediated adhesion processes, DC-SIGN also functions as an antigen receptor that captures and internalises antigens for presentation by DC. Strikingly, HIV-1 circumvents processing after binding DC-SIGN and remains infectious for several days after capture. A better understanding of the action of this novel HIV receptor in initial viral infection and subsequent transmission will provide a basis for the design of drugs that inhibit or alter interactions of DC-SIGN with gp120, interfering with HIV-1 dissemination and that may have a therapeutic value in both immunological diseases and/or HIV-1 infections.

1 The Dendritic Cell, a Trojan Horse for HIV-1

Dendritic cells (DCs) are professional antigen-presenting cells that are pivotal in eliciting an efficient immune response against invading pathogens, such as HIV-1. DCs sample foreign antigens in the periphery and migrate to lymphoid tissues, where they present processed antigens to naïve T cells, initiating an immune response (Banchereau and Steinman 1998). However, besides being responsible for eliciting an HIV-1-specific response (Buseyne et al. 2001), it is now becoming evident that DCs also participate in the dissemination of HIV-1 (reviewed in the chapter by Steinman et al., this volume).

Sexual transmission via oral, rectal and vaginal mucosa is the most common route for infection with HIV-1 worldwide. Although the nature of the immune response at these sites may differ, evidence suggests that dissemination of HIV-1 from these distinct sites in the body is facilitated by a common cell-type, the DC. Various groups have shown that DCs residing in blood, skin or mucosal surfaces play an important role in HIV-1 capture and dissemination (Cameron et al. 1992; Pope et al. 1994; Patterson et al. 1994; Spira et al. 1996). The positioning of DC in mucosal sites, together with their function as sentinels of the immune system, makes DCs crucial in the first contact with invading pathogens including HIV-1. However, DCs are not efficiently infected by HIV-1, and instead capture the virus with high efficiency and subsequently transport it from the periphery into the lymphoid tissues. Here, DC-bound HIV-1 is transmitted to T cells expressing CD4 and HIV-1 co-receptors, mediating viral entry and productive infection. Thus the minute amounts of HIV-1 initially encountered by DCs are very rapidly transported to the T cell compartment, leading to a full-blown infection in T cells. The molecular basis behind this process remained unclear until the DC-specific HIV-1 *trans*-receptor DC-SIGN was recently identified (Geijtenbeek et al. 2000a,b; Geijtenbeek and van Kooyk 2001).

2
DC-SIGN Is Important for an Efficient Immune Response

DC-SIGN is essential in several key functions throughout the life cycle of DCs (Fig. 1; Bleijs et al. 2001). DC-SIGN was recently identified as a cellular adhesion molecule on DCs by the generation of antibodies that inhibited the binding of DCs to ICAM-3 (Geijtenbeek et al. 2000b). Cloning of the adhesion receptor DC-SIGN on DCs revealed that it is identical to the previously cloned HIV-1 envelope-binding C-type lectin (Curtis et al. 1992). DC-SIGN (CD209) is a type II transmembrane protein that, on the basis of its structure, belongs to the C-type lectin family. DC-SIGN contains a short cytoplasmic N-terminal domain with several intracellular sorting motifs, an extracellular stalk consisting of seven complete and one partial tandem repeat that is required for tetramerisation, and a C-terminal lectin or carbohydrate-recognition domain (Curtis et al. 1992; Geijtenbeek et al. 2000b; Mitchell et al. 2001). The carbohydrate recognition domain mediates ligand binding, in which Ca^{2+}-coordinating amino acid residues mediate the ligand interaction (Feinberg et al. 2001; Geijtenbeek et al., submitted). The carbohydrate recognition pattern of DC-SIGN is largely distinct from that of other members of the C-type lectin family, indicating that DC-SIGN recognises unique carbohydrate-bearing ligands (Mitchell et al. 2001).

DC-SIGN has a high affinity for ICAM-3, an adhesion molecule that is abundantly expressed by naïve T cells, and the DC-SIGN-ICAM-3 interaction mediates the early transient DC-T cell clustering necessary for T cell activation (Fig. 1; Geijtenbeek et al. 2000b). This is further supported by the inhibition of DC-induced T cell activation by blocking anti-

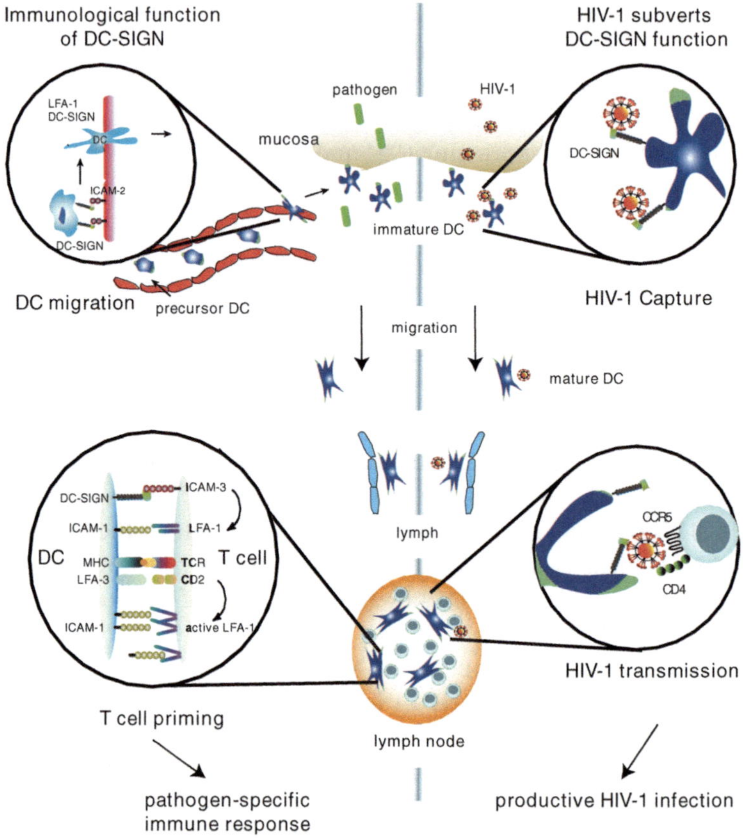

Fig. 1. The functions of DC-SIGN in the immune system and HIV-1 dissemination. DC-SIGN mediates rolling and tethering of DC precursors along ICAM-2 surfaces and migration across the endothelium in response to cytokines. Immature DCs capture pathogens and migrate to lymphoid tissues to present the processed antigens to T cells. DC-SIGN mediates the initial clustering of DCs with T cells necessary for an efficient immune response (Bleijs et al. 2001). HIV-1 subverts the trafficking capacity of DCs to gain access to the $CD4^+$ T cell compartment of lymphoid tissues. DC-SIGN captures HIV-1 at mucosal sites of entry and protects it during migration to the lymphoid tissues. Once arrived, DC-SIGN enhances the HIV-1 infection of T cells *in trans*, resulting in a productive HIV-1 infection

bodies against DC-SIGN (Geijtenbeek et al. 2000b). Thus DC-SIGN mediates the initial contact between DCs and naïve T cells by interacting with ICAM-3. This interaction may also provide positional stability to facilitate engagement of the T cell receptor (TCR) with peptide-MHC class II complexes at the immunological synapse. TCR triggering leads to the initiation of other strong DC-T cell adhesive interactions mediated by other adhesion molecules present on T cells (Fig. 1; Bleijs et al. 2001). This tighter DC-T cell contact enables secondary interactions of the co-stimulatory molecules CD80, CD83 and CD86 on DCs to provide the necessary signals for efficient activation of the T cell (Banchereau and Steinman 1998).

Another important feature of DCs is their capacity to migrate, and DC-SIGN also plays a role in this process by interacting with the endothelial adhesion molecule ICAM-2 (Geijtenbeek et al. 2000c). DC precursors are constitutively recruited from the blood to peripheral tissues to replenish the steady-state level of tissue DCs or in response to inflammation (Fig. 1). Tethering to the blood vessel endothelium is an essential prerequisite for the exit of DC precursors from the blood and is necessary for subsequent firm adhesion and extravasation. DC-SIGN mediates this initial tethering of DC-SIGN-positive cells along surfaces expressing ICAM-2, an endothelial ligand abundantly expressed on both blood and lymphatic vessels (Geijtenbeek et al. 2000c). The transendothelial migration of DC precursors increases on activation of endothelial cells and is mediated by DC-SIGN-ICAM-2 interactions, because antibodies against DC-SIGN inhibit transendothelial migration. The presence of DC-SIGN-positive DC precursors in blood further supports the hypothesis that, under physiological conditions, DC-SIGN-ICAM-2 interactions mediate rolling along endothelial linings and transmigration of DCs into the periphery (Fig. 1; Geijtenbeek et al. 2000c). These DC-SIGN-positive DC precursors could be poised to exit the blood at inflammatory sites, allowing rapid recruitment of these cells to sites where their surveillance function is needed. The high expression of DC-SIGN on immature DCs in the peripheral tissues, and the expression of ICAM-2 on lymphatic vessels, further support a central role for the DC-SIGN-ICAM-2 interaction in DC-specific migration from blood into the periphery, whether inflamed or not, and subsequently of immature DCs via the lymph into lymphatic tissues (Fig. 1; Geijtenbeek et al. 2000c; Bleijs et al. 2001).

Other C-type lectins besides DC-SIGN are also abundantly expressed by DCs, such as the macrophage mannose receptor (MMR), BDCA2 and DEC-205, which function as pathogen recognition receptors (Sallusto et al. 1995; Dzionek et al. 2001; Kato et al. 1998). This family of receptors contains at least one carbohydrate recognition domain, which recognises various carbohydrates with different specificities. Such C-type lectins are important in the innate immune system because they can recognise a broad repertoire of conserved pathogen-specific molecules (Drickamer 1995). Although DC-SIGN was identified primarily as an adhesion receptor, we recently demonstrated that DC-SIGN also functions as an antigen receptor (Engering et al. 2002). Many C-type lectins, such as MMR, DEC-205 and BDCA-2 on DCs mediate pathogen recognition, internalisation and processing of the antigen for subsequent presentation by DCs. Targeting of DC-SIGN with soluble antigen or cross-linking with anti-DC-SIGN antibodies leads to rapid internalisation of DC-SIGN-antigen complexes (Engering et al. 2002). A di-leucine motif within the cytoplasmic tail of DC-SIGN is important in targeting of these DC-SIGN-antigen complexes to lysosomal compartments. Internalisation of antigen through capture by DC-SIGN leads to presentation of peptides to T cells, indicating an important function in antigen capture and presentation (Engering et al. 2002). This function is not unique to DC-SIGN, as other C-type lectins on DCs also capture and process antigen; however, selectivity may exist in the antigen capture process, because the receptors may recognise several different antigens. The variety of C-type lectin receptors on DCs may be essential to elicit immune responses against a broad spectrum of pathogens.

3
DC-SIGN as a Novel HIV-1 Receptor

The identification of DC-SIGN as a DC-specific adhesion receptor (Geijtenbeek et al. 2000a) demonstrated that it is identical to the previously cloned HIV-1 envelope-binding C-type lectin (Curtis et al. 1992) and initiated a detailed investigation into the function of DC-SIGN as an HIV-1 receptor on DCs (Geijtenbeek et al. 2000a). Early work revealed that DCs pulsed with HIV-1 promote a robust infection of co-cultivated T cells (Cameron et al. 1992; Pope et al. 1994). This property of DCs to enhance infection of T cells was molecularly defined 8 years later when DC-SIGN was identified as the HIV-1 *trans*-receptor on DCs (Geijtenbeek et al. 2000a). The affinity of DC-SIGN for the HIV-1 envelope glycoprotein gp120 exceeds that of CD4 (Curtis et al. 1992), but, in contrast to CD4, DC-SIGN does not function as a classic HIV-1 entry receptor; co-expression of DC-SIGN with either CD4 or CCR5 does not enable HIV-1 entry into HIV-1 non-permissive cells (Geijtenbeek et al. 2000a). In contrast, DC-SIGN acts as an HIV-1 *trans*-receptor that binds HIV-1 and transmits it very efficiently to neighbouring permissive target cells (Fig. 1; Geijtenbeek et al. 2000a). DC-SIGN expressed on DCs, and on transfectants, binds both M- and T-tropic HIV-1, HIV-2 and SIV and transmits these viruses to recipient T cells to result in an efficient infection of T cells (Geijtenbeek et al. 2000a; Pohlmann et al. 2001b). So far, no differences in affinity of DC-SIGN for either R5- or X4-tropic strains have been demonstrated which could explain the selective infection by R5-tropic strains in sexual transmission of HIV-1 (Pohlmann et al. 2001b). Strikingly, DC-SIGN not only captures and transmits HIV-1, but also enhances infection; at low virus titres CD4/CCR5-expressing cells are not detectably infected without the assistance of DC-SIGN *in trans* (Geijtenbeek et al. 2000a). Conditions in which the number of HIV-1 particles is limiting are likely to resemble those found early during infection in vivo and thus suggest that DC-SIGN may be required not only for HIV-1 transmission from mucosa to lymphoid tissues but also for efficient infection of T cells. The binding of HIV-1 to DC-SIGN alone is not sufficient for transmission of HIV-1, because HIV-1 binding and transfer by DC-SIGN have been shown to be dissociated functions (Pohlmann et al. 2001c).

To date, DC-SIGN is unique in this function because it not only binds HIV-1 gp120 but also enhances the HIV-1 infection of T cells *in trans* (Geijtenbeek et al. 2000a). The process through which DC-SIGN promotes efficient infection *in trans* of cells through their CD4/chemokine receptor complex remains unclear to date. Binding of the viral envelope glycoprotein to DC-SIGN may induce a conformational change in gp120 that enables a more efficient interaction with CD4 and/or the chemokine receptor and subsequent membrane fusion with T cells. This is supported by our observation that the interaction of CD4 with gp120 increases on DC-SIGN binding to gp120 (T.G. Geijtenbeek, unpublished data). Alternatively, binding of viral particles to DC-SIGN may focus or concentrate the virus particles at the surface of the DCs and may thus increase the probability that entry will occur after they bind to the CD4 and co-receptor complex on target cells. The potency of DC-SIGN to capture and transmit HIV to T cells may largely depend on the membrane organisation of DC-SIGN in rafts or its capacity to multimerise. Future experiments will determine the molecular mechanism by which DC-SIGN enhances the infection of T cells, and will elucidate whether a diversity in multimerisation and membrane organisation of DC-SIGN are instrumental for its function as an HIV-1 *trans*-receptor.

4 HIV-1 Circumvents DC-SIGN-Mediated Endocytosis

Where the role of DC-SIGN in HIV-1 capture and transmission by DCs has been demonstrated, it is still largely unknown how virions are conveyed from the periphery to lymphoid organs by DCs. Our studies demonstrate that HIV-1 bound to DC-SIGN is remarkably stable and can thus retain infectivity for the prolonged periods required for DCs to traffic via the lymphatic from mucosa to regional lymph nodes (Geijtenbeek et al. 2000a). It is not clear to date whether infectious virions remain associated with the cell surface or, more likely, are internalised into intracellular vesicles (Blauvelt et al. 1997; Granelli-Piperno et al. 1999; Geijtenbeek et al. 2000a). but enzymatic digestion of cell surface-bound HIV-1 demonstrated that HIV-1 is protected and probably 'hides' within the cell close to the cell membrane, without being degraded. Indeed, the finding that DC-SIGN contains internalisation motifs in its cytoplasmic tail supports the idea of endocytosis of infectious virions into DCs (Fig. 2; Engering et al. 2002). DC-SIGN-ligand complexes are normally targeted to late endosomes/lysosomes of immature DCs, resulting in an efficient antigen presentation of the internalised ligands to $CD4^{+}$ T cells (Fig. 2; Engering et al. 2002). The finding that DC-SIGN-bound HIV-1 is protected suggests that it is not routed to the lysosomal compartment to induce an anti-HIV-1 immune response, and therefore HIV-1 may interfere with targeting of DC-SIGN-HIV-1 complexes to lysosomes and/or with lysosomal degradation (Fig. 2). In mature DCs, DC-SIGN is targeted to early endosomal compartments, where HIV-1 would be protected against degradation (Engering et al. 2002; Kwon et al. 2002).

Recent results demonstrate that the intracellular trafficking of DC-SIGN is affected by the Nef protein of HIV-1. Nef is required for efficient in vivo viral replication and pathogenesis (Cullen 1998). In immature DCs, Nef exerts a marginal role on the replication of R5 HIV strains only, but it is required for optimal virus production in DC-T cell mixtures (Petit et al. 2001). Expression of Nef in immature DCs results in a redistribution of DC-SIGN and a two- to threefold increase in its cell surface expression, leading in turn to enhanced formation of DC-T cell clusters and virus transmission (Sol-Foulon et al. 2002). The effect of Nef on DC-SIGN distribution could be mimicked by removing the di-leucine sorting motif from cytoplasmic domain of DC-SIGN, and it also depends on the di-leucine motif present in Nef (Sol-Foulon et al. 2002).

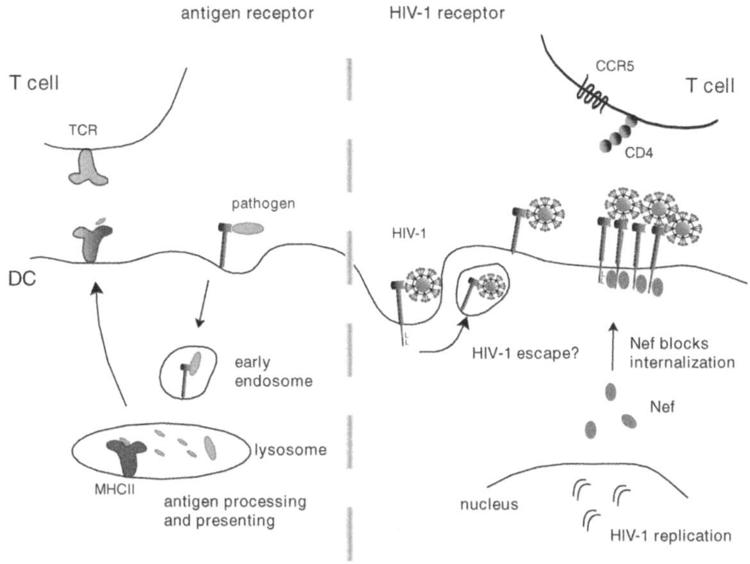

Fig. 2. HIV-1 interferes with the internalisation pathway of DC-SIGN. DC-SIGN functions as an antigen receptor. DC-SIGN rapidly internalises on soluble ligand binding and is targeted to late endosomes/lysosomes, where antigens are processed and presented on MHC class II molecules. DC-SIGN binding stabilises infectious HIV-1, indicating that HIV-1 escapes internalisation and processing either by blocking internalisation or by preventing targeting to late endosomes. Strikingly, HIV-1 Nef blocks internalisation of DC-SIGN, possibly by interacting with di-leucine motif in the cytoplasmic tail of DC-SIGN, resulting in both an increased cell surface expression and an HIV-1 transmission to T cells

This critical motif is known to mediate the direct interactions of Nef with adaptor complexes (Piguet et al. 1998). This leads to a model in which Nef, by interacting with adaptor complexes, affects the proper recognition and recycling of DC-SIGN by the cellular sorting machinery. Thus HIV-1 infection of DCs alters DC-SIGN internalisation to result in enhanced up-regulation of DC-SIGN and in increased virus capture and transmission to T cells (Fig. 2). These findings demonstrate that HIV-1 interferes with normal ligand-driven DC-SIGN internalisation, which may account for HIV-1 protection from degradation and processing by DCs. Identification of the mechanism by which HIV-1 prevents its own degradation may lead to the development of successful strategies to combat HIV-1 dissemination. Moreover, finding a way to stimulate DC-SIGN-mediated HIV-1 internalisation would greatly facilitate HIV-1 processing in DCs and would enhance specific anti-HIV-1 immune responses while reducing infection of T cells (Engering et al. 2002).

5
Other C-Type Lectins as HIV-1 gp120 Receptors on DC

DC-SIGN is a member of the C-type lectin superfamily and is specifically expressed at high levels by DCs (Geijtenbeek et al. 2000b). Other C-type lectins, such as MMR and DEC-205, are abundantly expressed by DCs and macrophages and function as pathogen recognition receptors (Sallusto et al. 1995; Kato et al. 1998). These C-type lectins consist of 8- to 10-lectin domains recognising specific carbohydrates. Such C-type lectins are important in the innate immune system because they can recognise a broad repertoire of pathogen-specific molecules (Drickamer 1995) and may potentially bind similarly to DC-SIGN the heavily glycosylated HIV-1 gp120. Indeed, some of these C-type lectins have been suggested to bind HIV-1 in a CD4-independent manner (Weissman and Fauci 1997; Blauvelt et al. 1997). However, binding of HIV-1 gp120 to monocyte-derived immature DCs, which express high levels of MMR and DEC-205, is mediated by DC-SIGN alone (Geijtenbeek et al. 2000a), because both gp120 binding and HIV-1 transmission by immature DCs are completely inhibited by specific antibodies against DC-SIGN (Geijtenbeek et al. 2000a). This block by antibodies against DC-SIGN implies that although other C-type lectins are expressed by immature DCs, they do not mediate high-affinity interactions with gp120, contrary to a recent suggestion (Turville et al. 2001). Other as yet unidentified HIV-1 *trans*-receptors could exist and may play a role, especially on DC subsets that do not express DC-SIGN, such as Langerhans cells and plasmacytoid DC (Patterson et al. 2001; Geijtenbeek et al. 2000a). However, HIV-1 gp120 binding alone is not sufficient for transmission of HIV-1 by DCs, because HIV-1 binding and transfer by DC-SIGN have been shown to be dissociable functions (Pohlmann et al. 2001c). Thus, to date, DC-SIGN is the only C-type lectin receptor expressed on DCs that is unique in this function because it not only binds HIV-1 gp120 but also enhances the HIV-1 infection of T cells *in trans* (Geijtenbeek et al, 2000a). Future identification of other putative C-type lectins that may function as HIV-1 *trans*-receptors, similarly to DC-SIGN, will need to be performed not only through HIV-1 binding studies but also through examination of their ability to mediate HIV-1 transfer and enhance infection.

6
DC-SIGN Interaction with gp120 Is Carbohydrate Independent

Detailed information about the mechanism of ligand binding is crucial for the development of specific DC-SIGN-based inhibitors and/or vaccines. The HIV-1 gp120 binding activity of DC-SIGN is calcium dependent, and DC-SIGN binds two Ca^{2+} ions (Curtis et al. 1992; Geijtenbeek et al. 2000b); one Ca^{2+} ion is essential for the tertiary structure and the other ion coordinates ligand binding (Geijtenbeek et al. 2000b; Feinberg et al. 2001), similarly to other C-type lectins (Drickamer 1995). The interaction of DC-SIGN with HIV-1, ICAM-2 and ICAM-3 is blocked by the polycarbohydrate mannan (Geijtenbeek et al. 2000a–c). The inhibition of DC-SIGN function by mannan suggests that its interaction with its ligands is mediated by mannose-like carbohydrates (Geijtenbeek et al. 2000b). Indeed, recent studies demonstrated that DC-SIGN is able to interact with carbohydrates (Geijtenbeek et al. 2000b; Mitchell et al. 2001). Its natural ligand ICAM-3 contains N-linked glycosylations consisting of high-mannose-type oligosaccharides (Funatsu et al. 2001), and, indeed, enzymatic removal of the N-linked carbohydrates from ICAM-3 completely abrogates binding to DC-SIGN (Geijtenbeek et al. 2002a). The HIV-1 envelope glycoprotein gp120 is also heavily glycosylated and contains high-mannose-type oligosaccharides (Leonard et al. 1990; Mizuochi et al. 1988). Strikingly, although glycosylation seems to enhance the affinity of gp120 binding, neither O- nor N-linked glycosylations are vital for the interaction of the HIV-1 gp120 with DC-SIGN, because DC-SIGN interacts with both enzymatically deglycosylated and non-glycosylated gp120 (Geijtenbeek et al. 2002a). Therefore, the interaction of DC-SIGN with gp120 differs from that with ICAM-3, indicating different binding sites for ICAM-3 and HIV-1 gp120 in DC-SIGN; the C-type lectin DC-SIGN interacts with its ligands either through carbohydrate interactions, such as mannan and ICAM-3, or through protein interactions, such as gp120. However, the polycarbohydrate mannan is able to inhibit both ICAM-3 and gp120 binding to DC-SIGN by occupying the binding site of DC-SIGN, suggesting that the carbohydrate and protein interactions to DC-SIGN are mediated by overlapping but distinct binding sites in DC-SIGN.

7 The Structure of DC-SIGN and Its Distinct Binding Sites for HIV-1 and ICAM

Recent elucidation of the three-dimensional structure of DC-SIGN co-crystallised with an oligosaccharide identified several important features of ligand binding by DC-SIGN (Feinberg et al. 2001). Part of the pentasaccharide ligand forms coordination bonds with the Ca^{2+} at the principal site 2 (Feinberg et al. 2001). Carbohydrate binding by DC-SIGN is mediated by the amino acid residues which are also in close contact with this Ca^{2+} and which form the core of the carbohydrate-binding site (Fig. 3). This type of binding is a hallmark of C-type lectin-carbohydrate interactions (Weis et al. 1998). However, the DC-SIGN interaction with carbohydrate-ligands has also some unique characteristics compared with that of other C-type lectins (Feinberg et al. 2001). Most C-type lectins, such as mannose binding protein-A, interact with a single terminal mannose or GlcNAc residue in an oligosaccharide ligand, and the rest of the oligosaccharide points away from the surface of the protein (Feinberg et al. 2001). In contrast, DC-SIGN binds an internal mannose of the oligosaccharide, and the external saccharides also interact with the surface of DC-SIGN (Feinberg et al. 2001).

Even though the nature of the interaction of DC-SIGN with gp120 differs from that with ICAM-3, site-directed mutagenesis demonstrates that both gp120 and ICAM-3 binding by DC-SIGN are mediated by the Ca^{2+} at site 2 and nearby amino acid residues which also serve as Ca^{2+} ligands (Fig. 3; Geijtenbeek et al. 2002a). The auxiliary Ca^{2+} bound at site 1 probably correctly orients the primary binding site, and, indeed, removal of this Ca^{2+} by mutagenesis completely abrogates both ICAM-3 and HIV-1 binding (Geijtenbeek et al. 2002a). The importance of the Ca^{2+} ions is underscored by the finding that the epitopes of the blocking antibodies against DC-SIGN, AZN-D1 and -D2 are located at the Ca^{2+} sites (Geijtenbeek et al. 2002a).

The differences in carbohydrate dependency suggest that gp120 binding by DC-SIGN differs from ICAM-3 binding. The three-dimensional structure of DC-SIGN with the pentasaccharide shows that the Val^{351} residue in DC-SIGN participates in carbohydrate binding through van der Waals interactions (Feinberg et al. 2001).We have demonstrated that this Val^{351} in DC-SIGN denotes the difference between the interaction of DC-SIGN with gp120 and ICAM-3 (Geijtenbeek et al. 2002a). The Val^{351} resi-

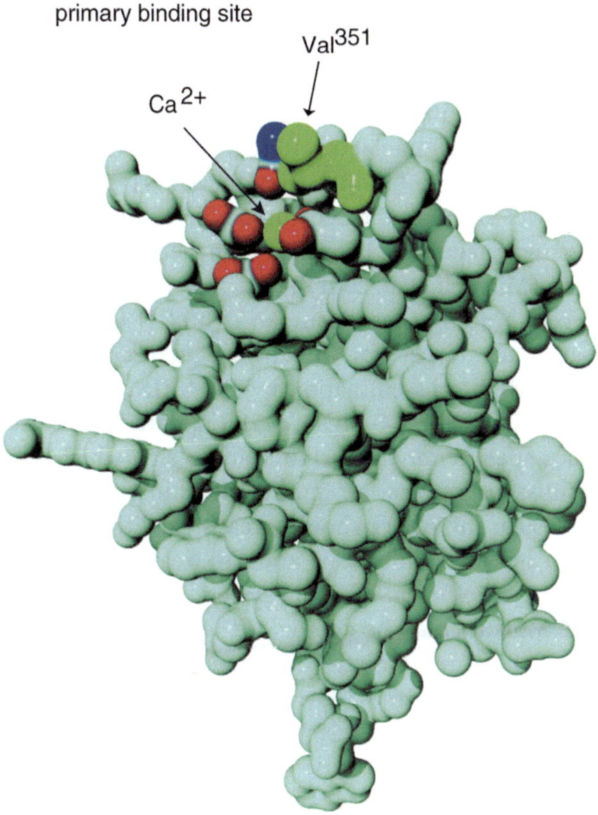

Fig. 3. The primary ligand binding site in DC-SIGN is located at the top of the C-type lectin domain and includes the Ca^{2+} at site 2. Molecular model of the ligand binding site, the Ca^{2+} site 2 and the Val^{351} amino acid residue in DC-SIGN. Nitrogen and oxygen atoms of the amino acid residues involved in both ligand and Ca^{2+} binding are shown are shown in *blue* and *red spheres*, respectively. The four amino acid residues Glu^{347}, Asn^{349}, Glu^{354}, Asn^{365} and Asp^{366} co-ordinate binding of the Ca^{2+} ion and, except for Asp^{366}, they all contribute to the binding of the ligand through hydrogen bonding via acidic or amide groups. The Val^{351} amino acid residue involved in ICAM-3 but not in gp120 binding is shown in *green*. (Geijtenbeek et al. 2002a)

due is important in binding of both carbohydrates and ICAM-3, but it is not essential for gp120 binding because the V351G mutant of DC-SIGN still interacts with gp120 but is unable to bind ICAM-3 (Geijtenbeek et al. 2002a). These data confirm that the interaction of DC-SIGN with ICAM-3, in contrast to that with gp120, is carbohydrate dependent. Thus

DC-SIGN has a distinct binding site for HIV-1 gp120 which is primarily mediated by protein-protein interactions (Geijtenbeek et al., submitted). These findings further indicate that, in addition to the ligands interacting with the primary binding site centred around the Ca^{2+} at site 2, the ligands form additional contacts with the surface of DC-SIGN, thereby creating different ligand binding sites for HIV-1 gp120 and ICAM-3. Clustering of DC-SIGN on the cell surface in a novel tetrameric coiled-coil motif is suggested to contribute to a further increase in ligand binding specificity (Mitchell et al. 2001).

The different binding sites in human DC-SIGN for both ICAM-3 and gp120 are attractive targets for therapeutic intervention of DC-induced immunity and HIV-1 dissemination by DC-SIGN, respectively. For example, the gp120-specific DC-SIGN V351G mutant can be used to generate antibodies specific for the gp120 binding site, which should inhibit HIV-1 transmission but not the immune function of DC-SIGN. The development of such inhibitors should greatly facilitate the prevention of dissemination of HIV-1 infections and the treatment of chronic HIV-1 infection, where an efficient immune system is vital. Moreover, the DC-SIGN V351G mutant could be conjugated to toxins to specifically target HIV-1-infected T cell reservoirs, which express surface gp120. Chimeric toxins such as CD4-PE40 and 3B3(Fv)-PE38 represent a complementary class of agents which already selectively kill productively infected cells, diminishing HIV reservoirs in patients with AIDS (Goldstein et al. 2000), but the higher affinity of DC-SIGN for gp120 compared with CD4-gp120 interactions (Curtis et al. 1992) suggests that DC-SIGN-toxin conjugates might be more effective. We anticipate that chimeric toxins based on the V351G mutant will target and kill only HIV-1-infected cells, in contrast to DC-SIGN wild-type-based chimeras which might also target healthy ICAM-3-expressing T cells.

8
DC-SIGN Expression in Relation to HIV-1 Transmission

DC-SIGN is abundantly expressed by both monocyte- and $CD34^+$-derived DCs, as well as in vivo by DCs present in mucosal tissues involved in sexual HIV-1 transmission, such as rectum, cervix and uterus (Table 1; Geijtenbeek et al. 2000a,b). DC-SIGN is also expressed in the skin by dermal DC (Geijtenbeek et al. 2000b). The presence of DC-SIGN in those

Table 1. Expression of DC-SIGN homologues in human, primate and mouse

Species	Receptor	Tissue expression	Specific expression	Function	References
Human	DC-SIGN	Lymphoid organs, dermis, mucosa, placenta, blood	DCs, Hofbauer cells	HIV-1, ICAM-2, ICAM-3	(Geijtenbeek et al. 2000a–c, 2001a; Pohlmann et al. 2001b)
	L-SIGN/ DC-SIGNR	Liver, lymph node	LSECs, MΦ in lymph node	HIV-1, ICAM-2, ICAM-3	(Bashirova et al. 2001; Soilleux et al. 2000; Pohlmann et al. 2001a; Engering 2002)
Primate [a]	DC-SIGN	Lymph, node, spleen, mucosa	DCs, MΦ	HIV-1, ICAM-2, ICAM-3	(Geijtenbeek et al. 2001b; Baribaud et al. 2001)
	L-SIGN	Liver	LSECs	nd	(Geijtenbeek et al. 2001b)
Mouse	mDC-SIGN/ CIRE	Spleen, lung(mRNA levels)	$CD8\alpha^-$ DCs >B cells (mRNA levels)	nd	(Park et al. 2001; Caminschi et al. 2001)
	SIGNR1	Spleen, liver, lymph node	LSECs, MZM, medullary and subcapsular MΦ	HIV-1, ICAM-2, ICAM-3	(Park et al. 2001; Baribaud et al. 2001; Geijtenbeek 2002b)
	SIGNR2–4	Spleen (R3, R4), testis (R2, R4) (mRNA levels)	DCs, B and T cells (R2, R3; mRNA levels)	nd	(Park et al. 2001)

MΦ, macrophages; nd, not determined.[a]Rhesus macaque and chimpanzee.

tissues involved in sexual transmission of HIV-1 further supports the importance of this receptor in HIV-1 dissemination (Table 1).

Peripheral blood DC subsets have also been implicated in HIV-1 transmission through blood (Weissman et al. 1995; Patterson et al. 1999; Patterson et al. 2001). Recent studies demonstrated that, by RT-PCR analysis, DC-SIGN is expressed neither by plasmacytoid DCs (Patterson et al. 2001) nor by myeloid blood precursor DCs (Turville et al. 2001), whereas DC-SIGN mRNA was detected in myeloid blood precursors (Patterson et al. 2001). This discrepancy could be due to differences in DC isolation procedures. These subsets were isolated by depletion of $CD14^+$ cells from peripheral blood. With the use of a novel isolation procedure which included $CD14^+$ cells, a $CD14^+$ $DC\text{-}SIGN^+$ DC precursor subset was recently identified in blood (Table 1; Geijtenbeek et al. 2000c) which readily acquires typical DC morphology and efficiently stimulates T cell proliferation (Engering 2002). Despite lower levels of DC-SIGN expression on this novel subset of blood $DC\text{-}SIGN^+$ DCs, they very efficiently transmit HIV-1 to T lymphocytes similarly to immature DCs (Engering 2002). This indicates that on HIV-1 infection through blood, either by injection or blood transfusion, the $DC\text{-}SIGN^+$ blood DCs may be the first target for HIV-1 and efficiently capture low amounts of virus. On migration into T cell areas of lymphoid tissues, $DC\text{-}SIGN^+$ blood DCs may facilitate infection of T cells *in trans*. Moreover, DC-SIGN is expressed on both DCs and Hofbauer cells within the placenta and may therefore also be involved in the vertical transmission of HIV-1 from mother to child (Geijtenbeek et al. 2001a). The DC-restricted expression of DC-SIGN in a wide range of tissues such as mucosal tissue, lymph node, blood and placenta implicates DC-SIGN as an initial target for HIV-1 not only during sexual transmission but also during vertical transmission, which may contribute to the persistence of T cell infections by HIV-1, because $DC\text{-}SIGN^+$ DCs are abundantly present in the lymph nodes. The dissemination of HIV-1 on infection through DC-SIGN may depend not only on the amount of virus present but also on the number of DC-SIGN molecules expressed per cell. So far, no individuals have been identified as having DCs lacking DC-SIGN on the cell surface, emphasising the importance of DC-SIGN in the immune response. However, differences in DC-SIGN expression levels on DCs have been found between healthy individuals (T.G. Geijtenbeek, unpublished data). It will be important to determine whether variability in levels of DC-SIGN expression between individuals affects HIV dissemination. Polymorphism within the promotor region of DC-SIGN may regulate expression levels of DC-SIGN and may determine viral transmission, disease progression to AIDS and clinical outcome of the disease.

9
A Role for L-SIGN in HIV-1 Infections

Initially, the analysis of DC-SIGN expression was confused by the presence of the DC-SIGN homologue L-SIGN (Bashirova et al. 2001), also called DC-SIGN-related (DC-SIGNR) (Soilleux et al. 2000; Table 1). L-SIGN functions as an HIV-1 *trans*-receptor similarly to DC-SIGN but is not expressed by DCs (Bashirova et al. 2001). RT-PCR analysis of DCs with L-SIGN-specific primers, as well as RNA blot analysis with L-SIGN-specific probes, failed to identify any L-SIGN-specific products (Bashirova et al. 2001). Moreover, no staining of L-SIGN on DCs was found in vitro or in vivo with L-SIGN-specific antibodies (A. Engering, personal communication). Recently, Mummidi et al. showed that mRNA encoding human DC-SIGN2, identical to L-SIGN, was present in DCs as detected by RT-PCR (Mummidi et al. 2001). This discrepancy could result from the presence of contaminating L-SIGN-positive cells in DC preparations, because L-SIGN expression can be induced on cells of myeloid origin (T.G. Geijtenbeek, unpublished results).

L-SIGN/DC-SIGNR is specifically expressed by liver sinusoidal endothelial cells (LSEC), a liver-resident antigen-presenting cell population of monocyte origin (Bashirova et al. 2001; Pohlmann et al. 2001a) and by a specific subpopulation of macrophages in lymph nodes (A. Engering, manuscript in preparation) (Table 1). The expression pattern of L-SIGN in liver sinusoids suggests that LSEC, which are in continual contact with circulating leukocytes, can capture HIV-1 from the blood and promote *trans*-infection of circulating T cells in the liver. Moreover, prior studies indicated that LSEC themselves may be susceptible to HIV-1 infection (Steffan et al. 1992). Thus it is possible that L-SIGN promotes infection of these cells, thereby establishing a reservoir for new viruses to be passed on to T cells trafficking through the liver sinusoid. Its expression on macrophages in lymph nodes (A. Engering, manuscript in preparation) suggests that L-SIGN may play an additional role in HIV-1 pathogenesis by promoting HIV-1 infection of T cells in lymph nodes and thus L-SIGN may be involved in the persistence of chronic HIV-1 infections. Additional functional studies are necessary to understand the physiological role of L-SIGN and its possible role in HIV-1 pathogenesis.

10
Primate and Murine In Vivo Models for DC-SIGN

The development of a vaccine that protects against HIV-1 infection remains one of the most pressing challenges facing modern medicine. Most current vaccine and inhibition studies target the binding of HIV-1 through CD4 and the chemokine receptors, as well as the replication of HIV-1 in infected T cells. The identification of DC-SIGN as the HIV-1 *trans*-receptor provides a new early target for preventing HIV-1 transmission to T cells and subsequent dissemination. The function of DC-SIGN can be blocked in vitro by antibodies against DC-SIGN which inhibit both HIV-1 binding and subsequent transmission to T cells. A model for in vivo studies must therefore be used to evaluate the effects of neutralising DC-SIGN antibodies on HIV-1 transmission and immune function. The in vivo rhesus macaque model has been valuable to the development and evaluation of HIV-1 vaccines and novel clinical therapeutics. SIV-rhesus macaque models have been used to advance our understanding of the involvement of DCs in HIV-1 dissemination, especially early after infection (Reimann et al, 1994; Hu et al. 1999; Spira et al. 1996). These data indicated that DCs represent the initial targets for SIV and that they transmit the virus from the periphery to T cells in lymphoid tissues. Both rhesus macaque and chimpanzee homologues of DC-SIGN are highly similar to human DC-SIGN (Geijtenbeek et al. 2001b). The primate homologues function similarly to human DC-SIGN, having a high affinity for HIV-1 gp120, ICAM-2 and ICAM-3 (Geijtenbeek et al. 2001b). The primate homologues of DC-SIGN are able to transmit both HIV-1 and SIV to recipient cells (Baribaud et al. 2001). The antibodies against human DC-SIGN, AZN-D1 and AZN-D2, cross-react with the primate homologues and immunohistochemical analyses demonstrate abundant expression of these homologues in lymphoid and mucosal tissues (Geijtenbeek et al. 2001b). The in vivo expression of DC-SIGN in the primates seems largely similar to the expression in humans, with the primate homologues being abundantly expressed by DCs present in lymphoid as well as mucosal tissues (Table 1). However, staining of macrophages in both rhesus macaque and chimpanzee lymphoid tissue sections is also observed. indicating a difference from the DC-restricted expression of human DC-SIGN (Table 1; Geijtenbeek et al. 2001b). Thus, dissemination of HIV-1 is likely to be facilitated by DC-SIGN in primates, and this in vivo model is suitable for investigating the function of DC-SIGN both in HIV/SIV infections and in the efficacy of the inhibiting antibodies against DC-SIGN, AZN-D1 and AZN-D2 in therapeutic strategies.

Similarly, a murine in vivo model would also be very useful to investigate the role of DC-SIGN in HIV-1 dissemination. Five murine DC-SIGN homologues were identified by RT-PCR, and one, called murine DC-SIGN, was expressed at high mRNA levels in $CD11c^+$ DCs (Park et al. 2001; Caminschi et al. 2001) (Table 1). In contrast, the other cDNAs, designated SIGNR1–4, were hardly detectable in DCs but were detected at various mRNA levels in B and T cells by RT-PCR (Park et al. 2001; Table 1). One of these homologues, SIGNR1, was shown to bind HIV-1. However, in contrast to human DC-SIGN, this murine homologue was not able to transfer the virus to recipient cells under the conditions tested (Baribaud et al. 2001). The reason for this is not clear but may be its expression in a human kidney cell line which could interfere with its function as a HIV-1 *trans*-receptor. We have demonstrated that this murine homologue is specifically expressed by liver sinusoidal endothelial cells and by macrophages in spleen and lymph node, but not by DCs (Geijtenbeek 2002b) (Table 1), and is thus likely the murine L-SIGN homologue. Therefore, the situation in mouse is much more complex than in human and primates, and the expression and function of the murine homologues of DC-SIGN must be further investigated before a murine model can be productively investigated.

11 Concluding Remarks

It is now becoming evident that, as sentinels of the immune system, DCs are exploited by HIV-1: HIV-1 uses DC-SIGN, which normally mediates antigen presentation, DC migration and T cell stimulation, as a receptor to ferry HIV-1 from the periphery directly to T cell-rich areas. HIV-1 has found a way to circumvent the normal DC-SIGN internalisation and presentation route and to remain infectious for several days, allowing time to be transported to T cell-rich lymphoid compartments. DC-SIGN functions as a novel class of HIV receptors, as it does not mediate HIV entry into DCs, like CD4 and the HIV-1 co-receptors, but enhances HIV-1 infection of $CD4^+$ T cells *in trans*.

The fact that expression of DC-SIGN is found on DCs at many sites in the body, including mucosal tissues, lymph node, blood, and placenta. strongly suggests a contribution to HIV-1 transmission during sexual transmission, contamination through blood and mother-to-child transfer. Because DC-SIGN also has a strong immune function in mediating DC migration and DC-T cell clustering through binding its cellular ligands ICAM-2 and ICAM-3, it is necessary to develop inhibitors that only block HIV-1 binding without affecting its immunological function. Structural analysis and binding studies have demonstrated that, fortunately. the DC-SIGN binding site for HIV-1 gp120 is different to that for ICAM-2/3, which opens up the possibility of designing drugs that specifically inhibit the interaction of DC-SIGN with HIV-1 without affecting its binding to cellular ligands. Although specific carbohydrate recognition by DC-SIGN has been observed, it is now becoming clear that DC-SIGN interactions with gp120 are not primarily mediated by glycosylations. Further research will be necessary to demonstrate the exact role of carbohydrates in mediating binding to gp120. To date, the function of DC-SIGN in HIV transmission is based only on in vitro experiments, and an in vivo validation is needed to confirm its importance in HIV-1 capture, transmission and immune evasion that will enable the successful development of both preventative and therapeutic DC-SIGN-based strategies in the treatment of HIV-1.

Acknowledgements. We thank S. van Vliet and G. van Duijnhoven for their technical support and A. Engering for providing unpublished data. We are grateful to E. Krieger for providing the three-dimensional structure of DC-SIGN and L. Colledge for helpful suggestions and editing of the manuscript.

References

Banchereau, J, Steinman, RM (1998) Dendritic cells and the control of immunity. Nature 392, 245–52.

Baribaud F, Pohlmann S, Sparwasser T, Kimata MT, Choi YK, Haggarty BS, Ahmad N, Macfarlan T, Edwards TG, Leslie GJ, Arnason J, Reinhart TA, Kimata JT, Littman DR, Hoxie JA, Doms RW (2001) Functional and antigenic characterization of human, rhesus macaque, pigtailed macaque, and murine DC-SIGN. J Virol 75, 10281–9

Bashirova AA, Geijtenbeek TBH, van Duijnhoven GCF, van Vliet SJ, Eilering JB, Martin MP, Wu L, Martin TD, Viebig N, Knolle PA, KewalRamani VN, van Kooyk Y, Carrington M (2001) A dendritic cell-specific intercellular adhesion molecule 3-grabbing nonintegrin (DC-SIGN)-related protein is highly expressed on human liver sinusoidal endothelial cells and promotes HIV-1 infection. J Exp Med 193, 671–8

Blauvelt A, Asada H, Saville MW, Klaus-Kovtun V, Altman DJ, Yarchoan R, Katz SI (1997) Productive infection of dendritic cells by HIV-1 and their ability to capture virus are mediated through separate pathways. J Clin Invest 100, 2043–53

Bleijs DA, Geijtenbeek TBH, Figdor CG, van Kooyk Y (2001) DC-SIGN and LFA-1: a battle for ligand. Trends Immunol 22, 457–63

Buseyne F, Le Gall S, Boccaccio C, Abastado JP, Lifson JD, Arthur LO, Riviere Y, Heard JM, Schwartz O (2001) MHC-I-restricted presentation of HIV-1 virion antigens without viral replication. Nat Med 7, 344–9

Cameron PU, Freudenthal PS, Barker JM, Gezelter S, Inaba K, Steinman RM (1992) Dendritic cells exposed to human immunodeficiency virus type-1 transmit a vigorous cytopathic infection to CD4+ T cells. Science 257, 383–7

Caminschi I, Lucas KM, O'Keeffe, MA Hochrein, H Laabi, Y Brodnicki, TC Lew, AM Shortman K, Wright, MD (2001) Molecular cloning of a C-type lectin superfamily protein differentially expressed by CD8alpha(-) splenic dendritic cells. Mol Immunol 38, 365–73

Cullen BR (1998) HIV-1 auxiliary proteins: making connections in a dying cell. Cell 93, 685–92

Curtis BM, Scharnowske S, Watson AJ (1992) Sequence and expression of a membrane-associated C-type lectin that exhibits CD4-independent binding of human immunodeficiency virus envelope glycoprotein gp120. Proc Natl Acad Sci USA 89, 8356–60

Drickamer K (1995) Increasing diversity of animal lectin structures. Curr Opin Struct Biol 5, 612–6

Dzionek A, Sohma Y, Nagafune J, Cella M, Colonna M, Facchetti F, Gunther G, Johnston I, Lanzavecchia A, Nagasaka T, Okada T, Vermi W, Winkels G, Yamamoto T, Zysk M, Yamaguchi Y, Schmitz J (2001) BDCA-2, a novel plasmacytoid dendritic cell-specific type II C-type lectin mediates antigen capture and is a potent inhibitor of interferon alpha/beta induction. J Exp Med 194, 1823–34

Engering A, Geijtenbeek TBH, van Vliet SJ, Wijers M, van Liempt E, Demaurex N, Lanzavecchia A, Fransen J, Figdor CG, Piguet V, van Kooyk Y (2002a) The dendritic cell-specific adhesion receptor DC-SIGN internalizes antigen for presentation to T cells. J Immunol 168, 2118–2126

Engering A, van Vliet SJ, Geijtenbeek TB, van Kooyk Y (2002b) Subset of DC-SIGN(+) dendritic cells in human blood transmit HIV to T lymphocytes. Blood 100, 1780–1786

Feinberg H, Mitchell DA, Drickamer K, Weis WI (2001) Structural basis for selective recognition of oligosaccharides by DC-SIGN and DC-SIGNR. Science 294, 2163–2166

Funatsu O, Sato T, Kotovuori P, Gahmberg CG, Ikekita M, Furukawa K (2001) Structural study of N-linked oligosaccharides of human intercellular adhesion molecule-3 (CD50). Eur J Biochem 268, 1020–9

Geijtenbeek TBH, van Kooyk Y (2001) Dendritic cells ferry HIV-1 from periphery into lymphoid tissues. In: Cossarizza A, Kaplan A (eds) Cellular aspects of HIV infection, Wiley-Liss Inc, New York, pp 229–247

Geijtenbeek TBH, Kwon DS, Torensma R, van Vliet SJ, van Duijnhoven GCF, Middel J, Cornelissen IL, Nottet HS, KewalRamani VN, Littman DR, Figdor CG, van Kooyk Y (2000a) DC-SIGN, a dendritic cell-specific HIV-1-binding protein that enhances trans-infection of T cells. Cell 100, 587–97

Geijtenbeek TBH, Torensma R, van Vliet SJ, van Duijnhoven GCF, Adema GJ, van Kooyk Y, Figdor CG (2000b) Identification of DC-SIGN, a novel dendritic cell-specific ICAM-3 receptor that supports primary immune responses. Cell 100, 575–85

Geijtenbeek TBH, Krooshoop DJEB, Bleijs DA, vanVliet SJ, van Duijnhoven GCF, Grabovsky V, Alon R, Figdor CG, van Kooyk Y (2000c) DC-SIGNICAM-2 interaction mediates dendritic cell trafficking. Nat Immunol 1, 353–357

Geijtenbeek TBH, van Vliet SJ, van Duijnhoven GCF, Figdor CG, van Kooyk Y (2001a) DC-SIGN, a dendritic cell-specific HIV-1 receptor present in placenta that infects T cells in trans – a review. Placenta 22, S19–23

Geijtenbeek TBH, Koopman G, van Duijnhoven GC, van Vliet SJ, van Schijndel AC, Engering A, Heeney JL, van Kooyk Y (2001b) Rhesus macaque and chimpanzee DC-SIGN act as HIV/SIV gp120 trans- receptors, similar to human DC-SIGN. Immunol Lett 79, 101–7

Geijtenbeek TBH, van Duijnhoven GCF, van Vliet SJ, Krieger E, Vriend G, Figdor CG, van Kooyk Y (2002a) Identification of different binding sites in the dendritic cell-specific receptor DC-SIGN for ICAM-3 and HIV-1. J Biol Chem 277, 11314–11320

Geijtenbeek TBH, Groot PC, Nolte MA, van Vliet SJ, Gangaram-Panday ST, van Duijnhoven GC, Kraal G, van Osterhout AJ, van Kooyk Y (2002b) Marginal zone macrophages express a murine homologue of DC-SIGN that captures blood-borne antigens in vivo. Blood 100, 2908–2916

Goldstein H, Pettoello-Mantovani M, Bera TK, Pastan IH, Berger EA (2000) Chimeric toxins targeted to the human immunodeficiency virus type 1 envelope glycoprotein augment the in vivo activity of combination antiretroviral therapy in thy/liv-SCID-Hu mice. J Infect Dis 181, 921–6

Granelli-Piperno A, Finkel V, Delgado E, Steinman RM (1999) Virus replication begins in dendritic cells during the transmission of HIV-1 from mature dendritic cells to T cells. Curr Biol 9, 21–9

Hu J, Miller CJ, O'Doherty U, Marx PA, Pope M (1999) The dendritic cell-T cell milieu of the lymphoid tissue of the tonsil provides a locale in which SIV can reside and propagate at chronic stages of infection. AIDS Res Hum Retroviruses 15, 1305–14

Kato M, Neil TK, Clark GJ, Morris CM, Sorg RV, Hart DN (1998) cDNA cloning of human DEC-205, a putative antigen-uptake receptor on dendritic cells. Immunogenetics 47, 442–50

Kwon DS, Gregorio G, Bitton N, Hendrickson WA, Littman DR (2002) DC-SIGN-mediated internalization of HIV is required for trans-enhancement of T cell infection. Immunity 16, 135–144

Leonard CK, Spellman MW, Riddle L, Harris RJ, Thomas JN, Gregory TJ (1990) Assignment of intrachain disulfide bonds and characterization of potential glycosylation sites of the type 1 recombinant human immunodeficiency virus envelope glycoprotein (gp120) expressed in Chinese hamster ovary cells. J Biol Chem 265, 10373–82

Mitchell DA, Fadden AJ, Drickamer K (2001) A novel mechanism of carbohydrate recognition by the C-type lectins DC-SIGN and DC-SIGNR subunit organization and binding to multivalent ligands. J Biol Chem 276, 28939–45

Mizuochi, T Spellman, MW Larkin, M Solomon, J Basa, LJ, Feizi, T (1988) Carbohydrate structures of the human-immunodeficiency-virus (HIV) recombinant envelope glycoprotein gp120 produced in Chinese-hamster ovary cells. Biochem J 254, 599–603

Mummidi S, Catano G, Lam L, Hoefle A, Telles V, Begum K, Jimenez F, Ahuja SS, Ahuja SK (2001) Extensive repertoire of membrane-bound and soluble DC-SIGN1 and DC- SIGN2 isoforms: Inter-individual variation in expression of DC-SIGN transcripts. J Biol Chem 276, 33196–212.

Park CG, Takahara K, Umemoto E, Yashima Y, Matsubara K, Matsuda Y, Clausen BE, Inaba K, Steinman RM (2001) Five mouse homologues of the human dendritic cell C-type lectin, DC- SIGN. Int Immunol 13, 1283–1290

Patterson S, Roberts MS, English NR, Macatonia SE, Gompels MN, Pinching AJ, Knight SC (1994) Detection of HIV DNA in peripheral blood dendritic cells of HIV-infected individuals. Res Virol 145, 171–6

Patterson S, Robinson SP, English NR, Knight SC (1999) Subpopulations of peripheral blood dendritic cells show differential susceptibility to infection with a lymphotropic strain of HIV-1. Immunol Lett 66, 111–6

Patterson S, Rae A, Hockey N, Gilmour J, Gotch F (2001) Plasmacytoid dendritic cells are highly susceptible to human immunodeficiency virus type 1 infection and release infectious virus. J Virol 75, 6710–3

Petit C, Buseyne F, Boccaccio C, Abastado JP, Heard JM, Schwartz O (2001) Nef is required for efficient HIV-1 replication in cocultures of dendritic cells and lymphocytes. Virology 286, 225–36

Piguet V, Chen YL, Mangasarian A, Foti M, Carpentier JL, Trono D (1998) Mechanism of Nef-induced CD4 endocytosis: Nef connects CD4 with the mu chain of adaptor complexes. EMBO J 17, 2472–81

Pohlmann S, Soilleux EJ, Baribaud F, Leslie GJ, Morris LS, Trowsdale J, Lee B, Coleman N, Doms RW (2001a) DC-SIGNR, a DC-SIGN homologue expressed in endothelial cells, binds to human and simian immunodeficiency viruses and activates infection in trans. Proc Natl Acad Sci USA 98, 2670–2675

Pohlmann S, Baribaud F, Lee B, Leslie GJ, Sanchez MD, Hiebenthal-Millow K, Munch J, Kirchhoff F, Doms RW (2001b) DC-SIGN interactions with human immunodeficiency virus type 1 and 2 and simian immunodeficiency virus. J Virol 75, 4664–72

Pohlmann S, Leslie GJ, Edwards TG, Macfarlan T, Reeves JD, Hiebenthal-Millow K, Kirchhoff F, Baribaud F, Doms RW (2001c) DC-SIGN interactions with human immunodeficiency virus: virus binding and transfer are dissociable functions. J Virol 75, 10523–6

Pope M, Betjes MG, Romani N, Hirmand H, Cameron PU, Hoffman L, Gezelter S, Schuler G, Steinman RM (1994) Conjugates of dendritic cells and memory T lymphocytes from skin facilitate productive infection with HIV-1. Cell 78, 389–98

Reimann KA, Tenner-Racz K, Racz P, Montefiori DC, Yasutomi Y, Lin W, Ransil BJ, Letvin NL (1994) Immunopathogenic events in acute infection of rhesus monkeys with simian immunodeficiency virus of macaques. J Virol 68, 2362–70

Sallusto F, Cella M, Danieli C, Lanzavecchia A (1995) Dendritic cells use macropinocytosis and the mannose receptor to concentrate macromolecules in the major histocompatibility complex class II compartment: downregulation by cytokines and bacterial products. J Exp Med 182, 389–400

Soilleux EJ, Barten R, Trowsdale J (2000) Cutting Edge: DC-SIGN; a Related Gene, DC-SIGNR; and CD23 Form a Cluster on 19p13. J Immunol 165, 2937–2942

Sol-Foulon N, Moris A, Nobile C, Boccaccio C, Engering A, Abastado P, Heard J-M, van Kooyk Y, Schwartz O (2002) HIV-1 nef induced up-regulation of DC-SIGN in dendritic cells promotes lymphocyte clustering and viral spread. Immunity 16, 145–155

Spira AI, Marx PA, Patterson BK, Mahoney J, Koup RA, Wolinsky SM, Ho DD (1996) Cellular targets of infection and route of viral dissemination after an intravaginal inoculation of simian immunodeficiency virus into rhesus macaques. J Exp Med 183, 215–25

Steffan AM, Lafon ME, Gendrault JL, Schweitzer C, Royer C, Jaeck D, Arnaud JP, Schmitt MP, Aubertin AM, Kirn A (1992) Primary cultures of endothelial cells from the human liver sinusoid are permissive for human immunodeficiency virus type 1. Proc Natl Acad Sci USA 89, 1582–6

Turville SG, Arthos J, Mac Donald K, Lynch G, Naif H, Clark G, Hart D, Cunningham AL (2001) HIV gp120 receptors on human dendritic cells. Blood 98, 2482–8

Weis WI, Taylor ME, Drickamer K (1998) The C-type lectin superfamily in the immune system. Immunol Rev 163, 19–34

Weissman D, Li Y, Ananworanich J, Zhou LJ, Adelsberger J, Tedder TF, Baseler M, Fauci AS (1995) Three populations of cells with dendritic morphology exist in peripheral blood, only one of which is infectable with human immunodeficiency virus type 1. Proc Natl Acad Sci USA 92, 826–30

Weissman D, Fauci AS (1997) Role of dendritic cells in immunopathogenesis of human immunodeficiency virus infection. Clin Microbiol Rev 10, 358–67

Chapter 3

Contrasting Roles of Dendritic Cells and B Cells in the Immune Control of Epstein-Barr Virus

K. Bickham, C. Münz

Abstract. The human γ-herpesvirus, Epstein-Barr virus (EBV), has growth-transforming potential in vivo and in vitro. Despite this, most healthy carriers remain free of EBV-associated malignancies because of effective T cell-mediated immune control of the virus. A better understanding of these highly efficient control mechanisms is important in the

K. Bickham, C. Münz (✉)
Laboratory of Cellular Physiology and Immunology, The Rockefeller University, New York, NY 10021, USA

development of new treatment strategies for EBV-associated malignancies.

A rational approach to EBV immunotherapy requires answering two questions about the initiation of the protective EBV-specific immune response. The first question is, what is the antigen-presenting cell responsible for priming EBV specific immunity? Second, which viral antigen is central to protective EBV adaptive immunity seen in healthy carriers of the virus?

We provide evidence in this review that dendritic cells rather than EBV-transformed B cells are responsible for orchestrating protective EBV immunity and that the EBV nuclear antigen 1 (EBNA1)-specific $CD4^+$ T cell response probably plays a role in resistance against all types of EBV-associated malignancies in healthy carriers. This implies that EBNA1 targeting to dendritic cells should be a component of vaccine and immunotherapy development against EBV-associated malignancies.

1
Introduction

Herpesviruses are some of the most successful pathogens colonizing the human host. For example, Epstein-Barr virus (EBV), a γ-herpesvirus, resides in more than 95% of the adult population. During the long coevolution of the virus family with humans, host and pathogen have developed a largely peaceful coexistence. Most infected individuals carry herpesviruses as latent infections without pathological manifestations. In contrast to lytic infection, herpesviruses do not produce infectious virions during latency but can persist through survival of the host cell. Latency is further characterized by a downregulation of expressed viral antigens in total number and copies per cell, further evading elimination by the immune system (Roizman 2001).

The various subfamilies of herpesviruses achieve latent infection differently. The α-herpesvirus herpes simplex hides during latency in peripheral neurons (Roizman and Kaipe 2001). This is a clever strategy, because these cells are long-lived, do not divide, and express low levels of MHC molecules, the scaffolds for antigen presentation to the adaptive immune system. Herpes simplex, therefore, does not need to express any latent antigens to persist.

EBV, in contrast, infects primarily B cells (Kieff and Rickinson 2001). Naïve and effector B cells are short-lived but can persist longer if they become memory cells. In addition, B cells express large numbers of MHC molecules on their surfaces, which allow efficient immune system surveillance and detection of pathogen-derived intracellular proteins. To overcome these obstacles, EBV has evolved a specialized set of latent antigens to transform infected B cells into memory cells, to actively sustain survival, and to evade immune detection. Once the EBV-infected B cell has reached the memory pool, EBV persists, not unlike herpes simplex, in a long-lived host cell with very low to undetectable expression of viral antigens (Babcock et al. 1998).

EBV ensures the transition of infected cells into the memory pool by constitutive activation and provision of survival signals for the B cell. These features, however, can lead to a massive B cell proliferation and malignant transformation both in vitro and in immunosuppressed hosts in vivo. The nonpathological coexistence of EBV and its human host is therefore achieved by continuous immune control of EBV-induced B cell lymphomas (Rickinson and Kieff 2001). On the following pages, we outline the immune response toward the Epstein-Barr nuclear antigen 1 (EBNA1) and its central role in the maintenance of this immune control. In addition, we discuss the evidence that EBNA1-specific T cells are not primed by the EBV-infected antigen-presenting cell (APC), the B cell, but rather by dendritic cells (DCs).

2 B Cells and Dendritic Cells Have Different Roles in the Induction of T Cells Specific for EBNA1

2.1 Differences in Antigen Presentation by B Cells and Dendritic Cells

Primary EBV infection generally occurs asymptomatically in early childhood. However, if infection is delayed until adolescence, it can be accompanied by fevers, pharyngeal inflammation, hepatosplenomegaly, and an abundance of atypical lymphocytes in the peripheral blood, from which the clinical syndrome, infectious mononucleosis (IM) derives its name (Rickinson and Kieff 2001). Most of the information known about the primary immune response to EBV is a consequence of studies in patients with IM. The atypical lymphocytes found in these patients are activated T cells, a majority of which are cytotoxic, EBV-specific $CD8^+$ lymphocytes (Sheldon et al. 1973). The exuberance of the cellular immune response to EBV is thought to contribute to the clinical presentation of IM. This is demonstrated by the failure of treatment with acyclovir to alleviate the severity of disease symptoms or diminish the number of circulating latently infected B cells, despite abrogating viral replication in the oropharynx (Lin et al. 1984).

The antigen-presenting cell responsible for the initiation of this strong cellular response is unknown. The B cell, preferentially infected by EBV, is a leading candidate as the architect of the primary immune reaction. B cells perform many functions in the initiation and maintenance of the adaptive response. They are capable of receptor-mediated uptake and presentation of specific antigen, such as bacterial toxins, for priming of naïve T cells (Fuhrman and Cebra 1981).

Once naïve $CD4^+$ T cells encounter APCs presenting antigen in the T cell-rich areas of secondary lymphoid organs (Jenkins et al. 2001), they then migrate to the borders between B and T cell areas for interaction with B cells (Jenkins et al. 2001; Linton et al. 2000). This may contribute to increased survival of the newly activated $CD4^+$ T cells after priming.

B cells also appear to have a unique role in the priming of naïve T cells during parasitic infections. Resolution of infections with *Trichuris muris*, a murine intestinal parasite, requires a predominant Th2 response (Bancroft et al. 1997; Else et al. 1994). B cell-deficient mice are susceptible to disease because they lack parasitic antigen-specific secretion of IL-4, IL-5, and IL-9; however, immunity can be restored by B cell infusion (Blackwell and Else 2001). Because earlier studies demonstrate that adoptive transfer of antigen-specific $CD4^+$ T cells into SCID mice was sufficient to eradicate the parasites (Else and Grencis 1996), it appears that the role of the B cell in this system is to prime the effector arm of the immune response. However, B cell-deficient mice are capable of developing a polarized Th2 response in other settings (Brown and Reiner 1999), thus arguing against the B cell as the sole determinant of the Th2-phenotype.

Apart from naïve T cell priming, B cells appear to be required for alteration of the phenotype of an established Th1 immune response to a Th2 response. In a murine malaria model, B cell-deficient mice are unable to eradicate blood stage parasites, because this requires an important switch to a Th2-driven immune response. The recurring parasitemia experienced by these mice can be resolved by infusion of B cells (Langhorne et al. 1998) and a subsequent switch of the immune phenotype to Th2. This function of B cells may be important in the loss of protective Th1 immunity to EBV and the development of EBV-associated malignancies as discussed in Sect. 3.

Dendritic cells are professional antigen-presenting cells that efficiently stimulate both memory and quiescent naïve T cells (reviewed in Banchereau and Steinman 1998). DCs exist in two developmental stages. In their immature form, they can capture large amounts of exogenous antigen (Koch et al. 1995; Romani et al. 1989). Mature DCs are potent stimulators of T cell activity as they express high levels of MHC-peptide complexes on their cell surface (Inaba et al. 1997) in conjunction with an abundance of accessory molecules that facilitate the adhesion and costimulation of T cells (Caux et al. 1994; Inaba et al. 1994; Larsen et al. 1994). DCs secrete large amounts of IL-12 (Koch et al. 1996), skewing the phenotype of the immune response to Th1 (Maldonado-Lopez et al. 1999), whereas B cells typically do not produce IL-12. Although B cells can efficiently present antigens cognate to their B cell receptor, only DCs have been found to effectively cross-present dying cells (Albert et al. 1998), including antigens from LCLs (Ferlazzo et al. 2000; Münz et al. 2000; Subklewe et al. 2001; Herr et al. 2000). Therefore, the two APC subsets, DCs and B cells, are specialized to address different needs of the adaptive immune response. These peculiarities must be exploited to further understand the APC responsible for initiating EBV immunity.

2.2 EBNA1-Specific $CD4^+$ Th1 Immunity Implicates Dendritic Cells in Its Generation

$CD4^+$ T cells likely play a vital role in the maintenance of the healthy carrier state in latent EBV infection, indicated by the increase of EBV-associated malignancies in AIDS patients (Levine 1992), after $CD4^+$ T cell depletion by HIV. One EBV latent antigen, EBNA1, is consistently recognized by $CD4^+$ T cells in healthy carriers (Münz et al. 2000). EBNA1 expression is crucial to EBV in proliferating cells. Because the EBV episome does not integrate into the host DNA, it must replicate independently before mitosis and copies of it must be distributed to the daughter cells during cell division. These functions are mediated by EBNA1, which initiates EBV genome replication and tethers the episomes to mitotic chromosomes during separation (Kieff 2001). Because of these crucial functions, EBNA1 is expressed in all EBV-associated malignancies (Khanna and Burrows 2000). Therefore, the EBNA1-specific $CD4^+$ T cell response provides an opportunity for immune prevention of EBV-associated malignancies and immune control of latent EBV infection.

There are a variety of known $CD4^+$ T cell functions whereby EBNA1-specific T cells could achieve these tasks. The phenotype of the activated $CD4^+$ T cell influences the outcome of the immune response (reviewed in O'Garra and Murphy 1994). Th1 $CD4^+$ T cells secrete IFNγ and help in the development of cellular immunity. Th2 $CD4^+$ T cells secrete IL-4 and IL-5, thereby stimulating eosinophils and antibody production. Th1 cells differ from Th2 cells in other ways such as expression of chemokine receptors (Sallusto et al. 1998) and a capacity to kill targets via Fas–FasL interactions (reviewed in Hahn et al. 1995). Th1 immunity has been shown to be protective in viral infection in mice (Maloy et al. 2000) as well as in humans (Rentenaar et al. 2000). Th1 cells appear to be more effective than Th2 cells in the maintenance of a memory response on rechallenge (Maloy et al. 2000; Nishimura et al. 1999; Rentenaar et al. 2000). Furthermore, Th1 cells express the chemokine receptors CCR2 and CXCR3, which allows them to migrate in response to inflammatory chemokines expressed in infected tissues (Maloy et al. 2000). In the absence of acute infection and inflammation, $CD4^+$ T cells are vital to $CD8^+$ T cell function in two ways. Activated $CD4^+$ T cells deliver survival and maturation stimuli to dendritic cells, which in turn induce and expand antigen-specific $CD8^+$ T lymphocytes (Ridge et al. 1998; Bennett et al. 1998; Schoenberger et al. 1998). Furthermore, they are required for the maintenance of $CD8^+$ T cell memory (Cardin et al. 1996; Zajac et al. 1998).

Studies in healthy EBV-seropositive adults demonstrate that EBNA1-specific $CD4^+$ T cells are predominantly of the Th1 phenotype and are consistently found across a number of HLA haplotypes (Bickham et al. 2001; Münz et al. 2000). Though both subdominant IL-4 and dominant IFNγ secreting EBNA1-specific T cells can be expanded from healthy donors, only the Th1-polarized cells are capable of lysing EBNA1 expressing cells (Bickham et al. 2001), thus suggesting that the Th1 response is protective in EBV latency. Indeed, EBV-specific $CD4^+$ T cells are now known to prevent the outgrowth of EBV-transformed B cells in freshly EBV-infected in vitro cultures from peripheral blood (Nikiforow et al. 2001).

Because B cells and B cell lines typically do not actively produce IL-12 or bias the $CD4^+$ T cell response toward Th1, DCs, which produce high levels of IL-12 (Cella et al. 1996; Koch et al. 1996), are likely candidates for the orchestration of the polarized Th1 response. It is now known that human DCs efficiently cross-present EBNA1 from dying EBV-infected B cells on MHC class II (Münz et al. 2000), thereby providing a mechanism whereby DCs, rather than infected B cells, can directly induce the Th1 response to EBNA1 in healthy carriers. Thus the phenotype of the immune response to EBV latency antigens, such as EBNA1, implicates a role for DCs in the initial priming of naïve T cells.

2.3 CD8$^+$ T Cells Against EBNA1 Indicate Cross-Priming In Vivo

Perhaps the strongest in vivo evidence for the involvement of an antigen-presenting cell nonpermissive for EBV infection in the induction of EBV immunity is the existence of EBNA1-specific CD8$^+$ T cells. EBNA1 prevents its endogenous MHC class I processing and presentation. Therefore, EBNA1 expression in EBV-transformed B cells is invisible to CD8$^+$ T cells (Blake et al. 1997). EBNA1 achieves this task by blocking proteasomal processing with its Gly/Ala domain (Levitskaya et al. 1995, 1997). Nevertheless, up to 5% of peripheral blood CD8$^+$ T cells in infectious mononucleosis patients are EBNA1-specific effector cells (Blake et al. 2000). The inability of these CTL clones to recognize autologous LCL indicates that EBV-transformed B cells are unable to present EBNA1 epitopes on MHC class I after endogenous (their own EBNA1) and exogenous (EBNA1 of dying LCL neighbors) processing (Blake et al. 1997, 2000). In contrast, they are capable of lysing DCs that have been loaded with EBNA1 protein (Blake et al. 2000). These data and the strong bias toward Th1 in the EBNA1-specific CD4$^+$ T cell response suggest a prominent role for DCs in the initiation of EBNA1-specific T cell immunity.

3 The Potential Role of EBNA1-Specific CD4$^+$ T Cell Responses in EBV-Associated Malignancies

3.1 Pattern of Latent EBV Antigen Expression In Vivo and Site of EBV Immunity Initiation

EBV immune control, however, fails in a variety of spontaneously developing EBV-associated tumors. To understand the etiology of EBV-associated malignancies, we must review the different expression patterns of the eight EBV latent antigens in healthy EBV carriers. EBV encodes six nuclear latent antigens (EBNA1, -2, -3A, -3B, -3C, and -LP) and two latent membrane proteins (LMP1 and -2). In persistent infection under immune control, three forms of EBV latency can be found (Thorley-Lawson 2001). In tonsils, IgD^+ naïve B cells express all EBV latent antigens (latency III), whereas $CD10^+$ germinal center B cells are positive only for EBNA1, LMP1, and LMP2 (latency II) (Babcock et al. 2000; Babcock and Thorley-Lawson 2000). IgD^-, $CD10^-$, $CD27^+$ memory B cells in peripheral blood lack any latent EBV antigen expression (Babcock et al. 1998; Joseph et al. 2000). In the absence of established EBV immune control in IM patients, only EBV latency III can be detected in tonsils (Kurth et al. 2000).

The only latent EBV antigen expression pattern that has not been found yet in healthy EBV carriers is latency I. It is characterized by the exclusive expression of EBNA1 and found, for example, in Burkitt lymphoma (BL) (Rickinson and Kieff 2001). The hypermutated variable regions of their immunoglobulin genes indicate that BL cells have passed through germinal centers (Harris et al. 2001; Küppers et al. 1999). The increased BL incidence rate during immune deficiency caused by the HIV virus argues further in favor of EBV latency I occurrence during normal EBV persistence (Levine 1992; Schulz et al. 1996; Shibata et al. 1993). Failing immune control in these patients might lead to the outgrowth of BL.

The immune system responds to EBV infection and limits EBV latent antigen expression, probably already in the tonsils. The tonsils are the primary site of EBV infection, and reactivation of lytic EBV replication in the tonsils leads to transmission of the virus via the saliva (Fig. 1; Kieff 2001). This suggests that the tonsils probably generate EBV immunity. It has indeed been shown that during asymptomatic seroconversion, the T cell pool in peripheral blood remains unperturbed (Biggar et al. 1978; Silins et al. 2001), and establishment of EBV immune control is most likely handled entirely in the tonsils.

With this description of EBV latency patterns and the putative site of EBV immunity generation in healthy EBV carriers, we have set the stage to discuss how EBV specific $CD4^+$ T cell responses, most consistently recognizing EBNA1, might fail in patients with EBV-associated malignancies.

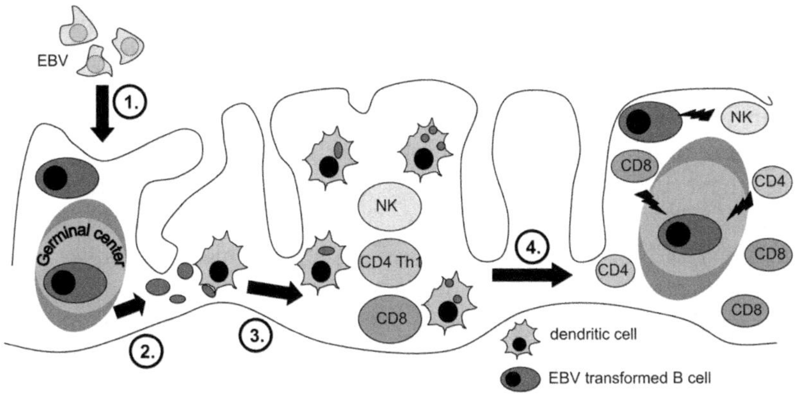

Fig. 1. EBV immunity is probably initiated in the tonsils. EBV infects B cells at this site after mucosal transmission (*1*). EBV lytic infection activates DCs and produces LCL fragments (*2*). Activated DCs loaded with LCL fragments activate NK cells and prime $CD4^+$ Th1 cells as well as $CD8^+$ CTL (*3*). These cells then lyse most EBV-transformed B cells and limit the EBV latent antigen expression to the low levels and characteristic expression patterns found in tonsils of healthy EBV carriers (*4*)

3.2
Hodgkin Disease Conditions Its Environment with Th2 Cytokines

Hodgkin lymphoma is one of the most common malignancies in the Western world with a bimodal age of distribution. An association with EBV was first entertained because the early onset form of the disease is seen two to four times more frequently in individuals with a history of IM in the 3 years before tumor development, in addition to the similarity in demographics between Hodgkin and IM patients (Rickinson and Kieff 2001). Overall, 40%–60% of cases of Hodgkin lymphoma are EBV positive (Jarrett et al. 1991) with 90% of the mixed cellularity and lymphocyte-depleted subtypes being EBV associated (Pallesen et al. 1993).

EBV-associated Hodgkin lymphoma Reed-Sternberg cells, like nasopharyngeal carcinoma cells (Deacon et al. 1993), demonstrate a latency II pattern of EBV protein expression. This is exemplified by the production of EBNA1, LMP1, and LMP2 concurrent with a downregulation of the more immunogenic EBV proteins, such as those of the EBNA3 family (Fahraeus et al. 1988; Young et al. 1988).

Histologically, Hodgkin lymphoma is characterized by the presence of a small number (approximately 1% of total tumor cells) of large, multinucleated cells termed Reed-Sternberg cells surrounded by a large population of T cells, eosinophils, B cells, plasma cells and histiocytes (Küppers and Rajewsky 1998). Reed-Sternberg cells produce the cytokine IL-13 in vivo (Kapp et al. 1999), which promotes immunoglobulin class switching to IgE and IgG4 (Zurawski and de Vries 1994) and may play a role in the induction of a polarized Th2 phenotype in newly primed $CD4^+$ T cells (Murphy et al. 2000). This helps explain the plethora of Th2-polarized $CD4^+$ T cells and eosinophils in biopsy samples from these patients. Patients whose tumor specimens reveal an abundance of eosinophils (von Wasielewski et al. 2000) or elevated tumor IL-10 levels (Herbst et al. 1996) have a poor prognosis, highlighting the clinical relevance of the phenotypic switch of the immune response in the microenvironment of the tumor.

Systemic effects of cytokine dysregulation are apparent in these patients as well. Patients with Hodgkin lymphoma often have a constellation of symptoms, which include weight loss, night sweats, and fever. These "B" symptoms are thought to be associated with increased levels of cytokines in the peripheral blood. Elevations of Th2 cytokines in the serum, such as IL-6 (Kurzrock 1997) and IL-10 (Sarris et al. 1999; Bohlen et al. 2000; Viviani et al. 2000; Kurzrock 1997), are known to be poor prognostic indicators in patients. Although this Th2 cytokine shift is present in both EBV-positive and -negative cases of Hodgkin lymphoma, it suggests that the defect in EBV immunity that allows for the outgrowth of tumor cells in EBV-associated Hodgkin lymphoma could be either the loss of a protective, cytotoxic Th1 response or the acquisition of a permissive Th2-polarized response to EBV latency antigens. Consequently, the presence of a Th2 cytokine milieu both systemically and in the tumor's microenvironment may well facilitate the escape of Hodgkin tumor cells from immune surveillance.

3.3
Nasopharyngeal Carcinoma Patients Carry Increased Levels of EBV-Specific Antibodies

Nasopharyngeal carcinoma (NPC) has an unusual predilection for certain ethnic groups, particularly southern Chinese, Eskimo, and northeast African populations. An EBV association is consistently demonstrated in cases of poorly differentiated (WHO classification II) and undifferentiated (WHO classification III) subtypes of NPC (Rickinson and Kieff 2001).

A link between NPC and EBV was first postulated because patients were noted to have unusually high EBV IgG antibody titers compared with control subjects (Henle et al. 1970; de Schryver et al. 1969). On further investigation, it was shown that the presence of IgA antibodies to viral capsid antigen (VCA) was so peculiar to patients with NPC (Henle and Henle 1976) that they could be used as a screening tool for the diagnosis of preclinical malignancy. Elevations in VCA IgA antibody titers can be seen as early as 8–30 months before the development of histological changes consistent with NPC and are therefore extremely useful in the detection of early-stage disease (Zeng 1985). Furthermore, levels of antibody titers are correlated with tumor burden and can be used to monitor for recurrence of disease (Sam et al. 1989). Because antibody production is a prominent feature in a Th2-polarized immune response, the appearance of these antibodies in NPC patients could mark a shift in the normal protective Th1 phenotype to a permissive Th2 response allowing for the outgrowth of tumor cells.

Moreover, LCL regression assays performed with PBMCs from patients with NPC demonstrate a defect in controlling the outgrowth of EBV-transformed B cells. In a study comparing 54 NPC patients with 38 healthy control subjects, only 6% of cultures from NPC patients showed complete regression compared with 34% of control subjects (Chan and Chew 1981). Further studies revealed that in NPC patients, separated into newly diagnosed versus long-term survivors, the PBMCs from newly diagnosed patients were significantly impaired in their ability to control outgrowth of LCL. This lack of regression was noted most consistently in newly diagnosed patients with IgA antibodies (Moss et al. 1983). Interestingly, the T cell deficiency that fails to cause LCL regression in the newly diagnosed patients appears to be rectified with the resolution of disease, as is seen in long-term survivors, thus arguing for tumor-derived factors responsible for this phenomenon. NPC patients, therefore, demonstrate perturbations in their EBV immune response that reflect a shift to a more Th2-dominant phenotype exemplified by the increase in EBV antibody titers of the IgA type, as well as the loss of cytolytic activity against EBV-infected cells.

3.4 Burkitt Lymphoma Occurs in Holoendemic Malaria Regions

The three forms of BL, endemic, sporadic, and AIDS-related, are associated with EBV to different degrees. Although 100% of endemic BL is EBV^+, only 30% of sporadic and AIDS-related BL carry EBV (Wright 1999). Endemic BL develops mainly in children of holoendemic malaria regions (Burkitt 1962; de-The et al. 1978). In children, the immune responses to both pathogens must develop concurrently. Continuous exposure to *Plasmodium falciparum* and active immune defense against it most likely influence the priming of EBV immunity. The clearance of the erythrocytic stage of the parasite is mediated by a humoral, Th2-based immune response (Good and Doolan 1999; Riley 1999). It is tempting to speculate that the Th2 cytokine milieu created by the systemic malaria-specific immune response polarizes or diverts EBV immunity that is just emerging to Th2. The outcome is a decrease in Th1-type, EBV-specific immune control in the malaria patients (Whittle et al. 1984).

This failure in the establishment of EBV immunity might originate in the inhibition of DCs by *Plasmodium falciparum* (Urban et al. 1999). Erythrocytes are the main host cell during systemic *Plasmodium falciparum* infection. Malaria-infected red blood cells adhere to DCs, prevent their maturation, and subsequently reduce their capacity to stimulate T cells (Urban et al. 1999). Instead of DCs, B cells are required for protection against systemic malaria infection, probably inducing a Th2-like immune response (Langhorne et al. 1998). A lack of EBV antigen presentation by IL-12-secreting DCs, combined with direct presentation by infected B cells, might lead to less efficient priming and a more Th2-polarized EBV-specific immune response. This immunity, less efficient against latently EBV-infected B cells (Whittle et al. 1984), might be incapable of controlling BL and, therefore, responsible for the outgrowth of this tumor in the affected children.

4
EBV Immunity Upstream and Downstream of EBNA1-Specific CD4$^+$ T Cell Responses

4.1
Natural Killer Cells

The knowledge about immunity diversion in EBV-associated malignancies places CD4$^+$ T cell responses, especially against EBNA1, at the checkpoint that decides whether immune control against EBV is successfully established. However, there are other lymphocyte subsets that are probably important in either slowing down EBV infection or augmenting Th1 immunity against EBV.

Natural killer (NK) cells are one arm of the innate immune response that can limit viral infections. Indeed, they constitute the main lymphocyte reactivity against EBV-transformed B cells in cord blood (Moretta et al. 1997). Another hint to their importance in EBV clearance comes from the finding that they cannot lyse LCL in patients with X-linked lymphoproliferative disease (XLP) (Parolini et al. 2000) and that XLP males frequently develop fatal lymphomas after EBV infection (Purtilo et al. 1974).

Interestingly, DCs are capable of activating this innate immune response (Ferlazzo et al. 2002; Fernandez et al. 1999). Furthermore, the IL-12-dependent pathway of NK activation seems to be crucial for the clearance of herpesvirus infections in mice, such as murine cytomegalovirus infection (Orange and Biron 1996; Orange et al. 1995). Therefore, DCs, as the main source of IL-12 in the immune system, are prime candidates for eliciting this innate antiviral immunity. The activated NK cells could in turn limit growth of EBV-transformed B cells by cytolysis, resulting in LCL fragments for DC-mediated cross-priming of EBV-specific T cells, including EBNA1-specific CD4$^+$ Th1 cells. DC-activated NK responses could be an important upstream event to EBV-specific Th1 immunity in the development of EBV immune control.

4.2
T Cell Responses Against Other EBV Latent Antigens

CD4$^+$ and CD8$^+$ T cells complement each other in the detection of EBV latent antigens and therefore constitute a comprehensive immune surveillance against EBV-associated malignancies. $CD8^+$ T cells primarily recognize EBNA3A, -3B, and -3C and in some MHC class I haplotype LMP2 (Murray et al. 1992; Steven et al. 1996). In contrast, $CD4^+$ T cell recognition shows a strong preference for EBNA1, EBNA3B, EBNA3C, and LMP1 (Leen et al. 2001; Münz et al. 2000; Rajnavölgyi et al. 2000; Steigerwald-Mullen et al. 2000).

All these EBV-specific T cell populations can be stimulated with dendritic cells. We have demonstrated that EBNA3B-, EBNA3C-, and LMP1-specific $CD4^+$ T cells can be expanded with DCs presenting EBV latent antigens in vitro (Münz et al. 2000). EBV-specific $CD8^+$ T cell responses can also be recalled with DCs, either after infection with recombinant vaccinia viruses expressing EBV latent antigens or after loading with EBV T cell epitopes (Redchenko and Rickinson 1999; Subklewe et al. 1999a,b). In addition, DCs can expand EBV-specific $CD8^+$ T cells after cross-presentation of dying LCL (Ferlazzo et al. 2000; Herr et al. 2000; Subklewe et al. 2001), specifically demonstrated for the EBNA3A and LMP2 antigens (Subklewe et al. 2001).

Therefore, DCs might be involved in the priming of all EBV-specific T cell responses, although the best evidence exists for DC priming of EBNA1-specific $CD4^+$ and $CD8^+$ T cell responses.

5
Summary

EBV mainly infects B cells. Despite the fact that these cells in their naïve state are short-lived and on activation can potentially become effective antigen-presenting cells, EBV manages to establish a lifelong persistent infection in most adult humans. This success is probably due to two viral strategies. First, EBV shows a remarkable flexibility in switching its antigens on and off. It not only shifts to a completely different antigen subset in latent infection but is also capable of successively downregulating these latent antigens to fit the activation requirements of the host B cell. This leads from eight expressed antigens in naïve B cells, to three in germinal center B cells and finally none in memory B cells, the safe haven of persistence. Second, EBV-infected B cells are probably biasing an EBV-specific immune response toward Th2. In contrast, virus-specific immunity has been shown to be most protective when based on Th1 responses, and all healthy EBV carriers harbor EBNA1-specific Th1 $CD4^+$ T cells. Dendritic cells are capable of inducing such Th1 immunity. They can cross-present EBNA1 on MHC class II for CD4$^+$ T cell recognition and probably cross-prime EBNA1-specific CD8$^+$ T cells in vivo. In addition, they are also good candidates for the activation of NK cells and priming of EBV-specific T cells against EBV latent antigens other than EBNA1. These events might contribute to EBV immune control upstream and downstream of EBNA1-specific CD4$^+$ T cell responses, respectively.

The evidence provided in this review emphasizes the need for DC-mediated cross-priming against antigens that are not expressed in DCs. In the case of EBV this even extends to a B lymphotrophic virus, thus suggesting that the important role of DCs in the initiation of cell-mediated adaptive immunity is not redundant and cannot be easily substituted by another APC, such as the B cell.

References

Albert M. L, Pearce S. F. A, Francisco L. M, Sauter B, Roy P, Silverstein R. L, and Bhardwaj N. (1998). Immature dendritic cells phagocytose apoptotic cells via $\alpha_v\beta_5$ and CD36 and cross-present antigens to cytotoxic T lymphocytes. J Exp Med 188 1359–1368

Arulanandam B. P, Mittler J. N, Lee W. T, O'Toole M, and Metzger D. W. (2000). Neonatal administration of IL-12 enhances the protective efficacy of antiviral vaccines. J Immunol 164 3698–704

Babcock G. J, Decker L. L, Volk M, and Thorley-Lawson D. A. (1998). EBV persistence in memory B cells in vivo. Immunity 9 395–404

Babcock G. J, Hochberg D, and Thorley-Lawson A. D. (2000). The expression pattern of Epstein-Barr virus latent genes in vivo is dependent upon the differentiation stage of the infected B cell. Immunity 13 497–506

Babcock G. J, and Thorley-Lawson D. A. (2000). Tonsillar memory B cells latently infected with Epstein-Barr virus express the restricted pattern of latent genes previously found only in Epstein-Barr virus-associated tumors. Proc Natl Acad Sci USA 97 12250–5

Banchereau J, and Steinman R. M. (1998). Dendritic cells and the control of immunity. Nature 392 245–52

Bancroft A. J, Else K. J, Sypek J. P, and Grencis R. K. (1997). Interleukin-12 promotes a chronic intestinal nematode infection. Eur J Immunol 27 866–70

Bennett S. R. M, Carbone F. R, Karamalis F, Flavell R. A, Miller J. F. A. P, and Heath W. R. (1998). Help for cytotoxic-T-cell responses is mediated by CD40 signalling. Nature 393 478–480

Bickham K, Münz C, Tsang M. L, Larsson M, Fonteneau J. F, Bhardwaj N, and Steinman R. (2001). EBNA1-specific CD4$^+$ T cells in healthy carriers of Epstein-Barr virus are primarily Th1 in function. J Clin Invest 107 121–30

Biggar R. J, Henle G, Bocker J, Lennette E. T, Fleisher G, and Henle W. (1978). Primary Epstein-Barr virus infections in African infants. II. Clinical and serological observations during seroconversion. Int J Cancer 22 244–50

Blackwell N. M, and Else K. J. (2001). B cells and antibodies are required for resistance to the parasitic gastrointestinal nematode *Trichuris muris*. Infect Immun 69 3860–8

Blake N, Haigh T, Shaka'a G, Croom-Carter D, and Rickinson A. (2000). The importance of exogenous antigen in priming the human $CD8^+$ T cell response: lessons from the EBV nuclear antigen EBNA1. J Immunol 165 7078–87

Blake N, Lee S, Redchenko I, Thomas W, Steven N, Leese A, Steigerwald-Mullen P, Kurilla M. G, Frappier L, and Rickinson A. (1997). Human $CD8^+$ T cell responses to EBV EBNA1: HLA class I presentation of the (Gly-Ala)-containing protein requires exogenous processing. Immunity 7 791–802

Bohlen H, Kessler M, Sextro M, Diehl V, and Tesch H. (2000). Poor clinical outcome of patients with Hodgkin's disease and elevated interleukin-10 serum levels. Clinical significance of interleukin-10 serum levels for Hodgkin's disease. Ann Hematol 79 110–3

Brown D. R, and Reiner S. L. (1999). Polarized helper-T-cell responses against *Leishmania major* in the absence of B cells. Infect Immun 67 266–70

Burkitt D. (1962). A children's cancer dependent on climatic factors. Nature 194 232–234

Cardin R. D, Brooks J. W, Sarawar S. R, and Doherty P. C. (1996). Progressive loss of $CD8^+$ T cell-mediated control of gamma-herpesvirus in the absence of $CD4^+$ T cells. J Exp Med 184 863–871

Caux C, Vanbervliet B, Massacrier C, Azuma M, Okumura K, Lanier L. L, and Banchereau J. (1994). B70/B7-2 is identical to CD86 and is the major functional ligand for CD28 expressed on human dendritic cells. J Exp Med 180 1841–1847

Cella M, Scheidegger D, Palmer-Lehmann K, Lane P, Lanzavecchia A, and Alber G. (1996). Ligation of CD40 on dendritic cells triggers production of high levels of interleukin-12 and enhances T cell stimulatory capacity: T-T help via APC activation. J Exp Med 184 747–752

Chan S. H, and Chew T. S. (1981). Lack of regression in Epstein-Barr virus infected leucocyte cultures of nasopharyngeal carcinoma patients. Lancet 2 1353

Deacon E. M, Pallesen G, Niedobitek G, Crocker J, Brooks L, Rickinson A. B, and Young L. S. (1993). Epstein-Barr virus and Hodgkin's disease: transcriptional analysis of virus latency in the malignant cells. J Exp Med 177 339–49

de Schryver A, Friberg S, Jr, Klein G, Henle W, Henle G, De-The G, Clifford P, and Ho H. C. (1969). Epstein-Barr virus-associated antibody patterns in carcinoma of the post-nasal space. Clin Exp Immunol 5 443–59

de-The G, Geser A, Day N. E, Tukei P. M, Williams E. H, Beri D. P, Smith P. G, Dean A. G, Bronkamm G. W, Feorino P, and Henle W. (1978). Epidemiological evidence for causal relationship between Epstein-Barr virus and Burkitt's lymphoma from Ugandan prospective study. Nature 274 756–61

Else K. J, Finkelman F. D, Maliszewski C. R, and Grencis R. K. (1994). Cytokine-mediated regulation of chronic intestinal helminth infection. J Exp Med 179 347–51

Else K. J, and Grencis R. K. (1996). Antibody-independent effector mechanisms in resistance to the intestinal nematode parasite *Trichuris muris*. Infect Immun 64 2950–4

Fahraeus R, Fu H. L, Ernberg I, Finke J, Rowe M, Klein G, Falk K, Nilsson E, Yadav M, Busson P, et al. (1988). Expression of Epstein-Barr virus-encoded proteins in nasopharyngeal carcinoma. Int J Cancer 42 329–38

Ferlazzo G, Semino C, Spaggiari G. M, Meta M, Mingari M. C, and Melioli G. (2000). Dendritic cells efficiently cross-prime HLA class I-restricted cytolytic T lymphocytes when pulsed with both apoptotic and necrotic cells but not with soluble cell-derived lysates. Int Immunol 12 1741–7

Ferlazzo G, Tsang M. L, Moretta A, Melioli G, Steinman R. M, and Münz C. (2002) Human dendritic cells activate resting NK cells and are recognized via the NKp30 receptor by activated NK cells. J Exp Med 195 343–351

Fernandez N. C, Lozier A, Flament C, Ricciardi-Castagnoli P, Bellet D, Suter M, Perricaudet M, Tursz T, Maraskovsky E, and Zitvogel L. (1999). Dendritic cells directly trigger NK cell functions: cross-talk relevant in innate anti-tumor immune responses in vivo. Nat Med 5 405–11

Fuhrman J. A, and Cebra J. J. (1981). Special features of the priming process for a secretory IgA response. B cell priming with cholera toxin. J Exp Med 153 534–44

Good M. F, and Doolan D. L. (1999). Immune effector mechanisms in malaria. Curr Opin Immunol 11 412–9

Hahn S, Gehri R, and Erb P. (1995). Mechanism and biological significance of CD4-mediated cytotoxicity. Immunol Rev 146 57–79

Harris R. S, Croom-Carter D. S, Rickinson A. B, and Neuberger M. S. (2001). Epstein-Barr virus and the somatic hypermutation of immunoglobulin genes in Burkitt's lymphoma cells. J Virol 75 10488–92

Henle G, and Henle W. (1976). Epstein-Barr virus-specific IgA serum antibodies as an outstanding feature of nasopharyngeal carcinoma. Int J Cancer 17 1–7

Henle W, Henle G, Ho H. C, Burtin P, Cachin Y, Clifford P, de Schryver A, de-The G, Diehl V, and Klein G. (1970). Antibodies to Epstein-Barr virus in nasopharyngeal carcinoma other head and neck neoplasms and control groups. J Natl Cancer Inst 44 225–31

Herbst H, Foss H. D, Samol J, Araujo I, Klotzbach H, Krause H, Agathanggelou A, Niedobitek G, and Stein H. (1996). Frequent expression of interleukin-10 by Epstein-Barr virus-harboring tumor cells of Hodgkin's disease. Blood 87 2918–29

Herr W, Ranieri E, Olson W, Zarour H, Gesualdo L, and Storkus W. J. (2000). Mature dendritic cells pulsed with freeze-thaw cell lysates define an effective in vitro vaccine designed to elicit EBV-specific $CD4^+$ and $CD8^+$ T lymphocyte responses. Blood 96 1857–64

Inaba K, Pack M, Inaba M, Sakuta H, Isdell F, and Steinman R. M. (1997). High levels of a major histocompatibility complex II – self peptide complex on dendritic cells from lymph node. J Exp Med 186 665–672

Inaba K, Witmer-Pack M, Inaba M, Hathcock K. S, Sakuta H, Azuma M, Yagita H, Okumura K, Linsley P. S, Ikehara S, et al. (1994). The tissue distribution of the B7-2 costimulator in mice: abundant expression on dendritic cells in situ and during maturation in vitro. J Exp Med 180 1849–1860

Jarrett R. F, Gallagher A, Jones D. B, Alexander F. E, Krajewski A. S, Kelsey A, Adams J, Angus B, Gledhill S, Wright D. H, et al. (1991). Detection of Epstein-Barr virus genomes in Hodgkin's disease: relation to age. J Clin Pathol 44 844–8

Jenkins M. K, Khoruts A, Ingulli E, Mueller D. L, McSorley S. J, Reinhardt R. L, Itano A, and Pape K. A. (2001). In vivo activation of antigen-specific CD4 T cells. Annu Rev Immunol 19 23–45

Joseph A. M, Babcock G. J, and Thorley-Lawson D. A. (2000). Cells expressing the Epstein-Barr virus growth program are present in and restricted to the naive B-cell subset of healthy tonsils. J Virol 74 9964–71

Kapp U, Yeh W. C, Patterson B, Elia A. J, Kagi D, Ho A, Hessel A, Tipsword M, Williams A, Mirtsos C, et al. (1999). Interleukin 13 is secreted by and stimulates the growth of Hodgkin and Reed-Sternberg cells. J Exp Med 189 1939–46

Khanna R, and Burrows S. R. (2000). Role of cytotoxic T lymphocytes in Epstein-Barr virus-associated diseases. Annu Rev Microbiol 54 19–48

Kieff E, and Rickinson A.B. (2001). Epstein-Barr virus and its replication. In Fields Virology D. M. Knipe and P. M. Howley eds. (Philadelphia Lippincott-Raven Publishers) pp. 2511–2573

Koch F, Stanzl U, Jennewien P, Janke K, Heufler C, Kämpgen E, Romani N, and Schuler G. (1996). High level IL-12 production by murine dendritic cells: upregulation via MHC class II and CD40 molecules and downregulation by IL-4 and IL-10. J Exp Med 184 741–746

Koch F, Trockenbacher B, Kämpgen E, Grauer O, Stössel H, Livingstone A. M, Schuler G, and Romani N. (1995). Antigen processing in populations of mature murine dendritic cells is caused by subsets of incompletely matured cells. J Immunol 155 93–100

Küppers R, Klein U, Hansmann M. L, and Rajewsky K. (1999). Cellular origin of human B-cell lymphomas. N Engl J Med 341 1520–9

Küppers R, and Rajewsky K. (1998). The origin of Hodgkin and Reed/Sternberg cells in Hodgkin's disease. Annu Rev Immunol 16 471–93

Kurth J, Spieker T, Wustrow J, Strickler G. J, Hansmann L. M, Rajewsky K, and Küppers R. (2000). EBV-infected B cells in infectious mononucleosis: viral strategies for spreading in the B cell compartment and establishing latency. Immunity 13 485–95

Kurzrock R. (1997). Cytokine deregulation in hematological malignancies: clinical and biological implications. Clin Cancer Res 3 2581–4

Langhorne J, Cross C, Seixas E, Li C, and von der Weid T. (1998). A role for B cells in the development of T cell helper function in a malaria infection in mice. Proc Natl Acad Sci USA 95 1730–4

Larsen C. P, Ritchie S. C, Hendrix R, Linsley P. S, Hathcock R. J, Lowry R. P, and Pearson T. C. (1994). Regulation of immunostimulatory function and costimulatory molecule [B7-1 and B7-2] expression on murine dendritic cells. J Immunol 152 5208–5219

Leen A, Meij P, Redchenko I, Middeldorp J, Bloemena E, Rickinson A, and Blake N. (2001). Differential immunogenicity of Epstein-Barr virus latent-cycle proteins for human $CD4^+$ T-helper 1 responses. J Virol 75 8649–59

Levine A. M. (1992). Acquired immunodeficiency syndrome-related lymphoma. Blood 80 8–20

Levitskaya J, Coram M, Levitsky V, Imreh S, Steigerwald-Mullen P. M, Klein G, Kurilla M. G, and Masucci M. G. (1995). Inhibition of antigen processing by the internal repeat region of the Epstein-Barr virus nuclear antigen-1. Nature 375 685–688

Levitskaya J, Sharipo A, Leonchiks A, Ciechanover A, and Masucci M. G. (1997). Inhibition of ubiquitin/proteasome-dependent protein degradation by the Gly-Ala repeat domain of the Epstein-Barr virus nuclear antigen 1. Proc Natl Acad Sci USA 94 12616–21

Lin J. C, Smith M. C, and Pagano J. S. (1984). Prolonged inhibitory effect of 9-(13-dihydroxy-2-propoxymethyl)guanine against replication of Epstein-Barr virus. J Virol 50 50–5

Linton P. J, Harbertson J, and Bradley L. M. (2000). A critical role for B cells in the development of memory CD4 cells. J Immunol 165 5558–65

Maldonado-Lopez R, De Smedt T, Michel P, Godfroid J, Pajak B, Heirman C, Thielemans K, Leo O, Urbain J, and Moser M. (1999). $CD8\alpha^+$ and $CD8\alpha^-$ subclasses of dendritic cells direct the development of distinct T helper cells in vivo. J Exp Med 189 587–592

Maloy K. J, Burkhart C, Junt T. M, Odermatt B, Oxenius A, Piali L, Zinkernagel R. M, and Hengartner H. (2000). $CD4^+$ T cell subsets during virus infection: protective capacity depends on effector cytokine secretion and on migratory capability. J Exp Med 191 2159

Moretta A, Comoli P, Montagna D, Gasparoni A, Percivalle E, Carena I, Revello M. G, Gerna G, Mingrat G, Locatelli F, et al. (1997). High frequency of Epstein-Barr virus (EBV) lymphoblastoid cell line-reactive lymphocytes in cord blood: evaluation of cytolytic activity and IL-2 production. Clin Exp Immunol 107 312–20

Moss D. J, Chan S. H, Burrows S. R, Chew T. S, Kane R. G, Staples J. A, and Kunaratnam N. (1983). Epstein-Barr virus specific T-cell response in nasopharyngeal carcinoma patients. Int J Cancer 32 301–5

Münz C, Bickham K. L, Subklewe M, Tsang M. L, Chahroudi A, Kurilla M. G, Zhang D, O'Donnell M, and Steinman R. M. (2000). Human $CD4^+$ T lymphocytes consistently respond to the latent Epstein-Barr Virus nuclear antigen EBNA1. J Exp Med 191 1649–1660

Murphy K. M, Ouyang W, Farrar J. D, Yang J, Ranganath S, Asnagli H, Afkarian M, and Murphy T. L. (2000). Signaling and transcription in T helper development. Annu Rev Immunol 18 451–94

Murray R. J, Kurilla M. G, Brooks J. M, Thomas W. A, Rowe M, Kieff E, and Rickinson A. B. (1992). Identification of target antigens for the human cytotoxic T cell response to Epstein-Barr virus (EBV): implications for the immune control of EBV-positive malignancies. J Exp Med 176 157–68

Nikiforow S, Bottomly K, and Miller G. (2001). $CD4^+$ T-cell effectors inhibit Epstein-Barr virus-induced B-cell proliferation. J Virol 75 3740–52

Nishimura T, Iwakabe K, Sekimoto M, Ohmi Y, Yahata T, Nakui M, Sato T, Habu S, Tashiro H, Sato M, and Ohta A. (1999). Distinct roles of antigen-specific T helper type 1 (Th1) and Th2 cells in tumor eradication in vivo. J Exp Med 190 617–628

O'Garra A, and Murphy K. (1994). Role of cytokines in determining T-lymphocyte function. Curr Opin Immunol 6 458–66

Orange J. S, and Biron C. A. (1996). An absolute and restricted requirement for IL-12 in natural killer cell IFN-γ production and antiviral defense. Studies of natural killer and T cell responses in contrasting viral infections. J Immunol 156 1138–42

Orange J. S, Wang B, Terhorst C, and Biron C. A. (1995). Requirement for natural killer cell-produced interferon gamma in defense against murine cytomegalovirus infection and enhancement of this defense pathway by interleukin 12 administration. J Exp Med 182 1045–56

Pallesen G, Hamilton-Dutoit S. J, and Zhou X. (1993). The association of Epstein-Barr virus (EBV) with T cell lymphoproliferations and Hodgkin's disease: two new developments in the EBV field. Adv Cancer Res 62 179–239

Parolini S, Bottino C, Falco M, Augugliaro R, Giliani S, Franceschini R, Ochs H. D, Wolf H, Bonnefoy J. Y, Biassoni R, et al. (2000). X-linked lymphoproliferative disease. 2B4 molecules displaying inhibitory rather than activating function are responsible for the inability of natural killer cells to kill Epstein-Barr virus-infected cells. J Exp Med 192 337–46

Purtilo D. T, Cassel C, and Yang J. P. (1974). Letter: Fatal infectious mononucleosis in familial lymphohistiocytosis. N Engl J Med 291 736

Rajnavölgyi E, Nagy N, Thuresson B, Dosztanyi Z, Simon A, Simon I, Karr R. W, Ernberg I, Klein E, and Falk K. I. (2000). A repetitive sequence of Epstein-Barr virus nuclear antigen 6 comprises overlapping T cell epitopes which induce HLA-DR-restricted $CD4^+$ T lymphocytes. Int Immunol 12 281–93

Redchenko I. V, and Rickinson A. B. (1999). Accessing Epstein-Barr virus-specific T-cell memory with peptide-loaded dendritic cells. J Virol 73 334–42

Rentenaar R. J, Gamadia L. E, van DerHoek N, van Diepen F. N, Boom R, Weel J. F, Wertheim-Van Dillen P. M, van Lier R. A, and ten Berge I. J. (2000). Development of virus-specific $CD4^+$ T cells during primary cytomegalovirus infection. J Clin Invest 105 541–8

Rickinson A. B, and Kieff E. (2001). Epstein-Barr virus. In Fields Virology P. M. Knipe and P. M. Howley eds. (Philadelphia Lippincott-Raven) pp. 2575–2627

Ridge J. P, Di Rosa F, and Matzinger P. (1998). A conditioned dendritic cell can be a temporal bridge between a $CD4^+$ T helper and a T-killer cell. Nature 393 474–478

Riley E. M. (1999). Is T-cell priming required for initiation of pathology in malaria infections? Immunol Today 20 228–33

Roizman B, and Pellett P. E. (2001). The Family Herpesviridae: A Brief Introduction. In Fields Virology D. M. Knipe and P. M. Howley eds. (Philadelphia Lippincott-Raven) pp. 2301–2307

Roizman B, and Knipe D. M. (2001). Herpes simplex viruses and their replication. In Fields Virology D. M. Knipe and P. M. Howley eds. (Philadelphia Lippincott-Raven) pp. 2399–2459

Romani N, Koide S, Crowley M, Witmer-Pack M, Livingstone A. M, Fathman C. G, Inaba K, and Steinman R. M. (1989). Presentation of exogenous protein antigens by dendritic cells to T cell clones: intact protein is presented best by immature epidermal Langerhans cells. J Exp Med 169 1169–1178

Sallusto F, Lanzavecchia A, and Mackay C. R. (1998). Chemokines and chemokine receptors in T-cell priming and Th1/Th2-mediated responses. Immunol Today 19 568–74

Sam C. K, Prasad U, and Pathmanathan R. (1989). Serological markers in the diagnosis of histopathological types of nasopharyngeal carcinoma. Eur J Surg Oncol 15 357–60

Sarris A. H, Kliche K. O, Pethambaram P, Preti A, Tucker S, Jackow C, Messina O, Pugh W, Hagemeister F. B, McLaughlin P, et al. (1999). Interleukin-10 levels are often elevated in serum of adults with Hodgkin's disease and are associated with inferior failure-free survival. Ann Oncol 10 433–40

Schoenberger S. P, Toes R. E. M, van der Voort E. I. H, Offringa R, and Melief C. J. M. (1998). T-cell help for cytotoxic T lymphocytes is mediated by CD40-CD40L interactions. Nature 393 480–483

Schulz T. F, Boshoff C. H, and Weiss R. A. (1996). HIV infection and neoplasia. Lancet 348 587–91

Sheldon P. J, Hemsted E. H, Papamichail M, and Holborow E. J. (1973). Thymic origin of atypical lymphoid cells in infectious mononucleosis. Lancet 1 1153–5

Shibata D, Weiss L. M, Hernandez A. M, Nathwani B. N, Bernstein L, and Levine A. M. (1993). Epstein-Barr virus-associated non-Hodgkin's lymphoma in patients infected with the human immunodeficiency virus. Blood 81 2102–9

Silins S. L, Sherritt M. A, Silleri J. M, Cross S. M, Elliott S. L, Bharadwaj M, Le T. T, Morrison L. E, Khanna R, Moss D. J, et al. (2001). Asymptomatic primary Epstein-Barr virus infection occurs in the absence of blood T-cell repertoire perturbations despite high levels of systemic viral load. Blood 98 3739–3744

Steigerwald-Mullen P, Kurilla M. G, and Braciale T. J. (2000). Type 2 cytokines predominate in the human $CD4^+$ T-lymphocyte response to Epstein-Barr virus nuclear antigen 1. J Virol 74 6748–59

Steven N. M, Leese A. M, Annels N. E, Lee S. P, and Rickinson A. B. (1996). Epitope focusing in the primary cytotoxic T cell response to Epstein- Barr virus and its relationship to T cell memory. J Exp Med 184 1801–13

Subklewe M, Chahroudi A, Bickham K, Larsson M, Kurilla M. G, Bhardwaj N, and Steinman R. M. (1999a). Presentation of Epstein-Barr virus latency antigens to $CD8^+$ interferon-γ-secreting T lymphocytes. Eur J Immunol 29 3995–4001

Subklewe M, Chahroudi A, Schmaljohn A, Kurilla M. G, Bhardwaj N, and Steinman R. M. (1999b). Induction of Epstein-Barr Virus-specific cytotoxic T-lymphocyte responses using dendritic cells pulsed with EBNA-3A peptides or UV-inactivated recombinant EBNA-3A vaccinia virus. Blood 94 1372–1381

Subklewe M, Paludan C, Tsang M. L, Mahnke K, Steinman R. M, and Münz C. (2001). Dendritic cells cross-present latency gene products from Epstein-Barr virus-transformed B cells and expand tumor-reactive $CD8^+$ killer T cells. J Exp Med 193 405–11

Thorley-Lawson D. A. (2001). Epstein-Barr virus: exploiting the immune system Nat Rev Immunol 1 75–82

Urban B. C, Ferguson D. J, Pain A, Willcox N, Plebanski M, Austyn J. M, and Roberts D. J. (1999). *Plasmodium falciparum*-infected erythrocytes modulate the maturation of dendritic cells. Nature 400 73–7

Viviani S, Notti P, Bonfante V, Verderio P, Valagussa P, and Bonadonna G. (2000). Elevated pretreatment serum levels of Il-10 are associated with a poor prognosis in Hodgkin's disease the Milan Cancer Institute Experience. Med Oncol 17 59–63

von Wasielewski R, Seth S, Franklin J, Fischer R, Hubner K, Hansmann M. L, Diehl V, and Georgii A. (2000). Tissue eosinophilia correlates strongly with poor prognosis in nodular sclerosing Hodgkin's disease allowing for known prognostic factors. Blood 95 1207–13

Whittle H. C, Brown J, Marsh K, Greenwood B. M, Seidelin P, Tighe H, and Wedderburn L. (1984). T-cell control of Epstein-Barr virus-infected B cells is lost during *P. falciparum* malaria. Nature 312 449–50

Wright D. H. (1999). What is Burkitt's lymphoma and when is it endemic? Blood 93 758

Young L. S, Dawson C. W, Clark D, Rupani H, Busson P, Tursz T, Johnson A, and Rickinson A. B. (1988). Epstein-Barr virus gene expression in nasopharyngeal carcinoma. J Gen Virol 69 1051–65

Zajac A. J, Blattman J. N, Murali-Krishna K, Sourdive D. J, Suresh M, Altman J. D, and Ahmed R. (1998). Viral immune evasion due to persistence of activated T cells without effector function. J Exp Med 188 2205–13

Zeng Y. (1985). Seroepidemiological studies on nasopharyngeal carcinoma in China. Adv Cancer Res 44 121–38

Zurawski G, and de Vries J. E. (1994). Interleukin 13 an interleukin 4-like cytokine that acts on monocytes and B cells but not on T cells. Immunol Today 15 19–26

Chapter 4

Dendritic Cells and Measles Virus Infection

S. Schneider-Schaulies, I. M. Klagge, V. ter Meulen

Abstract. Measles is a major cause of childhood mortality in developing countries which is mainly attributed to the ability of measles virus (MV) to suppress general immune responses. Paradoxically, virus-specific immunity is efficiently induced, which leads to viral clearance from the host and confers long-lasting protection against reinfection. As sensitisers of pathogen encounter and instructors of the adaptive immune response, dendritic cells (DCs) may play a decisive role in the induction and quality of the MV-specific immune activation. The ability of MV wild-type strains in particular to infect DCs in vitro is clearly established, and the receptor binding haemagglutinin protein of these viruses

S. Schneider-Schaulies (✉), I.M. Klagge, V. ter Meulen
Institute for Virology and Immunobiology, University of Würzburg, Versbacher Str. 7, 97078 Würzburg, Germany

essentially determines this particular tropism. DC maturation as induced early after MV infection is likely to be of crucial importance for the induction of MV-specific immunity. DCs may, however, be instrumental in MV-induced immunosuppression. (1) T cell depletion could be brought about by DC-T cell fusion or TRAIL-mediated induction of apoptosis. (2) Inhibition of stimulated IL-12 production from MV-infected DCs might affect T cell responses in qualitative terms in favouring Th2 and suppressing Th1 responses. (3) The viral glycoprotein complex expressed at high levels on infected DCs late in infection is able to directly inhibit T cell proliferation by surface contact-dependent negative signalling. This most likely accounts for the failure of infected DC cultures to stimulate allogeneic and inhibit mitogen-stimulated T cell proliferation in vitro and the pronounced proliferative unresponsiveness of T cell ex vivo to polyclonal and antigen-specific stimulation which is a central finding of MV-induced immunosuppression.

1 Introduction

Despite the existence of an effective live attenuated vaccine, measles virus (MV) remains among the most potent global pathogens, killing more than 1 million children annually. This is mainly because the virus induces a profound suppression of immune functions during and for weeks after the acute disease, which favours the establishment of and aggravates the course of secondary infections. At the same time, however, virus-specific immune responses are efficiently generated which essentially clear the virus from the organism and leave the patient with a long-lasting immunity. Linking the innate to the adaptive immune response, professional antigen-presenting cells such as dendritic cells (DCs) most likely play a key role in the induction and quality of the virus-specific immune response. Many findings of immune suppression associated with MV infection, however, are also compatible with modulation of or interference with APC maturation and function, which then result in qualitative and quantitative disturbances of T cell activation. This review focuses on the interaction of MV with DCs with special emphasis on viral tropism for these cells and on functional consequences.

2
Measles Pathogenesis, Immune Activation and Suppression

After primary infection in the respiratory tract, MV is most likely acquired by tissue-resident macrophages or DCs from the basolateral side of epithelial cells (Moll et al. 2001; Naim et al. 2000) and transported to local lymphatic tissues. From there, after a first round of replication, MV spreads by a cell-associated viraemia, and MV-specific RNA and proteins can be detected in a small proportion of lymphocytes and monocytes during and for few days after the rash (Esolen et al. 1993; Forthal et al. 1992; Hyypiae et al. 1985; Schneider-Schaulies et al. 1991). The onset of virus-specific immune responses is marked by the appearance of antiviral antibodies, the infiltration of mononuclear cells into local areas of virus replication and the appearance of virus-specific T cells in the blood. Activation of virus-specific T cells is documented by soluble forms of CD4, CD8, IL-2R and $\beta 2$ microglobulin and an initial Th1 cytokine profile which switches to a Th2 type as indicated by a rise in IL-4 plasma levels (reviewed in Griffin 1995). Whether or not generation of a Th1 responses is less efficient after vaccination than after natural infection has not been clearly resolved as yet (Gans et al. 1999; Schnorr et al. 2001; Ward and Griffin 1993).

Although MV-specific immunity is efficiently induced in the course of acute measles and after vaccination, there is a generalised suppression of immune responses to secondary infections. Although MV was the first pathogen recognised to cause immunosuppression, it was only recently that underlying molecular mechanisms were even partially elucidated. Characteristically, a marked lymphopenia affecting mainly the T cell population and a loss of delayed-type hypersensitivity reactions are observed. The impaired proliferative responses of lymphocytes to polyclonal and antigen-specific stimulation ex vivo is considered to be one of the major hallmarks of MV-induced immunosuppression (Borrow and Oldstone 1995; Schneider-Schaulies and ter Meulen 1998). This can be documented for several weeks after acute measles and also, albeit to a moderate extent, after vaccination (Hussey et al. 1996). In addition, a cytokine imbalance as seen by a predominant Th2 response and thereby suppression of cellular immunity has also been linked to immunosuppression (reviewed in Griffin 1995).

Complications of acute measles are frequent and mainly affect the respiratory and intestinal tract (reviewed in Katz 1995). In industrialised countries, about 10% of MV cases are associated with otitis, pneumonia, diarrhoea or encephalitis. CNS complications can occur early (postinfectious measles encephalitis, thought to be a virus-induced autoimmune disease) or, very infrequently, months to years after the acute disease. The late complications (subacute sclerosing panencephalitis, SSPE, or measles inclusion body encephalitis, MIBE) develop on the basis of a persistent MV infection of neuronal and macroglial cells. Remarkably, SSPE occurs in the face of a humoral hyperimmune reaction. Both diseases are characterised by restrictions of viral gene expression (reviewed in Schneider-Schaulies et al. 1995b) and are invariably fatal. In developing countries, MV infections have high rates of morbidity and mortality. Pneumonia and diarrhoea can be complicated by secondary viral, bacterial and parasitic infections established as a consequence of MV-induced immunosuppression (reviewed in Clements and Cutts 1995).

3
MV Strain Variations

Although measles is considered as monotypic, sequence analysis has shown that distinct lineages of wild-type viruses exist and cocirculate. Genetic characteristics of MVs were defined by sequencing the genes coding for the haemagglutinin (H) and the nucleocapsid (N) protein (see below) and, based on these sequences, MVs were assigned to eight clades (A to H). Within these clades, twenty genotypes were designated which reflect the genetic drift of MV strains (Rima et al. 1995, 1997). The availability of an extensive database for wild-type viruses enabled molecular epidemiological studies tracing the source and transmission pathways of individual viruses and documenting interruption of indigenous transmission in some areas. There is, however, no evidence that MV wild-type strains would differ in terms of pathogenicity or neurovirulence. Apparently, host rather than viral determinants are associated with MV pathogenicity such as immunocompetence. The basis for the attenuated phenotype of MV vaccine strains, which form a cluster within genotype A viruses, is unknown. It is, however, likely that, amongst other determinants, sequence motifs within the glycoprotein genes common to the vaccine strains are important. This is because these proteins essentially determine the ability of the virus to interact with surface receptors and thereby cellular tropism and spread. Moreover, detailed knowledge of the viral glycoproteins may also be important in understanding why the degree of both induction and suppression of immune responses by wild-type MV and vaccine differs.

4 The Virus: Protein Functions

MV is the type species of the morbillivirus subgroup of the mononegavirales and has a nonsegmented RNA genome of negative polarity. The pleomorphic virion particle consists of two structural subunits, the nucleocapsid core and the envelope (schematically depicted in Fig. 1a). The genome tightly associated with the nucleocapsid protein (N) is the only target for the viral polymerase complex [the large (L) protein together with its cofactor, the phosphoprotein (P)] for viral transcription and replication. The P gene of MV encodes for nonstructural proteins, the highly basic C protein and the V protein. Although these proteins are dispensable for viral replication in vitro (Radecke and Billeter 1996; Schneider et al. 1997), there is evidence that they may play a role in viral pathogenesis in vivo (Patterson et al. 2000; Tober et al. 1998; Valsamakis et al. 1998).

The matrix protein M (37 kDa) links the core to the host cell-derived lipid envelope and is thought to interact with the two glycosylated viral transmembrane proteins. This has been clearly revealed in functional terms. Recombinant viruses lacking the M reading frame show enhanced fusogenicity (Cathomen et al. 1997), and the M protein was found to specify apical virus release and sorting of the viral glycoproteins in epithelial cells (Naim et al. 2000). The haemagglutinin protein (H), a type II glycoprotein, has a dual role by mediating attachment of virions to cellular surface proteins and also providing a helper function for membrane fusion (Nussbaum et al. 1995). Viral and membrane fusion at neutral pH requires the MV glycoprotein complex consisting of an H tetramer and a fusion protein (F) trimer (Malvoisin and Wild 1993; Plemper et al. 2000). The latter protein, a type I glycoprotein, is synthesised as a precursor protein (F_0). Proteolytic activation into the disulfide bond linked F_1-F_2 heterodimer by furin occurs in the trans-Golgi compartment (Bolt and Pedersen 1999). Thereby, a first conformational change is induced which is indispensable for biologic activity of the protein (Kohama et al. 1981). Equally important, a second conformational change within the F_1-F_2 heterodimer occurs after receptor binding of the H protein, which leads to pairing of two amphipatic α-helical domains, thereby bringing

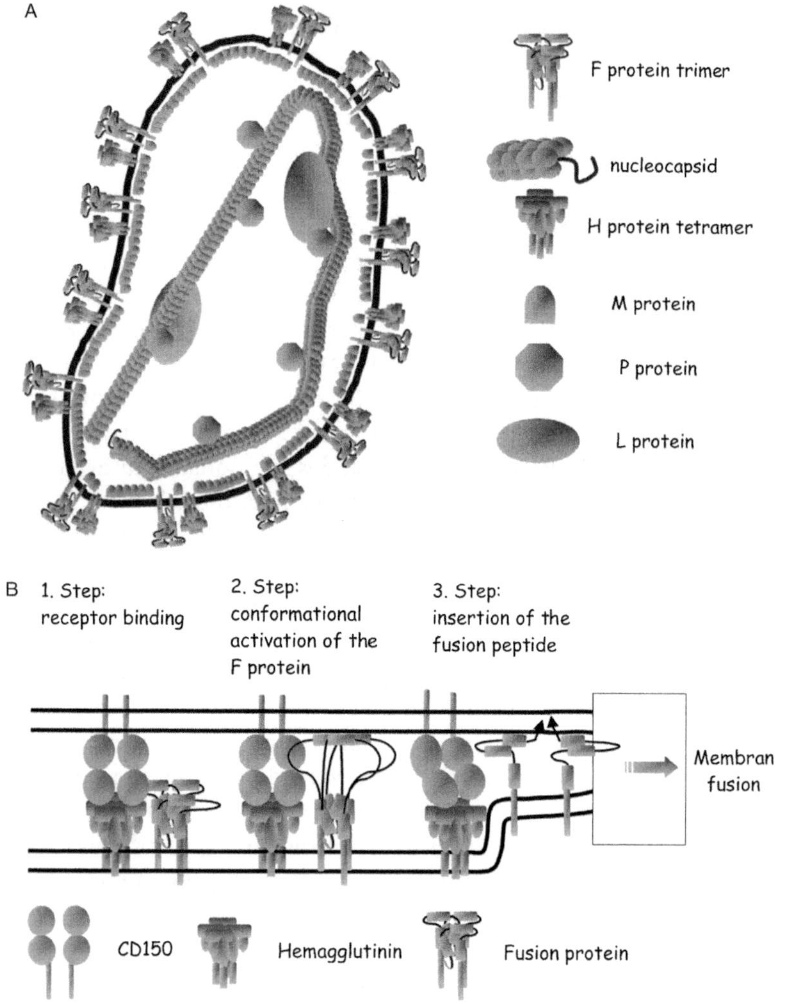

Fig. 1. **a** Schematic representation of the measles virus (MV) particle. **b** Model of MV induced membrane fusion and conformational changes within the F protein

the membranes to be fused in close proximity (Baker et al. 1999; Lamb 1993; Samuel and Shai 2001; Fig. 1b). Supporting the importance of these processes, abolishment of F_0 protein cleavage by mutation of the multibasic cleavage site, or inclusion of fusion inhibitory peptides which interfere with membrane insertion of the fusion domain or the second conformational change strongly reduces or completely abolishes infectivity (Buckland et al. 1992; Lambert et al. 1996; Maisner et al. 2000; Richardson and Choppin 1983).

5
MV Receptors

CD46 (membrane cofactor protein, MCP), which is ubiquitously expressed on human nucleated cells was the first identified cellular MV receptor (Dörig et al. 1993; Naniche et al. 1993; Fig. 2). Several splice variants of this protein are expressed in a tissue-specific manner, and all of them were found to support MV uptake after transgenic expression in certain rodent cell lines (Manchester et al. 1994) but not in transgenic rodents (Blixenkrone-Moller et al. 1998; Horvat et al. 1996; Niewiesk et al. 1997b). The ability to interact with CD46 could be assigned to a few amino acids within the MV H protein with amino acid 481 being most important (Bartz et al. 1996; Lecouturier et al. 1996; Nielsen et al. 2001). The MV H protein interacts with an extended binding site spanning the two most membrane-distal domains, short conserved domains (SCR1 and 2) of CD46, whereas for binding of the 'natural' ligands of CD46, the C3b/C4b complement components, mainly SCR3 and -4, are required (Casasnovas et al. 1999). High-affinity binding to CD46 is, however, confined to attenuated vaccine strains and particularly wild-type strains isolated on and adapted to Vero cells, whereas these strains, but also lymphotropic MV wild-type viruses, were recently found to use CD150 (SLAM; signalling lymphocyte activation molecule), a member of the CD2 subset of the Ig superfamily (Erlenhoefer et al. 2001; Hsu et al. 2001; Tatsuo et al. 2000; Fig. 2). The expression of CD150 is, however, confined to activated T and B cells, memory cells and immature thymocytes but is not found on unstimulated monocytes (Aversa et al. 1997). It is thus not clear how wild-type strains, which cannot use CD46, manage infection of CD150-negative cells such as brain cells, epithelial cells and particularly monocytes, which are important targets during acute measles.

Downregulation from the cell surface after contact with MV H protein has been described for CD150 (Erlenhoefer et al. 2001) and for CD46 (Schneider-Schaulies et al. 1995a,c). Whereas the consequences of CD150 downregulation remain speculative, it has been revealed that, in agreement with the natural function of this molecule to inactivate C3b/

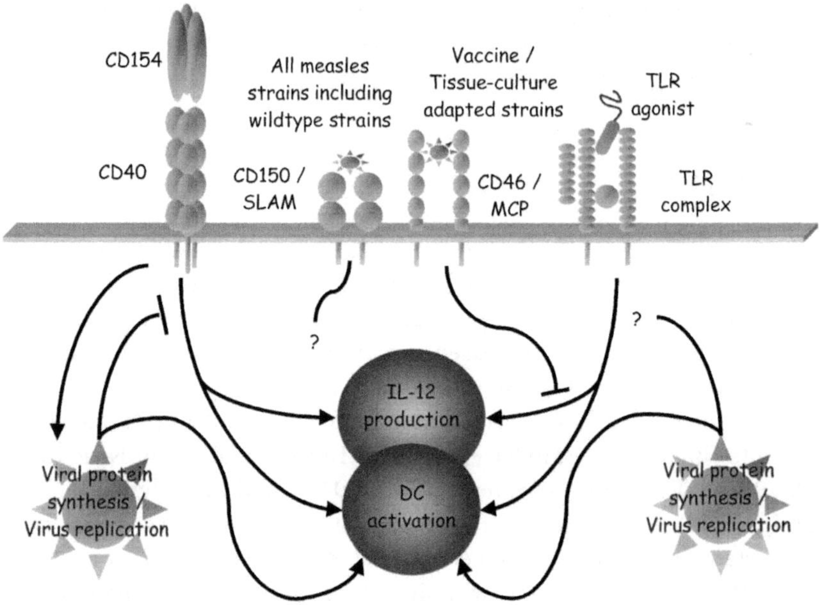

Fig. 2. Interaction of MV with cellular surface receptors and consequences for DC activation. Both CD150 (for all MV strains) and CD46 (for attenuated, Vero cell-adapted wild-type strains) can support viral entry into DCs. Whereas functional consequences of CD150 interaction with DCs are unknown, CD46 ligation can suppress stimulated IL-12 synthesis on monocytes. In DCs, this suppression apparently requires viral replication. Whereas CD46 ligation is able to enhance viral replication in DCs, this, in turn, was found to interfere with terminal, CD40-induced maturation of these cells. On the other hand, maturation of immature DCs by MV infection has also been documented, as indicated by upregulation of MHC and costimulatory molecules. Signalling pathways involved in these processes have not been unravelled as yet

C4b deposited on the membrane, MV-induced CD46 downregulation led to an enhanced sensitivity of lymphocytes to lysis by activated complement in vitro (Schneider-Schaulies et al. 1996; Schnorr et al. 1995).

6 MV Interaction with DCs

As 'nature's adjuvants' DCs play a major role in initiating and shaping the adaptive immune response to pathogens also including viruses.

There is, however, also evidence that viruses exploit DCs as transport vehicles into lymphatic tissues. Viruses are, however, also able to modify or interfere with DC maturation, migration and function, thereby inducing specific or generalised immunosuppression as intensely studied for HIV (for review, see Knight and Patterson 1997 and the chapter by Steinman et al., this volume). Although it is likely that uptake of MV in DCs and subsequent transport into the local lymphatics occurs after initial infection of the respiratory tract epithelium, this has not been documented directly. In experimentally infected rhesus macaques, syncytia formation, albeit to a very limited extent, was seen in lymph nodes (McChesney et al. 1997). Association of MV with FDC has also been documented in these animals. Whether FDC, as for HIV, could be considered as long-term repositories for MV is currently unknown. MV-infection of DCs in mice cannot be analysed because these animals are not permissive for MV infection. In genetically modified mice lacking the type I IFN receptor and transgenic for CD46 evidence for infection of DC was provided (Mrkic et al. 2000). However, studies in these animals exclude lymphotropic wild-type viruses and are restricted to MV strains which can use CD46 as receptors (see above).

6.1 Parameters of MV DC Infection

The interaction of MV with DC has been exclusively studied in tissue culture with DCs isolated from peripheral blood (P-DCs), Langerhans cells and DCs generated in vitro from monocyte precursors or $CD34^+$ stem cells. MV rapidly infects immature and mature DCs and causes extensive formation of syncytia both in pure DCs and mixed DC-T cell cultures (Fugier-Vivier et al. 1997; Grosjean et al. 1997; Klagge and Schneider-Schaulies 1999; Klagge et al. 2000; Schnorr et al. 1997b; Servet-Delprat et al. 2000b; Steineur et al. 1998; Fig. 3). Although DCs apparently support MV replication, production of infectious virus from pure DC cultures is usually low, but can be enhanced by CD40 ligation (Fugier-Vivier et al. 1997; Servet-Delprat et al. 2000b, and this volume; Fig. 2). Interestingly, transmission of virus from DCs to T cells only occurs to a limited extent in mixed DC-T cell cultures (Fugier-Vivier et al. 1997). It is currently unknown which cellular factors stimulated by CD40 ligation contribute to enhancement of viral replication. In contrast, a viral determinant favouring infection of DC by MV was recently defined. In both

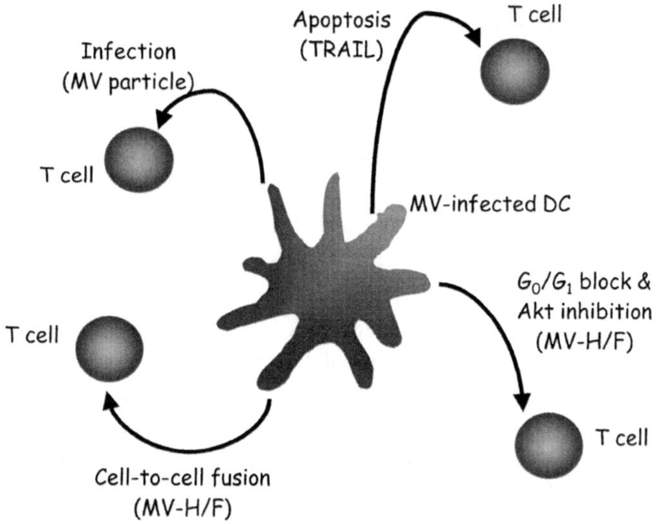

Fig. 3. Mechanisms that may account for suppression of T cell responses by MV-infected DCs. T cell depletion can be explained by DC/T cell fusion, transmission of infectious virus from DCs to T cells, or TRAIL-mediated T cell apoptosis. The MV glycoproteins expressed to high levels on the surface late in DC infection are, however, also able to inhibit allogenic and mitogen-driven T cell proliferation by a contact-mediated mechanism. In T cells, this F/H contact-mediated inhibition is characterised by a disruption of Akt kinase activation, and this is important in preventing S phase entry of these cells

immature and matured DC, infection with an MV wild-type virus (WTF) was found to be more efficient and to progress faster than that with the vaccine strain Edmonston (ED) (Ohgimoto et al. 2001; Schnorr et al. 1997b). As shown for LCMV (Sevilla et al. 2000), this particular targeting of WTF to DCs might be a property of the viral glycoproteins. Recombinant MVs were used in which the authentic MV-derived glycoproteins were singly or doubly swapped with those of the WTF strain (Johnston et al. 1999). Viruses expressing the ED H protein preferentially bound to primary lymphocytes, where they caused syncytia formation, whereas viruses expressing the H protein of WTF more efficiently entered into, replicated in and caused fusion of DCs generated from monocytes (M-DCs) (Ohgimoto et al. 2001; Table 1). In contrast to the observations made with lymphocytes and ED H-containing viruses, preferential infection of DCs by WTF H-containing viruses did not correlate with enhanced binding to these cells. Expression of CD46 was documented

Table 1. Receptor usage of measles virus strains in lymphocytes and dendritic cells (DC)

Measles virus strains		Receptor	Tropism
Vero-adapted strains	ED-B	CD46/CD150	PBL>>DC
	red ED-B (WTF-F)	CD46/CD150	PBL>>DC
Lymphocyte-adapted strains	WTF	CD150>>CD46	DC>>PBL
	ED-B (WTF-H)	CD150>>CD46	DC>>PBL
	rec ED-B (WTF-H+WTF-F)	CD150>>CD46	DC>>PBL

both on DCs isolated from peripheral blood (Per-DC) (Schnorr et al. 1997b) and on immature DC generated in vitro from monocytes (M-DC) (Ohgimoto et al. 2001). Although expression of CD150 was only detected after LPS- or CD40L-driven maturation of M-DC, and was found to be further upregulated by IL-1β (Bleharski et al. 2001; Kruse et al. 2001), we detected this protein, albeit at low levels, on immature M-DCs after a 7-day culture in GM-CSF/IL-4 (Ohgimoto et al. 2001). Because in our cultures d7 M-DCs might have been already partially matured as also indicated by the expression of intermediate levels of CD80 and CD86, CD150 might be considered as an activation marker on M-DC. Whether upregulation of CD150 also occurs after activation of monocytes is currently unknown. In agreement with the CD150 surface expression in our M-DC cultures, WTF H-containing viruses were able to infect and fuse these cells, and this could be efficiently prevented by CD150- but interestingly also, albeit less efficiently, CD46-specific antibodies (Ohgimoto et al. 2001). It is, however, also evident that the ability of WTF to replicate in DCs is also determined by viral proteins other than the glycoprotein. Thus virus production from these cells was higher and more prolonged with WTF per se than with the recombinants containing the WTF glycoproteins (Ohgimoto et al. 2001).

6.2 Impact of MV Infection on DC Maturation

Infection of immature DC with MV induces phenotypic maturation of these cells as indicated by enhancement of MHC class II and CD40 expression and induction of costimulatory molecules such as CD80, CD86, CD83 and CD25 (Klagge et al. 2000; Schnorr et al. 1997b; Servet-Delprat

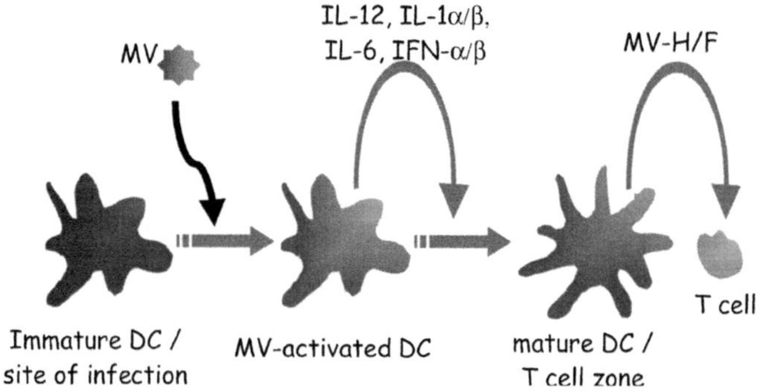

Fig. 4. Model for MV-induced alterations of DC maturation and function. MV infection of mucosa-resident immature DCs induces maturation as documented by the induction of cytokine expression and costimulatory molecules. Soluble mediators such IFN-α/β may play an important role in this maturation process. MV-induced maturation most likely also triggers DC migration to secondary lymphatic tissues, where they interact with T cells. There, early in infection, MV-specific immune responses are activated because DCs, if at all, only produce low levels of infectious virus and express low amounts of viral glycoproteins on their surface. Once they are further activated by T cells, viral replication and expression of the MV glycoproteins is enhanced, and both phenomena most likely contribute to generalised suppression of immune responses

et al. 2000b; Fig. 4). There is, however, also evidence that CD40L-dependent terminal maturation of MV-infected DC is impaired (Servet-Delprat et al. 2000b; Fig. 2). Induction of DC maturation by MV occurred with all virus strains tested. Whereas it was observed that WTF again was more potent in inducing DC maturation than ED (Schnorr et al. 1997b), this difference was not seen with another MV wild-type strain, LYS-1 (Grosjean et al. 1997; Steineur et al. 1998). Supernatants of MV-infected DC cultures were found to promote maturation of immature DCs, indicating the release of soluble maturation factors (Klagge et al. 2000). Production of type I IFN, a known DC maturation factor (Luft et al. 1998; Paquette et al. 1998), by MV has been shown by the induction of MxA protein in DCs (Fig. 4). The importance of this cytokine for DC maturation has been revealed by inclusion of neutralising antibodies. Unlike for influenza (Cella et al. 1999), the presence of MxA protein did not prevent viral spread in these cultures although the ability of this protein to interfere with MV replication has been clearly established in monocytic and neural cells (Schneider-Schaulies et al. 1994; Schnorr et al. 1993).

Thus levels of MxA might be insufficient to protect DC, or, alternatively, this protein does not inhibit MV replication in these cells. The mechanism of how MV induces type I IFN induction is unknown. According to the classic theory, dsRNAs as accidentally formed during viral transcription and replication or present in UV-inactivated virions should be the intracellular inducer structures. More recently, activation of monocytes and dendritic cells by dsRNA via Toll-like receptor 3 (TLR3) was described, which led to the induction of type I IFN (Alexopoulou et al. 2001), and this was suggested to occur after release of viral dsRNAs from necrotic or apoptotic cells. Of note, expression of TLR3 is confined to M-DCs (Kadowaki et al. 2001). Surprisingly then, a recent study seems to suggest that only MV vaccine strains efficiently induce type I IFN in PBMC cultures, whereas wild-type strains largely failed to do so and even actively suppressed induction of this cytokine (Naniche et al. 2000). Underlying mechanisms have, however, not been resolved, and it is unknown whether this is a general property of wild-type MV and, if so, whether differential induction of type I IFN also occurs in DCs.

In addition to type I IFN, production of TNF-α, another well-known DC maturation factor, was found stimulated in MV-infected DCs cultures (I.M. Klagge, unpublished data). In contrast to mediators promoting DC maturation, soluble factors inhibiting DC functions are obviously not induced by MV infection. This is because uninfected DCs matured in the presence of virus-free supernatants of infected cultures retained their activity to stimulate allogenic T cell proliferation in vitro (Klagge et al. 2000).

Indicative of DC activation and maturation, induction of low levels of IL-12p40-, IL-12p35-, IL-1 α/β-, IL-1RA- and IL-6-specific transcripts were detected in MV-infected DCs (Servet-Delprat et al. 2000b, and the chapter by Servet-Delprat et al., this volume). The mechanisms underlying this particular activation by MV are unclear. Conventional maturation stimuli such as LPS and SAC operate via TLRs to trigger signalling cascades which ultimately converge in activation of NF-kB and thereby to promote the expression of costimulatory molecules and cytokines such as IL-6, IL-1α/β, TNF-α and IL-12 (Akira et al. 2001; Fig. 2). Recently, the F protein of respiratory syncytial virus was found to act as an agonist for TLR4 on monocytes (Kurt-Jones et al. 2000). It is unknown as yet whether this is an exception, or whether MV is also able to interact with a member of the TLR family. It also remains to be determined whether these monokines are produced after MV infection in detectable amounts. At least production of bioactive IL-12p70 has not been seen after infection of P-DCs with ED or WTF (Schnorr et al. 1997b).

6.3 Impairments of DC Functions by MV

6.3.1 Regulation of Stimulated IL-12 Production

Although MV infection fails to alter stimulated production of IL-6, TNF-α and MIP-1 in monocytes, interference of MV with IL-12 synthesis was found in primary monocytes and M-DCs stimulated by CD40L or bacterial cell wall components (Fugier-Vivier et al. 1997; Karp et al. 1996; Servet-Delprat et al. 2000b, and the chapter by Servet-Delprat et al., this volume; Fig. 2). In contrast, the low levels of IL-12 released from P-DCs after LPS or SAC stimulation were unaffected by MV (Schnorr et al. 1997b). For monocytes, ablation of this cytokine was induced by cross-linking of CD46 by specific antibodies, multimeric C3b or MV, indicating that viral replication was not required (Karp et al. 1996). Downregulation of IL-12 after CD40 ligation in DC was also observed, but only with replication-competent and not UV-inactivated MV (Servet-Delprat et al. 2000b). Direct evidence for interference of MV infection with signalling pathways activated by CD40 ligation was obtained in this study, although the intracellular targets were not identified. It is also enigmatic how MV could modulate LPS- or SAC-driven activation of IL-12. The most attractive hypothesis, interference of MV with TLR signalling, must be addressed as well as the role of CD46 in this process as the signalling properties of CD46 are established in both T cells and monocytes (Astier et al. 2000; Katayama et al. 2000; Kurita-Taniguchi et al. 2000). The ability of MV wild-type strains, which do not interact with CD46, to suppress stimulated IL-12 release from M-DCs has not been addressed as yet. Their receptor, CD150, is probably only expressed in the course of DC maturation, and signalling via CD150 requires an integral regulatory protein, SAP (Latour et al. 2001), which has not been detected in these cells as yet. Thus, for MV wild-type strains, interference with stimulated IL-12 production by binding to their receptor remains speculative.

6.3.2
Loss of Allostimulatory Activity and Acquisition of an Inhibitory Phenotype

Despite their mature phenotype, MV-infected DCs fail to stimulate proliferation of allogenic T cells and even actively suppress stimulated T cell proliferation in vitro (Fugier-Vivier et al. 1997; Grosjean et al. 1997; Klagge et al. 2000; Schnorr et al. 1997b; Servet-Delprat et al. 2000b; Steineur et al. 1998). The inhibitory activity can be documented over a wide range of dilutions and can only be partially overcome on addition of mature, uninfected DCs (Klagge et al. 2000; Schnorr et al. 1997b). Although it is well established that MV infection of T cells is associated with proliferative arrest (McChesney et al. 1988; Naniche et al. 1999; Yanagi et al. 1992), this is not likely to account for inhibition of allogenic T cell proliferation in these systems, because transmission of infectious MV to T cells by DCs occurs only to a very limited extent (Grosjean et al. 1997). Loss of both DCs and T cells by fusion or apoptosis during the co-culture may, however contribute to a certain extent (Servet-Delprat et al. 2000a; Vidalain et al. 2000, 2001; Fig. 3). Nevertheless, it is likely that the MV glycoproteins expressed on the surface of infected DCs to high levels are accountable to a large extent for the inhibition of T cell proliferation. This is because the inhibitory activity of DCs directly correlates with the expression levels of these proteins, and T cell proliferation is not stimulated by uninfected DCs in the presence of UV-inactivated MV. Furthermore, DCs infected with a recombinant MV which contains the G protein of vesicular stomatitis virus instead of the MV glycoproteins reveal a high allostimulatory activity and do not inhibit mitogen-driven T cell proliferation (Klagge et al. 2000). In mice transgenic for CD46, evidence for a direct effect on DC functions by the MV N protein, but also the glycoproteins of CD46 adapted MV strains, has been obtained (Marie et al. 2001; the chapter by Servet-Delprat et al., this volume).

Although not directly verified in DCs, transient expression of the MV F/H complex on fibroblasts was found necessary and sufficient to induce arrest of stimulated T cell proliferation in vitro and ex vivo (Niewiesk et al. 1997a; Schlender et al. 1996; Weidmann et al. 2000b) (Fig. 4). In these systems, inhibition was independent of soluble mediators, complex glycosylation and the fusogenic activity of the effector F/H complex; however,proteolytic processing of the F protein is required (Weidmann et al. 2000a,b). In contrast to the observations made in CD46 transgenic mice

(Marie et al. 2001), induction of T cell proliferative arrest occurred independent of CD46 and also CD150 (Erlenhoefer et al. 2001; Niewiesk et al. 1999; Schlender et al. 1996). Apoptosis is not induced by the surface contact between the effector complex and an as-yet unknown receptor structure on lymphocytes which rather accumulate at the G_1/S-phase restriction point (Niewiesk et al. 1999; Schnorr et al. 1997a). On a molecular level, MV F/H contact-induced cell cycle retardation was associated with restrictions of the accumulation and activity of important S phase entry control proteins such as the regulatory and catalytic subunits of the cyclin-dependent kinases (CDKs), as well as their inhibitor, $p27^{Kip1}$ (Engelking et al. 1999). To gain insight into the signalling pathways targeted by the viral F/H effector complex, we recently analysed IL-2R signalling in T cells. This is because the expression levels of the IL-2R subunits were unaffected, stimulated levels of IL-2 release were, if at all, only slightly reduced after contact with the effector structure and exogenously added IL-2 did not restore their proliferative activity of T cells (Avota et al. 2001; Niewiesk et al. 1999). Corroborating our observation that T cell activation as indicated by upregulation of early activation markers and cytokine release was not affected, the F/H complex did not interfere with IL-2-dependent activation of the JAK1/3/STAT3/5 pathway (Avota et al. 2001). In contrast, IL-2-dependent activation of the Akt kinase, which plays a key role in transmitting survival signals in a variety of cell types and, particularly in T cells, in controlling CDK activity and S-phase entry (Brennan et al. 1997), proved to be disrupted after MV contact with these cells (Avota et al. 2001). In line with our observations that apoptosis is not induced by MV interaction, IL-2-stimulated phosphorylation and thereby inactivation of the proapoptotic protein Bad did occur, which is a well-known downstream target of Akt in some cell types. Supporting the importance of MV interference with Akt signalling, inhibition of this pathway could also be documented in T cells isolated from experimentally infected cotton rats. Additionally, transgenic expression of a constitutively active Akt kinase (Myr-Akt) strongly reduced the sensitivity of T cells to MV contact-mediated proliferative inhibition (Avota et al. 2001).

7
Concluding Remarks and Outlook

The interaction of dendritic cells with MV may be central to the induction of the MV-specific immune response but also for that of immunosuppression by this virus. Because MV causes maturation of these cells, albeit with different kinetics for wild-type and vaccine strains, and with low levels of MV glycoproteins expressed on these cells early after infection, immune responses might be efficiently primed and shaped. In support of this concept, the allostimulatory activity of MV-infected DC cultures was perfectly retained early after infection (Klagge et al. 2000). Later, when the expression levels of the viral glycoproteins exceed a certain threshold, both fusion of DCs or DCs and T cells and active T cell inhibition by DCs most likely occur and contribute to immunosuppression. Whether and on which level MV is able to directly interfere with signalling pathways in DC and thereby modulates their viability and function must be studied. As seen in T cells, MV can, by a mere surface contact, ablate activation of the phosphatidylinositol-3/Akt kinase pathway. Because this pathway is also involved in the survival and maturation of LPS-stimulated M-DCs (Ardeshna et al. 2000), and particularly TRANCE signalling in these cells (Wong et al. 1999), it is tempting to speculate that MV could also interfere with DC survival via this mechanism. Also by a CD46-dependent surface contact, certain MV strains could be able to interfere with TLR signalling as evidenced by suppression of IL-12 release by this mechanism in monocytes (Karp et al. 1996). Whether this also applies to DCs is unknown as yet, and it is also unknown whether this would also be specific for IL-12 and, as seen in monocytes, would not affect IL-6 and TNF-α release which are also induced after TLR activation. Most likely dependent on viral replication, MV obviously targets other intracellular pathways in DCs such as CD40 signalling, and it will be interesting to define the viral effector protein and the cellular targets involved in this interference.

The consequences of MV infection of DCs for the quality of the T cell response also requires direct assessment. Given the central role IL-12 on the generation of Th1 responses, downregulation of this cytokine could obviously have a significant impact on the quality of cellular immune responses. It should, however, be kept in mind that this downregulation is confined to MV strains strongly interacting with CD46 and is not exerted by lymphotropic MV wild-type strains. Furthermore, in the human system type I IFN can also induce production of type II IFN from T cells (O'Shea and Visconti 2000). The observation that type II IFN is produced in vivo early during infection seems to argue for an at least initially efficient Th1 activation.

Finally, and probably most importantly, as with many other viruses, in which interaction with DC has been mainly studied in vitro, infection of DCs by MV in vivo must be documented. For obvious reasons, this can only be done in animal models aided by fluorochrome-tagged MV. The animal models ideally suited for this purpose are rhesus macaques and cotton rats, because in these animals infection via the respiratory tract is possible and both animals are permissive for infection with wild-type and vaccine MV strains. As a small animal model with inbred strains available, cotton rats appear particularly attractive for these studies, and once the necessary reagents are at hand, they will give exiting and conclusive information on the tropism of MV strains for DCs, the type of DC infected, and their trafficking to lymphatic compartments.

Note added in proof. While this manuscript was under review, we established that the H protein of wild-type but not that of vaccine MV strains is an agonist for Toll-like receptor 2 on monocytes. As a typical TLR2 agonist, wild-type MV H protein activates production of cytokines from murine and human monocytes. Moreover, surface expression of CD150, which is absent from freshly isolated human monocytes and immature DCs, is also induced by this interaction, and this might explain how wild-type MV strains gain entry into these cells (Bieback et al. 2002, J Virol 76:8729–8736).

Acknowledgements. The authors thank the Deutsche Forschungsgemeinschaft, the Robert Pfleger Stiftung, the Bundesministerium für Bildung and Forschung, the World Health Organisation and the Humboldt foundation for financial support of their laboratory work.

References

Akira S, Takeda K, Kaisho T (2001) Toll-like receptors: critical proteins linking innate and acquired immunity. Nat Immunol 2:675–680

Alexopoulou L, Holt A, Medzhitov R, Flavell RA (2001) Recognition of double-stranded RNA and activation of NF-kB by Toll-like receptor 3. Nature 413:732–738

Ardeshna KM, Pizzey AR, Devereux S, Khwaja A (2000) The PI3 kinase, p38 SAP kinase, and NF-kB signal transduction pathways are involved in the survival and maturation of lipopolysaccharide-stimulated human monocyte-derived dendritic cells. Blood 96:1039–1046

Astier A, Trescol-Biemont MC, Azocar O, Lamouille B, Rabourdin-Combe C. (2000) CD46, a new costimulatory molecule for T cells, that induces p120^{CBL} and LAT phosphorylation. J Immunol 164:6091–6095

Aversa G, Chang CCJ, Carballido JM, Cocks BG, de Vries JE (1997) Engagement of the signalling lymphocytic activation molecule (SLAM) on activated T cells results in IL-2-independent, cyclosporin A-sensitive T cell proliferation and IFN-gamma production. J Immunol 158:4036–4044

Avota E, Avots A, Niewiesk N, Kane LP, Bommhardt U, ter Meulen V, Schneider-Schaulies S (2001) Disruption of Akt kinase activation is important for immunosuppression induced by measles virus. Nat Med 7:725–731

Baker KA, Dutch RE, Lamb RA, Jardetzky TS (1999) Structural basis for paramyxovirus mediated membrane fusion. Mol Cell 3:309–319

Bartz R, Brinckmann U, Dunster LM, Rima B, ter Meulen V, Schneider-Schaulies J (1996) Mapping amino acids of the measles virus hemagglutinin responsible for receptor (CD46) downregulation. Virology 224:334–337.

Bleharski JR, Niazi KR, Sieling PA, Cheng G, Modlin RL. (2001) Signaling lymphocytic activation molecule is expressed on CD40 ligand-activated dendritic cells and directly augments production of inflammatory cytokines. J Immunol 167:3174–3181

Blixenkrone-Moller M, Bernard A, Bencsik A, Sixt N, Diamond L, Logan JS, Wild F (1998) Role of CD46 in measles virus infection in CD46 transgenic mice. Virology 249:238–248

Bolt G, Pedersen IR (1999) The role of subtilisin-like proprotein convertases for cleavage of the measles virus fusion glycoprotein in different cell types. Virology 252:387–398

Borrow P, Oldstone MBA (1995) Measles virus- mononuclear cell interactions. In "Measles Virus" (Billeter MA, ter Meulen V, eds.), Vol. 191, pp. 85–100. Springer-Verlag, Berlin, Heidelberg, New York.

Brennan P, Babbage JW, Burgering BMT, Groner B, Reif K, Cantrell DA (1997) Phosphatidylinositol 3-kinase couples the interleukin-2 receptor to the cell cycle regulator E2F. Immunity 7:679–689

Buckland R, Malvoisin E, Beauverger P, Wild T (1992) A leucine zipper structure present in the measles virus fusion protein is not required for its tetramerization but is essential for fusion. J Gen Virol 73:1703–1707

Casasnovas JM, Larvie M, Stehle T (1999) Crystal structure of two CD46 domains reveals an extended measles virus-binding surface. EMBO J. 18:2911–2922

Cathomen T, Mrkic B, Spehner D, Drillien R, Naef R, Pavlovic J, Aguzzi A, Billeter MA, Cattaneo R (1997) A matrix-less measles virus is infectious and elicits extensive cell fusion: consequences for propagation in the brain. EMBO J 17:3899–3908

Cella M, Salio M, Sakakibara Y, Langen H, Julkunen I, Lanzavecchia A (1999) Maturation, activation and protection of dendritic cells induced by double stranded RNA. J Exp Med 189:821–829

Clements CJ, Cutts FT (1995) The epidemiology of measles: thirty years of vaccination. In "Measles Virus" (Billeter MA, ter Meulen V, eds.), Vol. 191, pp. 13–34. Springer-Verlag, Berlin, Heidelberg, New York.

Dörig RE, Marcil A, Chopra A, Richardson CD (1993) The human CD46 molecule is a receptor for measles virus (Edmonston strain). Cell 75:295–305

Engelking O, Fedorov LM, Lilischkis R, ter Meulen V, Schneider-Schaulies S (1999) Measles virus-induced immunosuppression in vitro is associated with deregulation of G1 cell cycle control proteins. J Gen Virol 80:1599–1608

Erlenhoefer C, Wurzer W, Loeffler S, Schneider-Schaulies S, ter Meulen V, Schneider-Schaulies J (2001) CD150 (SLAM) is a receptor for measles virus but is not involved in contact-mediated proliferation inhibition of lymphocytes. J Virol 75:4499–4505.

Esolen LM, Ward BJ, Moench TR, Griffin DE (1993) Infection of monocytes during measles. J Infect Dis 168:47–52

Forthal DN, Aarnaes S, Blanding J, de la Maza L, Tilles JG (1992) Degree and length of viremia in adults with measles. J Infect Dis 166:421–424

Fugier-Vivier I, Servet-Delprat C, Rivailler P, Rissoan M, Liu Y, Rabourdin-Combe C (1997) Measles virus suppresses cell-mediated immunity by interfering with the survival and function of dendritic cells. J Exp Med 186:813–823

Gans HA, Maldonado Y, Yusakawa LL, Beeler J, Audet S, Rinki MM, DeHovitz R, Arvin AA (1999) IL-12, IFN-γ and T cell proliferation to measles in immunised infants. J Immunol 162:5569–5575

Griffin DE (1995) Immune responses during measles virus infection. In "Measles Virus" (Billeter MA, ter Meulen V, eds.), Vol. 191, pp. 117–134. Springer-Verlag, Berlin, Heidelberg, New York.

Grosjean I, Caux C, Bella C, Berger I, Wild F, Banchereau J, Kaiserlian D (1997) Measles virus infects human dendritic cells and blocks their allostimulatory properties for CD4+ T cells. J Exp Med 186:801–812

Horvat B, Rivailler P, Varior-Krishnan G, Cardoso A, Gerlier D, Rarourdin-Combe C (1996) Transgenic mice expressing human measles virus (MV) receptor CD46 provide cells exhibiting different permissivities to MV infections. J Virol 70:6673–6681

Hsu EC, Iorio C, Sarangi F, Khine AA, Richardson CD (2001) CDw150 (SLAM) is a receptor for a lymphotropic strain of measles virus and may account for the immunosuppressive properties of this virus. Virology 279:9–21

Hussey GD, Goddard EA, Hughes J, Ryon JJ, Kerran M, Carelse E, Strebel PM, Markowitz LE, Moodie J, Barron P, Zaninub L, Sayed R, Beatty D, Griffin DE (1996) The effect of Edmonston-Zagreb and Schwarz measles vaccines on immune responses in infants. J Infect Dis 173:1320–1326

Hyypiae T, Korkiamaki P, Vanionpaa R (1985) Replication of measles virus in human lymphocytes. J Exp Med 161:1261–1271

Johnston ICD, ter Meulen V, Schneider-Schaulies J, Schneider-Schaulies S (1999) A recombinant measles vaccine virus expressing wild-type glycoproteins: consequences for viral spread and cell tropism. J Virol 73:6903–6915

Kadowaki N, Ho S, Antonenko S, de Waal Malefyt R, Kastelein RA, Bazan F, Liu Y (2001) Subsets of human dendritic cell precursors express different Toll-like receptors and respond to different microbial antigens. J Exp Med 194:863–869

Karp CL, Wysocka M, Wahl LM, Ahearn JM, Cuomo PJ, Sherry B, Trinchieri G, Griffin, DE (1996) Mechanism of suppression of cell-mediated immunity by measles virus. Science 273:228–231

Katayama Y, Hirano A, Wong TC (2000) Human receptor for measles virus (CD46) enhances nitric oxide production and restricts virus replication in mouse macrophages by modulating the production of alpha/beta interferon. J Virol 74:1252–1257

Katz M (1995). Clinical spectrum of measles. In "Measles Virus" (Billeter MA, ter Meulen V, eds.), Vol. 191, pp. 1–12. Springer-Verlag, Berlin, Heidelberg, New York.

Klagge IM, Schneider-Schaulies S. (1999) Virus interactions with dendritic cells. J Gen Virol 80:823–833

Klagge IM, ter Meulen V, Schneider-Schaulies S (2000) Measles virus-induced promotion of dendritic cell maturation by soluble mediators does not overcome the immunosuppressive activity of viral glycoproteins on the cell surface. Eur J Immunol 30:2741–2750

Knight SC, Patterson P (1997) Bone-marrow derived dendritic cells, infection with human immunodeficiency virus and immunopathology. Annu Rev Immunol 15:593–615

Kohama T, Garten W, Klenk HD (1981) Changes in conformation and charge paralleling proteolytic activation of Newcastle disease virus glycoproteins. Virology 111:364–376

Kruse M, Meinl E, Henning G, Kuhnt C, Berchtold S, Berger T, Schuler G, Steinkasserer A (2001) Signaling lymphocyte activation molecule is expressed on mature CD83+ dendritic cells and is upregulated by IL-1β. J Immunol 167:1989–1995

Kurita-Taniguchi M, Fukui A, Hazeki K, Hirano A, Tsuji S, Matsumoto M, Watanabe M, Ueda S, Seya T (2000) Functional modulation of human macrophages through CD46 (measles virus receptor): production of IL-12 p40 and nitric oxide in association with recruitment of protein-tyrosine phosphatase SHIP-1 to CD46. J Immunol 165:5143–5152

Kurt-Jones EA, Popova L, Kwinn L, Haynes LM, Jones LP, Tripp RA, Walsh EE, Freeman MW, Golenbock DT, Anderson LJ, Finberg RW (2000) Pattern recognition receptors TLR4 and CD14 mediate response to respiratory syncytial virus. Nat Med 1:398–401

Lamb R (1993) Paramyxovirus fusion: a hypothesis for changes. Virology 197:1–11

Lambert DM, Barney S, Lambert AL, Guthrie K, Medinas R, Davis D, Bucy T, Erickson J, Merutka G, Petteway SR (1996) Peptides from conserved regions of paramyxovirus fusion proteins are potent inhibitors of viral fusion. Proc Natl Acad Sci USA 93:2186–2191

Latour S, Gish G, Helgason CD, Humphries RK, Pawson T, Veillette A (2001) Regulation of SLAM-mediated signal transduction by SAP, the X-linked lymphoproliferative gene product. Nat Immunol 2:681–690

Lecouturier V, Fayolle J, Caballero M, Carabana J, Celma ML, Fernandez-Munoz R, Wild TF, Buckland R (1996) Identification of two amino acids in the hemagglutinin glycoprotein of measles virus (MV) that govern hemadsorption, HeLa cell fusion, and CD46 downregulation: phenotypic markers that differentiate vaccine and wild-type MV strains. J Virol 70:4200–4204

Luft T, Pang KC, Thomas E, Hertzog P, Hart DNJ, Trapani JCJ (1998) Type I interferons enhance the terminal differentiation of dendritic cells. J Immunol 161:1947–1953

Maisner A, Mrkic B, Herrler G, Moll M, Billeter MA, Cattaneo R, Klenk HD (2000) Recombinant measles virus requiring an exogenous protease for activation of infectivity. J Gen Virol 81:441–449

Malvoisin E, Wild T (1993) Measles virus glycoproteins: studies on the structure and interaction of the haemagglutinin and fusion proteins. J Gen Virol 74:2365–2372

Manchester M, Liszewski MK, Atkinson JP, Oldstone MB (1994) Multiple isoforms of CD46 (membrane cofactor protein) serve as receptors for measles virus. Proc Natl Acad Sci USA 91:2161–2165

Marie JC, Kehren J, Trescol-Biemont MC, Evlashev A, Valentin H, Walzer T, Tedone R, Loveland B, Nicolas JF, Rabourdin-Combe C, Horvat B (2001) Mechanisms of measles virus-induced suppression of inflammatory immune responses. Immunity 14:69–79

McChesney MB, Altmann A, Oldstone MBA (1988) Suppression of T lymphocyte function by measles virus is due to cell cycle arrest in G1. J Virol 140:1269–1273

McChesney MB, Rota, PA, Zhu YD, Antipa L, Lerche NW, Ahmed R, Bellini WJ (1997) Experimental measles. I. Pathogenesis in the normal and the immunized host. Virology 233:74–84

Moll M, Klenk HD, Herrler G, Maisner A (2001) A single amino acid change in the cytoplasmic domains of measles glycoproteins H and F alters targeting, endocytosis and cell fusion in polarised Madin-Darby canine kidney cells. J Biol Chem 276:17887–17894

Mrkic B, Odermatt B, Klein MA, Billeter MA, Pavlovic J, Cattaneo R (2000) Lymphatic dissemination and comparative pathology of recombinant measles viruses in genetically modified mice. J Virol 74:1364–1372

Naim HY, Ehler E, Billeter MA (2000) Measles virus matrix protein specifies apical virus release and glycoprotein sorting in epithelial cells. EMBO J 19:3576–3585

Naniche D, Varior-Krishnan G, Cervoni F, Wild TF, Rossi B, Rabourdin-Combe C, Gerlier D (1993) Human membrane cofactor protein (CD46) acts as a cellular receptor for measles virus. J Virol 67:6025–6032

Naniche D, Reed SI, Oldstone MBA (1999) Cell cycle arrest during measles virus infection: a G0-like block leads to suppression of Retinoblastoma protein expression. J Virol 73:1894–1901

Naniche D, Yeh A, Eto D, Manchester M, Friedman RM, Oldstone MBA (2000) Evasion of host defenses by measles virus: wildtype measles virus infection interferes with induction of alpha/beta interferon production. J Virol 74:7478–7484

Nielsen L, Blixenkrone-Moller M, Thylstrup M, Hansen NJV, Bolt G (2001) Adaptation of wild-type measles virus to CD46 receptor usage. Arch Virol 146:197–208

Niewiesk S, Eisenhuth I, Fooks A, Clegg JC, Schnorr JJ, Schneider-Schaulies S, ter Meulen V (1997a) Measles virus-induced immune suppression in the cotton rat (*Sigmodon hispidus*) model depends on viral glycoproteins. J Virol 71:7214–7219

Niewiesk S, Schneider-Schaulies J, Ohnimus H, Jassoy C, Schneider-Schaulies S, Diamond L, Logan J. ter Meulen V (1997b) CD46 expression does not overcome the intracellular block of measles virus replication in transgenic rats. J Virol 71:7969–7973

Niewiesk S, Ohnimus H, Schnorr JJ, Götzelmann M, Schneider-Schaulies S, Jassoy C, ter Meulen V (1999) Measles virus-induced immunosuppression in cotton rats is associated with cell cycle retardation in uninfected lymphocytes. J Gen Virol 80:2023–2029

Nussbaum O, Broder CC, Moss B, Stern LB, Rozenblatt S, Berger EA (1995) Functional and structural interactions between measles virus hemagglutinin and CD46. J Virol 69:3341–3349

O'Shea JJ, Visconti R (2000) Type 1 IFNs and regulation of TH1 responses: enigmas both resolved and emerge. Nat Immunol 1:17–19

Ohgimoto S, Ohgimoto K, Niewiesk S, Klagge IM, Pfeuffer J, Johnston ICD, Schneider-Schaulies J, Weidmann A, ter Meulen V, Schneider-Schaulies S (2001) The hemagglutinin protein is an important determinant for measles virus tropism for dendritic cells in vitro and immunosuppression in vivo. J Gen Virol 82:1835–1844

Paquette RL, Hsu NC, Kiertschner SM, Park AN, Tran L, Roth MD, Glapsy JA (1998) Interferon alpha and granulocyte-macrophage stimulating factor differentiate peripheral blood monocytes into potent antigen-presenting cells. J Leukoc Biol 64:358–367

Patterson JB, Thomas D, Lewicki H, Billeter MA, Oldstone MBA (2000) V and C proteins of measles virus function as virulence factors in vivo. Virology 267:80–89

Plemper RK, Hammond L, Cattaneo R (2000) Characterisation of a region of the measles virus hemagglutinin sufficient for its dimerisation. J Virol 74:6485–6493

Radecke F, Billeter MA (1996) The nonstructural C protein is not essential for multiplication of Edmonston B strain measles virus in cultured cells. Virology 217:418–421

Richardson CD, Choppin PW (1983) Oligopeptides that specifically inhibit membrane fusion by paramyxoviruses: studies on the site of action. Virology 131:518–532

Rima BK, Earle JAP, Yeo RP, Herlihy L, Baczko K, ter Meulen, V, Carabana J, Caballero M, Celma ML, Fernandez-Munoz R (1995) Temporal and geographical distribution of measles virus genotypes. J Gen Virol 76:1173–1180

Rima BK, Earle JAP, Baczko K, ter Meulen V, Carabana J, Caballero M, Celma ML, Fernandez-Munoz R (1997) Sequence divergence of measles virus haemagglutinin during natural evolution and adaptation to cell culture. J Gen Virol 78:97–106

Samuel O, Shai Y (2001) Participation of two fusion peptides in measles virus-induced membrane fusion: emerging similarity with other paramyxoviruses. Biochem 40:1340–1349

Schlender J, Schnorr JJ, Spielhofer P, Cathomen T, Cattaneo R, Billeter MA, ter Meulen V, Schneider-Schaulies S (1996) Interaction of measles virus glycoproteins with the surface of uninfected peripheral blood lymphocytes induces immunosuppression in vitro. Proc Natl Acad Sci USA 93:13194–13199

Schneider H, Kaelin K, Billeter MA (1997) Recombinant measles viruses defective for RNA editing and V protein synthesis are viable in cultured cells. Virology 227:314–322

Schneider-Schaulies J, Dunster LM, Kobune F, Rima B, ter Meulen V (1995a) Differential downregulation of CD46 by measles virus strains. J Virol 69:7257–7259

Schneider-Schaulies J, Dunster LM, Schneider-Schaulies S, ter Meulen V (1995b) Pathogenetic aspects of measles virus infections. Vet Microbiol 44:113–125

Schneider-Schaulies J, Schnorr JJ, Brinckmann U, Dunster LM, Baczko K, Liebert UG, Schneider-Schaulies S, ter Meulen V (1995c). Receptor usage and differential downregulation of CD46 by measles virus wild-type and vaccine strains. Proc Natl Acad Sci USA 92:3943–3947

Schneider-Schaulies J, Schnorr JJ, Schlender J, Dunster LM, Schneider-Schaulies S, ter Meulen V (1996) Receptor (CD46) modulation and complement-mediated lysis of uninfected cells after contact with measles virus-infected cells. J Virol 70:255–263

Schneider-Schaulies S, Kreth HW, Hofmann G, Billeter M, ter Meulen V (1991) Expression of measles virus RNA in peripheral blood mononuclear cells of patients with measles, SSPE, and autoimmune diseases. Virology 182:703–711

Schneider-Schaulies S, Schneider-Schaulies J, Schuster A, Bayer M, Pavlovic J, ter Meulen V (1994) Cell type-specific MxA-mediated inhibition of measles virus transcription in human brain cells. J Virol 68:6910–6917

Schneider-Schaulies S, ter Meulen V (1998) Measles virus induced immunosuppression. Nova Acta Leopoldina 307:1–13

Schnorr JJ, Schneider-Schaulies S, Simon-Jodicke A, Pavlovic J, Horisberger MA, ter Meulen V (1993) MxA-dependent inhibition of measles virus glycoprotein synthesis in a stably transfected human monocytic cell line. J Virol 67:4760–4768

Schnorr JJ, Dunster LM, Nanan R, Schneider-Schaulies J, Schneider-Schaulies S, ter Meulen V (1995) Measles virus-induced down-regulation of CD46 is associated with enhanced sensitivity to complement-mediated lysis of infected cells. Eur J Immunol 25:976–984

Schnorr JJ, Seufert M, Schlender J, Borst J, Johnston ICD, ter Meulen V, Schneider-Schaulies S (1997a) Cell cycle arrest rather than apoptosis is associated with measles virus contact-mediated immunosuppression in vitro. J Gen Virol 78:3217–3226

Schnorr JJ, Xanthakos S, Keikavoussi P, Kampgen E, ter Meulen V, Schneider-Schaulies S (1997b) Induction of maturation of human blood dendritic cell precursors by measles virus is associated with immunosuppression. Proc Natl Acad Sci USA 94:5326–5331

Schnorr JJ, Cutts FT, Wheeler JG, Zaman SMA, Alam S, Azim T, Schneider-Schaulies S, ter Meulen V (2001). Immune activation after measles vaccination of 6–9 months old Bangladeshi infants. Vaccine 19:1503–1510

Servet-Delprat C, Vidalain O, Azocar O, Le Deist F, Fischer A, Rabourdin-Combe C (2000a) Consequences of Fas-mediated human dendritic cell apoptosis induced by measles virus. J Virol 74:4387–4393

Servet-Delprat C, Vidalain O, Bausinger H, Manie O, Le Deist F, Azocar O, Fischer A, Rabourdin-Combe C (2000b) Measles virus induces abnormal differentiation of CD40-ligand activated human dendritic cells. J Immunol 164:1753–1760

Sevilla N, Kunz S, Holz A, Lewicki H, Homann D, Yamada H, Campbell KP, de la Torre JC, Oldstone MBA (2000) Immunosuppression and resultant viral persistence by specific targeting of dendritic cells. J Exp Med 192:1249–1260

Steineur M, Grosjean I, Bella C, Kaiserlian D (1998) Langerhans cells are susceptible to measles virus infection and actively suppress T cell proliferation. Eur J Dermatol 8:413–420

Tatsuo H, Ono N, Yanagi Y (2000) SLAM (CDw150) is a cellular receptor for measles virus. Nature 406:893–897

Tober C, Seufert M, Schneider H, Billeter MA, Johnston ICD, Niewiesk S, ter Meulen V, Schneider-Schaulies S (1998) Expression of measles virus V protein is associated with pathogenicity and control of viral RNA synthesis. J Virol 72:8124–8132

Valsamakis A, Schneider H, Auwaerter PG, Kaneshima H, Billeter MA, Griffin DE (1998) Recombinant measles viruses with mutations in the C, V or F reading gene have altered growth phenotypes in vivo. J Virol 72:7754–7761

Vidalain O, Azocar. Lamouille B, Astier A, Rabourdin-Combe C, Servet-Delprat C (2000) Measles virus induces functional TRAIL production by human dendritic cells. J Virol 74:556–559

Vidalain O, Azocar O, Yagita H, Rabourdin-Combe C, Servet-Delprat C (2001) Cytotoxic activity of human dendritic cells is differentially regulated by double stranded RNA and CD40 ligand. J Immunol 167:3765–3772

Ward BJ, Griffin DE (1993) Changes in cytokine production after measles virus vaccination: predominant production of IL-4 suggests induction of a Th2 response. Clin Immunol Immunopathol 67:171–177

Weidmann A, Fischer C, Ohgimoto S, Rüth C, ter Meulen V, Schneider-Schaulies S (2000a) Measles virus-induced immunosuppression in vitro is independent of complex glycosylation of viral glycoproteins and hemifusion. J Virol 74:7548–7553

Weidmann A, Maisner A, Garten W, Seufert M, ter Meulen V, Schneider-Schaulies S (2000b) Proteolytic cleavage of the fusion protein but not membrane fusion is required for measles virus-induced immunosuppression in vitro. J Virol 74:1985–1993

Wong BR, Besser D, Kim N, Arron JR, Vologodskaia M, Hanafusa H, Choi Y (1999) TRANCE, a TNF family member, activates Akt/protein kinase B through a signaling complex involving TRAF6 and c-Src. Mol Cell 4:1041–1049

Yanagi Y, Cubitt BA, Oldstone MB (1992) Measles virus inhibits mitogen-induced T cell proliferation but does not directly perturb the T cell activation process inside the cell. Virology 187:280–289

Chapter 5

Measles Virus and Dendritic Cell Functions: How Specific Response Cohabits with Immunosuppression

C. Servet-Delprat, P.-O. Vidalain, H. Valentin, C. Rabourdin-Combe

Abstract. Measles virus (MV) infection induces both an efficient MV-specific immune response and a transient but profound immunosuppression characterised by a panlymphopenia that occasionally results in opportunistic infections responsible for a high rate of mortality in children. On the basis of in vitro studies, the putative roles of dendritic cells

C. Servet-Delprat (✉), P.-O. Vidalain, H. Valentin, C. Rabourdin-Combe
Immunobiologie Fondamentale et Clinique, CERVI-INSERM U503,
21 avenue Tony Garnier, 69007 Lyon, France

(DCs) in MV infection are discussed. (1) DCs could participate in anti-MV innate immunity because MV turns on TNF-related apoptosis-inducing ligand (TRAIL)-mediated DC cytotoxicity. (2) Cross-priming by non-infected DCs might be the route of MV adaptive immune response. (3) After CD40-ligand activation in secondary lymphoid organs, MV-infected DCs could initiate the formation of Warthin-Finkeldey multinucleated giant cells, replicating MV and responsible for in vivo spreading of MV. (4) We review how integrated viral attack of the host immune system also targets DCs: Progress in understanding the immunobiology of MV-infected DCs that could account for MV-induced immunosuppression observed in vivo is presented and their potential role in lymphopenia is underlined. In conclusion, future research directions are proposed.

Abbreviations. AIDS, acquired immunodeficiency syndrom; APC, antigen presenting cells; CD40L, CD40 ligand; DC(s), dendritic cell(s); dsRNA, double strand RNA; H, haemaglutinin; HIV, human immunodeficiency virus; IFN, interferon; IL, interleukin; LPS, lipopolysaccharide; MHC-I/MHC-II, major histocompatibility complex class I/II; MV, measles virus; NP, nucleoprotein; TLR, toll receptor; TRAIL, TNF-related apoptosis-inducing ligand; WFC, Warthin-Finkeldey cells

1 Global Presentation of Measles

Measles virus (MV) (see Sect. 5 for structure and cycle) first infects the human respiratory tract, but the initial target cells are not entirely characterised (Fig. 1, A): As yet, we do not know whether human respiratory tract dendritic cells (DCs) are infected or not in vivo. In vitro, immature DCs weakly replicate MV (Fugier-Vivier et al. 1997). After a primo-replication at this entry site, anti-viral innate immunity does not eliminate MV that reaches the draining lymph nodes. We postulate that either virion may go alone through the lymphatic vessels or migrating infected DCs may transport replicating virus to these secondary lymphoid organs. Intensive replication occurs there, and then the virus disseminates through the whole body, infecting blood, spleen, lymphatic tissue, lung, thymus, liver and finally skin with a typical rash about 15 days after infection. Two paradoxical features accompany measles infection: On one hand, the immune system initiates an efficient adaptive immune re-

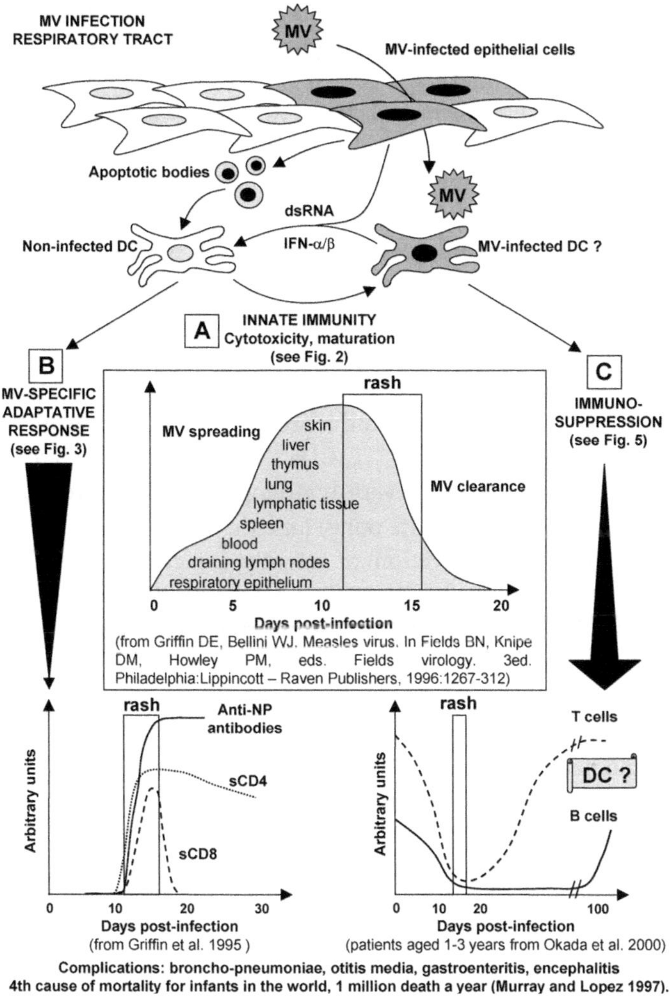

Fig. 1. Model of measles infection. Among various cell types, measles virus (*MV*) could infect epithelial and dendritic cells in respiratory tract. After a first phase of innate immunity in the periphery, which could involve DCs (*A*) besides NK cells, MV reach draining lymph nodes where a boost of MV replication occurs before dissemination. Non-infected migrating DCs could initiate MV-specific response (*B*), whereas infected DCs could participate in MV-induced immunosuppression (*C*). After spreading into the whole body, MV is eliminated, whereas immunosuppression persists for 2–3 months in patients aged 1–3 years

sponse (Fig. 1, B) that succeeds in virus clearance after around 20 days. Production of anti-MV antibodies (Graves et al. 1984; Norrby and Gollmar 1972) and differentiation of MV-specific $CD4^+$ or $CD8^+$ T cell clones (Ilonen et al. 1990; Jacobson et al. 1989; van Binnendijk et al. 1989) have been documented. In vivo, detections of soluble CD4 and soluble CD8 attest to the TCR engagement of $CD4^+$ and $CD8^+$ T cells (Furukawa et al. 1991; Griffin and Ward 1993). On the other hand, MV induces a transient but profound immunosuppression (Fig. 1, C) that takes place until 2 months after infection, depending on sanitary status and age of the patient. Consequently, MV presents the interesting and unique case of inducing a prolonged immunosuppression when there is no more virus. Measles patients were found to be in a severe panleukopenic state after onset of rash, where monocytes and a broad spectrum of various subsets of lymphocytes (including $CD4^+$ T, $CD8^+$ T and B cells) were depressed (Okada et al. 2000). Such a severe lymphopenia, similar to that of the terminal stage of AIDS, permits opportunistic infections with exogenous infective agents or re-activation of latently infecting agents. This explains why MV infection is responsible for an acute childhood disease that remains the fourth greatest cause of infant mortality in the world (Murray and Lopez 1997).

After this global presentation, we focus on the putative roles of DCs during measles. First, the TNF-related apoptosis-inducing ligand (TRAIL)-mediated cytotoxic function of DCs during anti-MV innate immunity is discussed. Second, the cross-priming ability of DCs as a route for MV adaptive immune response is discussed. Next, the CD40-activation of MV-infected DCs as a molecular key to boost MV replication is outlined, and, finally, the DCs are presented as a target of the integrated MV-induced immunosuppression.

2
Anti-Viral Innate Immunity: MV Infection Induces Maturation and Turns on TRAIL-Mediated Cytotoxic Functions in Dendritic Cells

Immature DCs form a network within all epithelia and mucosa, such as the Langerhans cells in the skin or the respiratory tract DCs in the respiratory mucosa. Since 1997, the human DC has been known to be susceptible to MV infection in vitro (Fugier-Vivier et al. 1997; Grosjean et al. 1997; Schnorr et al. 1997; Servet-Delprat et al. 2000b; Steineur et al. 1998). Even when few DCs are infected, the whole DC population undergoes maturation as demonstrated by (1) upregulation of CD40, CD80, major histocompatibility complex class II (MHC II) and MHC I, (2) induction of CD25, CD69, CD71, CD83 and CD86, (3) down-regulation of CD1a and CD32 and (4) weak induction of IL-12, IL-1α/β and IL-6 mRNA synthesis. When triggered by pathogens, the pattern recognition receptors expressed by immature DCs cause them to mature to immunogenic DCs.

Recently, double-strand RNA (dsRNA) was demonstrated to be the ligand for Toll-like receptor-3 (TLR-3) expressed by DCs (Alexopoulou et al. 2001). Such a dsRNA-TLR-3 interaction might be involved in the maturation of DCs observed after MV infection (Fig. 2, A). Interestingly, dsRNA or Edmonston strain of MV induces IFN-β production by DCs (Fig. 2, B) (Vidalain et al. 2001). IFN-α/β limit MV replication (Leopardi et al. 1992) and induce the synthesis of TRAIL by DCs (Vidalain et al. 2000) (Fig. 2, C). TRAIL is a type II transmembrane protein that belongs to the TNF family closely related to three other death-inducing ligands: FasL, TNF-α and TWEAK/Apo3L (Pitti et al. 1996; Wiley et al. 1995). It has been involved in apoptosis of a tumoural cell line (Pitti et al. 1996) and virus-infected cells (Sedger et al. 1999). After a 12-h infection with MV, MV-infected DCs are able to kill MDA231, a TRAIL-sensitive carcinoma cell line. As the death of MDA231 was specifically blocked by a TRAIL receptor:Fc chimera, cytotoxic activity of DCs was clearly mediated by TRAIL (Vidalain et al. 2000). Furthermore, monocytes have also been described to express and produce functional TRAIL after IFN-α treatment (Griffith et al. 1999).

Thus IFN-α/β-activated DC and monocytes may exert an innate TRAIL-mediated cytotoxic activity towards surrounding MV-infected cells (Fig. 2, D). However, some wild-type strains of MV limit induction of IFN-α/β (Naniche et al. 2000). On one hand, this could partially explain why innate immunity is unable to control MV infection, but on the other hand, natural killer NK cells, which also support anti-viral innate immunity, were increased in number and activated during measles (Okada et al. 2000).

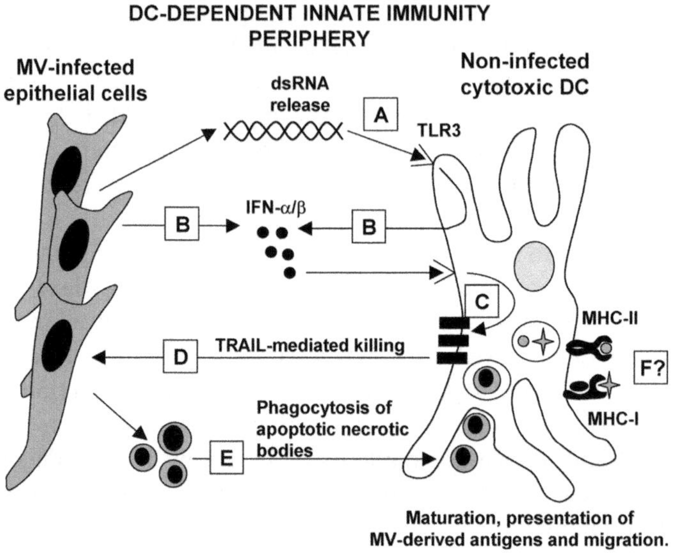

Fig. 2. Natural killer activity of DC: a link between innate and adaptive immunity. MV-infected epithelial cells in the respiratory tract replicate MV and consequently may produce dsRNA (*A*) and IFN-α/β (*B*), two molecular signals able to induce TRAIL synthesis by DCs (*C*). DCs are able to kill TRAIL-sensitive target (*D*) and subsequently phagocytise resulting apoptotic and/or necrotic bodies (*E*). Cross-priming may lead to MHC I and MHC II MV peptide presentation at the DC surface (*F*)

3
Cross-Priming by Non-Infected Dendritic Cells: The Route of MV Adaptive Immune Response?

Elimination of MV requires the initiation of adaptive immune response. DCs possess a specialised set of functions which allow them to act as professional antigen-presenting cells (APC), including pathogen recognition, antigen capturing, antigen processing, migratory capacity from the periphery to the draining secondary lymphoid organs and costimulatory molecule expression. As discussed further below, MV-infected DCs undergo an abnormal CD40-dependent maturation, as they are no longer able to exert their APC functions. We propose that the adaptive immune response might be initiated by non-infected DCs that have engulfed apoptotic bodies of MV-infected cells.

In ds-RNA-stimulated DCs, it has been demonstrated that cytotoxic DCs are able to phagocyte their target (Vidalain et al. 2002; Fig. 2, E). Indeed, cytotoxic DCs that have killed green fluorescence-labelled target cells with TRAIL, were able to phagocytise resulting apoptotic bodies as demonstrated by emergence of a new CMH-II$^+$/ green$^+$ DC population. Non-infected DCs were shown to undergo maturation after engulfment of MV-infected apoptotic cells (Servet-Delprat et al. 2000b). Maturing DCs upregulate CCR7, which is essential to drive their migration first to the lymphatics and then to the T cell areas, where the cognate ligands CCL19/ELC and CCL21/SLC are produced (Forster et al. 1999; Gunn et al. 1999). Cross-priming might occur: DCs which have processed MV-derived antigens from apoptotic or necrotic bodies might present MV-derived peptides both on MHC II and MHC I molecules (Albert et al. 1998; Fig. 2, F). In the absence of direct infection of DCs, or when infected DCs are unable to prime the immune response (as MV-infected DCs), cross-priming is an essential component of the immune response against pathogens, but this is not yet firmly established for MV.

Mature DCs produce chemokines that attract naïve T cells (Adema et al. 1997). Providing MHC-peptide complexes, co-stimulatory molecules and a cytokine pattern related to the pathogen, DCs prime naïve T cells. A sustained TCR stimulation is required to commit naïve T cells to proliferate and differentiate into effector cells (Iezzi et al. 2001). As documented by Lanzavecchia and colleagues, the kinetics of DC activation influences IL-12 production. Thus recently activated DCs can efficiently prime Th1 responses, whereas post-24h-activated DCs preferentially prime Th2 response (Langenkamp et al. 2000). Such a model could explain inductions of MV-specific cellular and humoral immune responses observed in vivo. Recently activated DCs produce IL-12 and may induce differentiation of CD4$^+$ Th0 MV-specific naïve T cells into Th1 effectors able to help differentiation of CD8$^+$ naïve MV-specific T cells into cytotoxic T cells (CTL) (Fig. 3, A); then post-24h-activated DCs may induce differentiation of Th2 effectors able to sustain differentiation of naïve B cells into plasma cells that secrete MV-specific antibodies (Fig. 3, B).

The transport of native antigen to lymph nodes is an absolute requirement for the induction of antibody responses (Zinkernagel 2000). Either free virions may enter in the draining lymph nodes or, as DC-SIGN for HIV (Geijtenbeek et al. 2000) (see Sect. 3), DCs may express a receptor not involved in viral entry but promoting transport to secondary lym-

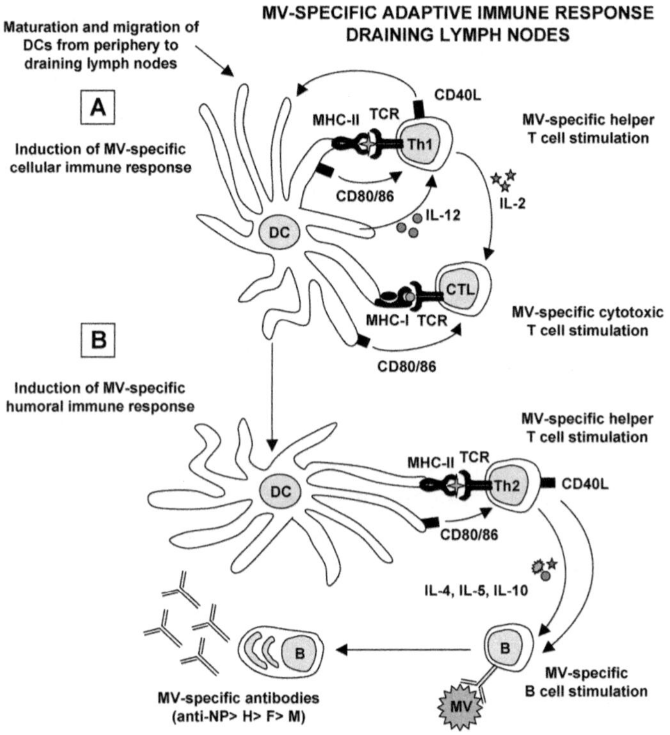

Fig. 3. Priming of MV-specific immune response by non-infected DCs. After maturation induced in the periphery after engagement of TLR3 and/or phagocytosis of necrotic bodies, DCs migrate to T cell area of draining lymph nodes. IL-12-secreting DCs, presenting MV-peptides-MHC complexes, may differentiate $CD4^+$ Th0-type T cells into Thl effector cells. These latter, providing CD40L activation, may participate in DC-dependent differentiation of $CD8^+$ pre-cytotoxic T cells into cytotoxic (CTL) effectors. Differentiation of MV-specific Thl and CTL effectors corresponds to induction of MV-specific cellular immune response (*A*). Then, as proposed by Lanzavecchia and colleagues, post-24-h activated DCs may initiate Th2 effectors able to cooperate with MV-activated B cells to organise MV-specific humoral immune response (*B*), in the B cell area

phoid organs; these virions may activate specific B cells and/or infect in trans cells that express CD46 or SLAM, two receptors used by wild-type MV strains (Manchester et al. 2000; Naniche et al. 1993; Tatsuo et al. 2000).

4
CD40 Ligand: The Signal Able to Boost Syncytia Formation and MV Replication in Infected Dendritic Cells

In secondary lymphoid organs, MV replication leads to syncytia formation. For a long time, giant multinucleated cells named Warthin-Finkeldey cells (WFCs) have been characterised in secondary lymphoid organs of measles patients (Tajima and Kudow 1976). WFCs have been also recently characterised in HIV infection, where they appear to represent a multinucleated form of follicular dendritic cells (Orenstein 1998). In vitro, syncytia formation is observed in MV-infected DC cultures and in DC-T cell cocultures (Fugier-Vivier et al. 1997). In this latter case, mixed DC-T cell syncytia have been detected (Fig. 4A). The presence of CD40 ligand (CD40L)-transfected fibroblasts or activated T cells increases both

A

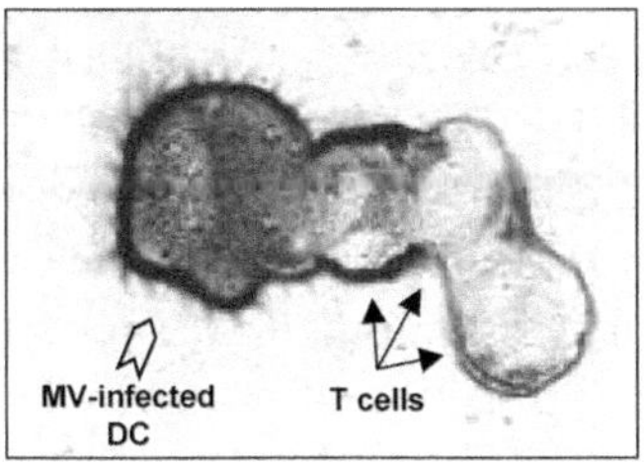

B

Number of syncytia / 10^6 DC	none	Inactivated UV-MV	MV	MV + anti-CD46 antibodies
DC	**0**	**0**	**31** +/- 8	**9** +/- 3
DC + $CD40L^+$ fibroblast or DC + activated T cells	**0**	**0**	**337** +/- 55	**31** +/- 4

Fig. 4A, B. DCs initiate syncytia formation. Immunochemical and morphologic analyses of MV-infected DC cultured alone, with CD40L-transfected fibroblasts or with activated T cells were performed as previously described (Fugier-Vivier et al. 1997), at day 3 after infection. **A** Anti-HLA-DR (MHC II) staining revealed formation of mixed DC-T cell syncytia. Here, originally infected DC (*left*) fused with three uninfected T cells identified by weak or absent MHC-II staining, reduced cell diameter and elevated nuclear/cytoplasm ratio (*right*). MHC II was progressing around the syncytia membrane from *left* to *right*. **B** Total syncytia were counted under microscope in 24-well plates containing 10^6 DCs/well. Experiments were performed with non-infected DCs, UV-inactivated MV-pulsed DCs and MV-infected DCs with or without CD46-blocking antibodies. Representative of three experiments

the size (5–50 nuclei) and the number (7- to 15-fold more) of multinuclear syncytia (Fig. 4B). All syncytia contain MV particles. Syncytia formation is blocked by anti-CD46 antibodies but does not occur in MV-infected monocyte cultures, although they express CD46. Consequently, CD46 is necessary but not sufficient to induce syncytia, and DCs express other proteins, which are absent from monocytes but essential for syncytia formation. We postulate that migrating MV-infected DCs may initiate syncytia formation in secondary lymphoid organs; the WFCs described above probably result from MV-induced syncytia including different cell types such as DCs, T cells and perhaps B cells and follicular dendritic cells. WFCs were morphologically and immunohistochemically divided into two groups. The first type was WFCs appearing in the germinal centres of lymphoid tissue, revealing many large nuclei and B cell markers. The second type was WFCs appearing in the interfollicular areas and thymus, showing many small hyperchromatic nuclei and T cell markers. The data indicate that there might be morphologic and immunophenotypic heterogeneity in WFCs (Nozawa et al. 1994). It would be interesting to further investigate MHC II, CD3, CD19 and CD21 expression in these giant multinucleated cells.

Syncytia formation is correlated with MV replication efficacy. Because nucleoprotein (NP) is the earliest and most abundant MV protein transcribed during the viral cell cycle, quantification of intracellular NP reflects the intensity of viral replication. As shown by detailed studies of intracellular NP synthesis, CD40 activation is necessary and sufficient to boost (10- to 100-fold) MV replication in infected DCs. In vitro, intracellular NP quantities increase proportionally to the amount of CD40L signal delivered to MV-infected DCs (Servet-Delprat et al. 2000a). As CD40L is a physiological signal that DCs receive after migration into T cell area of draining lymph nodes, we propose that CD40L expressed by activated T cells may enhance MV replication in DCs (Fig. 5, A). Consecutively, haemagglutinin (H) and fusion protein (F) transmembrane viral glycoproteins highly expressed by infected DCs may favour syncytia formation.

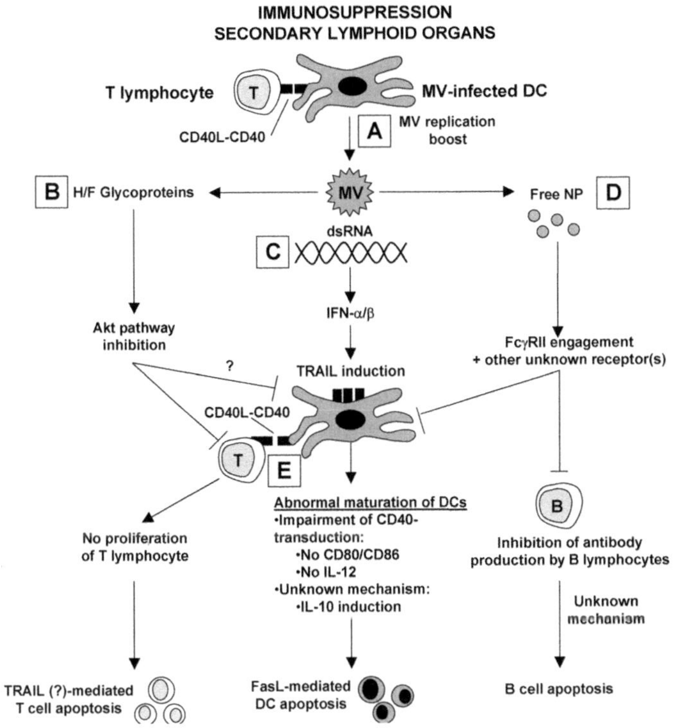

Fig. 5. MV-infected DCs, target of MV-induced immunosuppression. (*A*) CD40-activation of MV-infected DCs enhances MV replication, accumulating viral proteins and dsRNA. (*B*) H/F viral glycoproteins expressed at the surface of MV-infected cells inhibit Akt pathway, preventing T cell proliferation. Disruption of Akt pathway may also modify CD40-transduction in DC. (*C*) Like MV infection, dsRNA alone induces IFN-β production that mediates TRAIL induction by DCs. This death factor may be responsible for T cell apoptosis. (*D*) As demonstrated by anti-NP antibody production in vivo, and NP quantified by ELISA in UV-inactivated MV preparation in vitro (Marie et al. 2001), MV infection produces free extracellular NP. Binding of NP to FcγRII inhibits IL-12 production by DCs and antibody production by B cells. Unknown mechanism leads to extensive B cell lymphopenia observed in vivo. (*E*) When CD40-activated, MV-infected DCs show low IL-12 secretion and defective co-signal membrane molecules that render them unable to prime naïve T cells. They inhibit T cell proliferation and induce the death of activated T cells. Furthermore, IL-10 production is induced, which may differentiate Tr1 regulatory inhibitory effector T cells. Finally DCs undergo Fas-dependent apoptosis

5
MV-Induced Immunosuppression: An Integrated Viral Attack of the Host Immune System That Also Targets Dendritic Cells

5.1
Lymphopenia Induced by Cytolysis of Non-Infected Cells

Few peripheral blood mononuclear cells of measles patients (0.01%–0.1%) are infected, and they disappear just after the onset of rash (Esolen et al. 1993), indicating that indirect mechanisms rather than a direct effect of MV replication in lymphoid cells cause suppression. Similar to the observation of HIV-infected cells in AIDS patients (Gougeon et al. 1996), thousands times more non-infected lymphocytes were destroyed than virus-infected cells. This strongly suggests that the lymphopenia in measles patients was caused mostly by cytolysis of non-infected cells. In vivo, the mechanisms of this cytolysis, which is now believed to be the major cause of immunosuppression, remain to be elucidated. Apoptosis-associated molecules such as CD95 (Fas) and TRAIL on the cell surface of non-infected lymphocytes or plasma IL-18, which is responsible for inducing apoptosis in several infectious diseases, were highly expressed (Okada et al. 2000; Okada et al. 2001). As homeostasis of lymphocyte populations results from signals delivered in primary and secondary lymphoid organs, it would be interesting to study the mechanisms of MV-induced apoptosis in these environments.

5.2
Measles Infection in Primary Lymphoid Organs

Few data are available concerning MV infection of thymus and bone marrow. In vitro, MV infection induces terminal differentiation of human thymic epithelial cells (TECs) that leads to early apoptosis (Valentin et al. 1999). In vivo, thymic stromal cells have been described as a target for MV in the SCID-hu mouse model after direct intrathymic inoculation of MV leads to thymocyte apoptosis (Auwaerter et al. 1996). Because TECs are required for the generation of immunocompetent T lymphocytes, providing survival signal during positive selection, MV-induced terminal differentiation of TECs may contribute to immunosuppression. The last lymphocyte population to recover at the end of immunosuppression phase is composed of B cells (Okada et al. 2000), but no data are available concerning MV infection of B cell progenitors.

5.3 Measles Virus Infection in Secondary Lymphoid Organs: Role of H/F, NP and dsRNA

A number of mechanisms occur in secondary lymphoid organs. These involve viral proteins and dsRNA (Fig. 5).

- Membrane of MV-infected cells express H and F glycoproteins. H and F glycoproteins have been involved in direct inhibition of T cell activation by disruption of Akt pathway (Avota et al. 2001) (Fig. 5, B; see Sect. 5). H-CD46 interaction significantly suppresses inflammation in murine contact hypersensitivity model in vivo (Marie et al. 2001).
- dsRNA may be released during MV infection, as MV is a lytic virus with a ssRNA genome exhibiting a dsRNA structure during replication. However, it has not been demonstrated whether free dsRNA without NP exists. dsRNA induces IFN-α/β synthesis that activates TRAIL synthesis by DCs and monocytes (Griffith et al. 1999; Liu et al. 2001; Vidalain et al. 2001; Fig. 5, C). Besides its cytotoxic activity, TRAIL was recently shown to inhibit autoimmune arthritis and inflammation, possibly because it also acts as an inhibitor of activated T cell proliferation and cytokine production in some cases (Song et al. 2000).
- Cell-free NP is released during MV infection as demonstrated by anti-NP antibody production in vivo (Graves et al. 1984) and NP quantified by ELISA in UV-inactivated MV preparation in vitro (Marie et al. 2001). NP inhibits antibody production by B lymphocytes through engagement of their FcγRII (McChesney et al. 1986; Ravanel et al. 1997; Fig. 5D). In a murine delayed-type hypersensitivity model, NP was shown to abolish specific CD8$^+$ T cell proliferation and to decrease murine IL-12-secreting DCs in regional lymph nodes through engagement of their FcγR (Marie et al. 2001). Inhibition of human IL-12 production by MV has been documented in LPS-activated monocytes (Karp et al. 1996) and CD40-activated DCs (Fugier-Vivier et al. 1997). In both cases, inhibition could be obtained with UV-inactivated MV in the absence of MV replication, suggesting that structural viral proteins are directly involved. CD46-dependent (Karp et al. 1996)

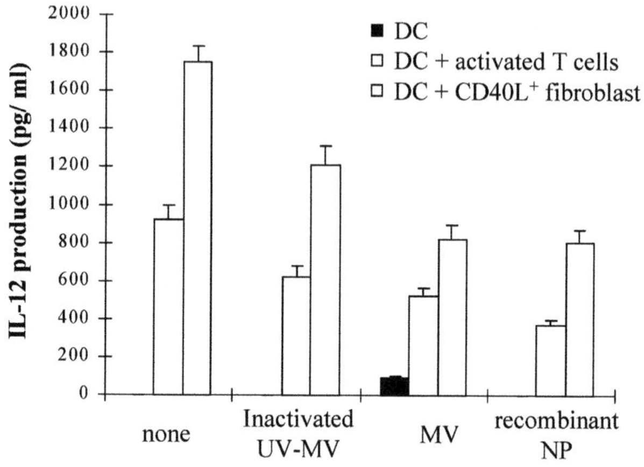

Fig. 6. Recombinant NP inhibits IL-12 production by human DCs. 10^6 DCs were cultured either alone or with activated T cells expressing CD40L or with CD40L-transfected fibroblasts, as previously described (Fugier-Vivier et al. 1997). Experiments were performed with non-infected DCs, UV-inactivated MV-pulsed DCs, MV-infected DCs or 1 μg/ml recombinant NP-incubated DCs. After 7 days, IL-12 production was quantified by ELISA. Representative of three experiments

and FcγR-dependent (Grazia Cappiello et al. 2001) mechanisms of the inhibition of IL-12 production were demonstrated in human monocytes and macrophages, respectively. High IL-12 production by human DCs can be obtained by co-cultivating DCs either with activated T cells or with CD40L-transfected fibroblasts in vitro. Recombinant NP is able to reproduce the inhibition of human IL-12 production obtained with MV infection without any death of DCs (Fig. 6). Thus binding of H and NP to CD46 and FcγR, respectively, on monocytes/ macrophages and DCs may represent key factors contributing to IL-12 inhibition and consecutive Th2 polarisation observed during and after measles, with increased IL-4 and decreased IL-2 and IFN-γ production (Ward and Griffin 1993).

5.4
Outcome of MV-Infected DC-T cell Interactions: Impairment of CD40 Transduction

The outcome of the crosstalk between MV-infected DCs and T cells in vitro has demonstrated that MV infection modifies CD40 transduction in DCs. Physiologically, CD40L supports two roles in DC-T cell interaction: (1) CD40L activates DC metabolism, performing terminal maturation, a phase required for the proliferation and generation of $CD8^+$ cytotoxic T cells against certain pathogens (Bennett et al. 1998; Ridge et al. 1998; Schoenberger et al. 1998); furthermore, (2) CD40L gives a survival signal, which protects DCs from Fas-dependent apoptosis, a crucial step when DCs face FasL-expressing activated T cells (Bjorck et al. 1997; Koppi et al. 1997; Ludewig et al. 1995). In the case of MV infection, biochemical analysis revealed that MV infection strongly modifies CD40-induced tyrosine phosphorylation in DCs (Servet-Delprat et al. 2000b) (Fig. 5, E). However, neither MV protein(s) nor a component of CD40 transduction has been identified to explain abnormal CD40 transduction into MV-infected DCs at a molecular level. As Akt is involved downstream of CD40 (Andjelic et al. 2000; Arron et al. 2001) and impairment of CD40 transduction is a dominant-negative effect observed in the whole DC population even though only 50% are infected (Servet-Delprat et al. 2000b), H/F glycoproteins able to disrupt Akt kinase activation (Avota et al. 2001) are good candidates.

5.5
Abnormal CD40-Dependent DC Maturation: Induction of Regulatory T Cells?

Impairment of CD40 transduction in DCs by MV induces abnormal DC differentiation with impaired phenotype and cytokine secretions as demonstrated by down-regulation of CD25, CD69, CD71, CD40, CD80, CD86 and CD83 expression, inhibition of IL-12 and IL-1α/β and induction of IL-10 mRNA synthesis (Servet-Delprat et al. 2000b).

IL-10 has recently been implicated in differentiation of regulatory T cells named Tr1. These cells display a cytokine profile distinct from that of the classic TH0, TH1 or TH2 phenotypes. They produce high levels of IL-10, low levels of IL-2 and no IL-4. When they were generated in vitro, exogenous IL-2 was found to be critical for the growth of these Tr1

(Groux 2001). Thus inhibition of T cell proliferation observed in MV-infected DC-T cell co-cultures could also result from induction of Tr1 effector T cells unable to proliferate in the absence of IL-2.

5.6 MV-Induced DC and T Cell Apoptosis

MV-infected DCs do not survive more than 3 days when co-cultured with FasL-expressing T cells; indeed, MV was shown to sensitise DCs to Fas-mediated apoptosis by up-regulating Fas, Flice/caspase-8 and FADD mRNAs (Servet-Delprat et al. 2000a). Addition of Fas-blocking antibodies, which rescue DCs in MV-infected DC-T cell co-cultures, established that death of DCs was Fas mediated (Servet-Delprat et al. 2000a). In vivo, abnormal differentiation could explain the inability of MV-infected DCs to induce naïve T cell proliferation in the absence of CD80/86 co-signals (Grosjean et al. 1997) and early apoptosis of Fas-sensitive MV-infected DCs would prevent a follow-up T cell activation. Finally, MV-infected DCs deliver death factor(s) able to kill facing activated T cells. In MV-infected DC-T cell co-cultures, 100% of activated T cells were dead after 3 days, although less than 10% were infected (Fugier-Vivier et al. 1997). Fas-blocking antibodies did not rescue the T cells in co-cultures of MV-infected DCs with activated T cells (Servet-Delprat et al. 2000a). Although it has not yet been demonstrated, TRAIL is a good candidate to explain massive apoptosis of non-infected T lymphocytes because TRAIL-receptors are increased in measles patients in vivo (Okada et al. 2001). Such a mechanism has been recently demonstrated in an in vivo murine SCID-hu model of HIV infection (Miura et al. 2001).

Thus H/F-mediated inhibition of T cell proliferation, NP-mediated inhibition of antibody secretion by B lymphocytes, NP-mediated inhibition of IL-12 production by DCs, FasL-mediated apoptosis of DCs and TRAIL-mediated apoptosis of T cells may account for MV-induced immunosuppression in vivo.

6 Conclusion and Perspectives

In conclusion, DCs could be implicated in all steps of measles infection: as effectors of innate cytotoxicity mediated by TRAIL, as APCs (non-infected DCs) to initiate an adaptive immune response through cross-priming mechanisms, as targets (infected DCs) for viral replication in conjunction with CD40-activation and as targets for immunosuppression subsequent to impairment of CD40 transduction. However, it is important to note that all the data concerning MV-infected human DCs have been generated in vitro, except NP-mediated inhibition of IL-12 production by DCs, which was also observed in vivo, in murine models. In vitro studies could be performed concerning interactions between MV and innate immune response: Is MV able to counteract innate response or do invasive properties of MV and its ability to rapidly reach secondary lymphoid organs explain that innate immunity is overcome? Studies are necessary to delineate initiation of an adaptive immune response against MV. Is cross-priming the route of MV adaptive immune response, as we propose? This issue may permit improvement of measles vaccination. Furthermore, it is now necessary to investigate whether in vitro-generated results concerning immunosuppression are relevant in vivo. It would be interesting to know whether respiratory tract DCs are infected, what cell types form the WFC multinucleated cells and what cell type is infected in the skin: does the rash correspond to infection of Langerhans cells, the skin DCs? Finally, the most important study to perform is probably one aimed at understanding the mechanism of in vivo depletion of immune cells. The role of TRAIL must be evaluated in vivo, and the consequence of measles infection in primary lymphoid organs should be addressed to discover the mechanism of such a long B cell lymphopenia.

Beyond measles virus, an understanding of host-pathogen interaction, particularly in the case of immunosuppressive agents, helps to investigate how the immune system acts, suggests possible therapeutic applications for immunosuppressive pathogenic proteins and may help to improve vaccines or imagine new strategy for designing vaccines.

Acknowledgements. This work was supported by institutional grants from the INSERM and from MENRT and by additional support from ARC (CRC 6108), Programme PRFMMIP and Region Rhone-Alpes.

References

Adema GJ, Hartgers F, Verstraten R, de Vries E, Marland G, Menon S, Foster J, Xu Y, Nooyen P, McClanahan T et al. 1997. A dendritic-cell-derived C-C chemokine that preferentially attracts naïve T cells. Nature 387(6634):713–7.

Albert ML, Sauter B, Bhardwaj N. 1998. Dendritic cells acquire antigen from apoptotic cells and induce class I-restricted CTLs. Nature 392(6671):86–9.

Alexopoulou L, Holt AC, Medzhitov R, Flavell RA. 2001. Recognition of double-stranded RNA and activation of NF-kappaB by Toll-like receptor 3. Nature 413(6857):732–8.

Andjelic S, Hsia C, Suzuki H, Kadowaki T, Koyasu S, Liou HC. 2000. Phosphatidylinositol 3-kinase and NF-kappa B/Rel are at the divergence of CD40-mediated proliferation and survival pathways. J Immunol 165(7):3860–7.

Arron JR, Vologodskaia M, Wong BR, Naramura M, Kim N, Gu H, Choi Y. 2001. A positive regulatory role for Cbl family proteins in tumor necrosis factor-related activation-induced cytokine (trance) and CD40L-mediated Akt activation. J Biol Chem 276(32):30011–7.

Auwaerter PG, Kaneshima H, McCune JM, Wiegand G, Griffin DE. 1996. Measles virus infection of thymic epithelium in the SCID-hu mouse leads to thymocyte apoptosis. J Virol 70(6):3734–3740.

Avota E, Avots A, Niewiesk S, Kane LP, Bommhardt U, ter Meulen V, Schneider-Schaulies S. 2001. Disruption of Akt kinase activation is important for immunosuppression induced by measles virus. Nat Med 7(6):725–31.

Bennett SR, Carbone FR, Karamalis F, Flavell RA, Miller JF, Heath WR. 1998. Help for cytotoxic-T-cell responses is mediated by CD40 signalling. Nature 393(6684):478–80.

Bjorck P, Banchereau J, Flores-Romo L. 1997. CD40 ligation counteracts Fas-induced apoptosis of human dendritic cells. Int Immunol 9(3):365–72.

Esolen LM, Ward BJ, Moench TR, Griffin DE. 1993. Infection of monocytes during measles. J Infect Dis 168(1):47–52.

Forster R, Schubel A, Breitfeld D, Kremmer E, Renner-Muller I, Wolf E, Lipp M. 1999. CCR7 coordinates the primary immune response by establishing functional microenvironments in secondary lymphoid organs. Cell 99(1):23–33.

Fugier-Vivier I, Servet-Delprat C, Rivailler P, Rissoan MC, Liu YJ, Rabourdin-Combe C. 1997. Measles virus suppresses cell-mediated immunity by interfering with the survival and functions of dendritic and T cells. J Exp Med 186(6):813–23.

Furukawa S, Matsubara T, Tsuji K, Motohashi T, Okumura K, Yabuta K. 1991. Serum soluble CD4 and CD8 levels in Kawasaki disease. Clin Exp Immunol 86(1):134–9.

Geijtenbeek TB, Kwon DS, Torensma R, van Vliet SJ, van Duijnhoven GC, Middel J, Cornelissen IL, Nottet HS, KewalRamani VN, Littman DR et al. 2000. DC-SIGN, a dendritic cell-specific HIV-1-binding protein that enhances trans-infection of T cells. Cell 100(5):587–97.

Gougeon ML, Lecoeur H, Dulioust A, Enouf MG, Crouvoiser M, Goujard C, Debord T, Montagnier L. 1996. Programmed cell death in peripheral lymphocytes from HIV-infected persons: increased susceptibility to apoptosis of CD4 and CD8 T cells correlates with lymphocyte activation and with disease progression. J Immunol 156(9):3509–20.

Graves M, Griffin DE, Johnson RT, Hirsch RL, de Soriano IL, Roedenbeck S, Vaisberg A. 1984. Development of antibody to measles virus polypeptides during complicated and uncomplicated measles virus infections. J Virol 49(2):409–12.

Grazia Cappiello M, Sutterwala FS, Trinchieri G, Mosser DM, Ma X. 2001. Suppression of Il-12 transcription in macrophages following Fc gamma receptor ligation. J Immunol 166(7):4498–506.

Griffin DE, Ward BJ. 1993. Differential CD4 T cell activation in measles. J Infect Dis 168(2):275–81.

Griffith TS, Wiley SR, Kubin MZ, Sedger LM, Maliszewski CR, Fanger NA. 1999. Monocyte-mediated tumoricidal activity via the tumor necrosis factor-related cytokine, TRAIL. J Exp Med 189(8):1343–54.

Grosjean I, Caux C, Bella C, Berger I, Wild F, Banchereau J, Kaiserlian D. 1997. Measles virus infects human dendritic cells and blocks their allostimulatory properties for CD4+ T cells. J Exp Med 186(6):801–12.

Groux H. 2001. An overview of regulatory T cells. Microbes Infect 3(11):883–9.

Gunn MD, Kyuwa S, Tam C, Kakiuchi T, Matsuzawa A, Williams LT, Nakano H. 1999. Mice lacking expression of secondary lymphoid organ chemokine have defects in lymphocyte homing and dendritic cell localization. J Exp Med 189(3):451–60.

Iezzi G, Scheidegger D, Lanzavecchia A. 2001. Migration and function of antigen-primed nonpolarized T lymphocytes in vivo. J Exp Med 193(8):987–93.

Ilonen J, Makela MJ, Ziola B, Salmi AA. 1990. Cloning of human T cells specific for measles virus haemagglutinin and nucleocapsid. Clin Exp Immunol 81(2):212–7.

Jacobson S, Sekaly RP, Jacobson CL, McFarland HF, Long EO. 1989. HLA class II-restricted presentation of cytoplasmic measles virus antigens to cytotoxic T cells. J Virol 63(4):1756–62.

Karp CL, Wysocka M, Wahl LM, Ahearn JM, Cuomo PJ, Sherry B, Trinchieri G, Griffin DE. 1996. Mechanism of suppression of cell-mediated immunity by measles virus. Science 273:228–31.

Koppi TA, Tough-Bement T, Lewinsohn DM, Lynch DH, Alderson MR. 1997. CD40 ligand inhibits Fas/CD95-mediated apoptosis of human blood-derived dendritic cells. Eur J Immunol 27(12):3161–5.

Langenkamp A, Messi M, Lanzavecchia A, Sallusto F. 2000. Kinetics of dendritic cell activation: impact on priming of TH1, TH2 and nonpolarized T cells. Nat Immunol 1(4):311–6.

Leopardi R, Hyypia T, Vainionpaa R. 1992. Effect of interferon-alpha on measles virus replication in human peripheral blood mononuclear cells. Apmis 100(2):125–31.

Liu S, Yu Y, Zhang M, Wang W, Cao X. 2001. The involvement of TNF-alpha-related apoptosis-inducing ligand in the enhanced cytotoxicity of IFN-beta-stimulated human dendritic cells to tumor cells. J Immunol 166(9):5407–15.

Ludewig B, Graf D, Gelderblom HR, Becker Y, Kroczek RA, Pauli G. 1995. Spontaneous apoptosis of dendritic cells is efficiently inhibited by TRAP (CD40-ligand) and TNF-alpha, but strongly enhanced by interleukin- 10. Eur J Immunol 25(7):1943–50.

Manchester M, Eto DS, Valsamakis A, Liton PB, Fernandez-Munoz R, Rota PA, Bellini WJ, Forthal DN, Oldstone MB. 2000. Clinical isolates of measles virus use CD46 as a cellular receptor. J Virol 74(9):3967–74.

Marie JC, Kehren J, Trescol-Biemont MC, Evlashev A, Valentin H, Walzer T, Tedone R, Loveland B, Nicolas JF, Rabourdin-Combe C et al. 2001. Mechanism of measles virus-induced suppression of inflammatory immune responses. Immunity 14(1):69–79.

McChesney MB, Fujinami RS, Lampert PW, Oldstone MB. 1986. Viruses disrupt functions of human lymphocytes. II. Measles virus suppresses antibody production by acting on B lymphocytes. J Exp Med 163(5):1331–6.

Miura Y, Misawa N, Maeda N, Inagaki Y, Tanaka Y, Ito M, Kayagaki N, Yamamoto N, Yagita H, Mizusawa H et al. 2001. Critical contribution of tumor necrosis factor-related apoptosis-inducing ligand (TRAIL) to apoptosis of human CD4+ T cells in HIV-1-infected hu-PBL-NOD-SCID mice. J Exp Med 193(5):651–60.

Murray CJ, Lopez AD. 1997. Mortality by cause for eight regions of the world: Global Burden of Disease Study. Lancet 349(9061):1269–76.

Naniche D, Varior-Krishnan G, Cervoni F, Wild TF, Rossi B, Rabourdin-Combe C, Gerlier D. 1993. Human membrane cofactor protein (CD46) acts as a cellular receptor for measles virus. J Virol 67(10):6025–32.

Naniche D, Yeh A, Eto D, Manchester M, Friedman RM, Oldstone MB. 2000. Evasion of host defenses by measles virus: wild-type measles virus infection interferes with induction of Alpha/Beta interferon production. J Virol 74(16):7478–84.

Norrby E, Gollmar Y. 1972. Appearance and persistence of antibodies against different virus components after regular measles infections. Infect Immun 6(3):240–7.

Nozawa Y, Ono N, Abe M, Sakuma H, Wakasa H. 1994. An immunohistochemical study of Warthin-Finkeldey cells in measles. Pathol Int 44(6):442–47.

Okada H, Kobune F, Sato TA, Kohama T, Takeuchi Y, Abe T, Takayama N, Tsuchiya T, Tashiro M. 2000. Extensive lymphopenia due to apoptosis of uninfected lymphocytes in acute measles patients. Arch Virol 145(5):905–20.

Okada H, Sato TA, Katayama A, Higuchi K, Shichijo K, Tsuchiya T, Takayama N, Takeuchi Y, Abe T, Okabe N et al. 2001. Comparative analysis of host responses related to immunosuppression between measles patients and vaccine recipients with live attenuated measles vaccines. Arch Virol 146(5):859–74.

Orenstein JM. 1998. The Warthin-Finkeldey-type giant cell in HIV infection, what is it? Ultrastruct Pathol 22(4):293–303.

Pitti RM, Marsters SA, Ruppert S, Donahue CJ, Moore A, Ashkenazi A. 1996. Induction of apoptosis by Apo-2 ligand, a new member of the tumor necrosis factor cytokine family. J Biol Chem 271(22):12687–90.

Ravanel K, Castelle C, Defrance T, Wild TF, Charron D, Lotteau V, Rabourdin-Combe C. 1997. Measles virus nucleocapsid protein binds to FcgammaRII and inhibits human B cell antibody production. J Exp Med 186(2):269–78.

Ridge JP, Di Rosa F, Matzinger P. 1998. A conditioned dendritic cell can be a temporal bridge between a CD4+ T-helper and a T-killer cell. Nature 393(6684):474–78.

Schnorr JJ, Xanthakos S, Keikavoussi P, Kampgen E, Ter Meulen V, Schneider-Schaulies S. 1997. Induction of maturation of human blood dendritic cell precursors by measles virus is associated with immunosuppression. Proc Natl Acad Sci USA 94(10):5326–31.

Schoenberger SP, Toes RE, van der Voort EI, Offringa R, Melief CJ. 1998. T-cell help for cytotoxic T lymphocytes is mediated by CD40-CD40L interactions. Nature 393(6684):480–83.

Sedger LM, Shows DM, Blanton RA, Peschon JJ, Goodwin RG, Cosman D, Wiley SR. 1999. IFN-gamma mediates a novel antiviral activity through dynamic modulation of TRAIL and TRAIL receptor expression. J Immunol 163(2):920–6.

Servet-Delprat C, Vidalain PO, Azocar O, Le Deist F, Fischer A, Rabourdin-Combe C. 2000a. Consequences of Fas-mediated human dendritic cell apoptosis induced by measles virus. J Virol 74(9):4387–93.

Servet-Delprat C, Vidalain PO, Bausinger H, Manie S, Le Deist F, Azocar O, Hanau D, Fischer A, Rabourdin-Combe C. 2000b. Measles virus induces abnormal differentiation of CD40 ligand-activated human dendritic cells. J Immunol 164(4):1753–60.

Song K, Chen Y, Goke R, Wilmen A, Seidel C, Goke A, Hilliard B. 2000. Tumor necrosis factor-related apoptosis-inducing ligand (TRAIL) is an inhibitor of autoimmune inflammation and cell cycle progression. J Exp Med 191(7):1095–104.

Steineur MP, Grosjean I, Bella C, Kaiserlian D. 1998. Langerhans cells are susceptible to measles virus infection and actively suppress T cell proliferation. Eur J Dermatol 8(6):413–20.

Tajima M, Kudow S. 1976. Morphology of the Warthin-Finkeldey giant cells in monkeys with experimentally induced measles. Acta Pathol Jpn 26(3):367–80.

Tatsuo H, Ono N, Tanaka K, Yanagi Y. 2000. SLAM (CDw150) is a cellular receptor for measles virus. Nature 406(6798):893–7.

Valentin H, Azocar O, Horvat B, Williems R, Garrone R, Evlashev A, Toribio ML, Rabourdin-Combe C. 1999. Measles virus infection induces terminal differentiation of human thymic epithelial cells. J Virol 73(3):2212–21.

van Binnendijk RS, Poelen MC, de Vries P, Voorma HO, Osterhaus AD, Uytdehaag FG. 1989. Measles virus-specific human T cell clones. Characterization of specificity and function of CD4+ helper/cytotoxic and CD8+ cytotoxic T cell clones. J Immunol 142(8):2847–54.

Vidalain PO, Azocar O, Lamouille B, Astier A, Rabourdin-Combe C, Servet-Delprat C. 2000. Measles virus induces functional TRAIL production by human dendritic cells. J.Virol. 74(1):556–9.

Vidalain PO, Azocar O, Rabourdin-Combe C, Servet-Delprat C. 2001. measles virus-infected dendritic cells develop immunosuppressive and cytotoxic activities. Immunobiol, 204:629–638.

Vidalain PO, Azocar O, Yagita H, Rabourdin-Combe C, Servet-Delprat C. 2001. Cytotoxic activity of human dendritic cells is differentially regulated by double-stranded rna and cd40 ligand. J Immunol 167(7):3765–72.

Ward BJ, Griffin DE. 1993. Changes in cytokine production after measles virus vaccination: predominant production of IL-4 suggests induction of a Th2 response. Clin Immunol Immunopathol 67(2):171–177.

Wiley SR, Schooley K, Smolak PJ, Din WS, Huang CP, Nicholl JK, Sutherland GR, Smith TD, Rauch C, Smith CA et al. 1995. Identification and characterization of a new member of the TNF family that induces apoptosis. Immunity 3(6):673–82.

Zinkernagel RM. 2000. Localization dose and time of antigens determine immune reactivity. Semin Immunol 12(3):163–71; discussion 257–344.

Chapter 6

Infection of Dendritic Cells by Lymphocytic Choriomeningitis Virus

N. Sevilla, S. Kunz, D. McGavern, M. B. A. Oldstone

Abstract. Dendritic cells (DCs) comprise the major antigen-presenting cells (APCs) of the host, uniquely programmed to stimulate immunologically naïve T lymphocytes. Viruses that can target and disorder the function of these cells enjoy a selective advantage. The cellular receptor for lymphocytic choriomeningitis virus (LCMV), Lassa fever virus (LFV), and several other arenaviruses is α-dystroglycan (α-DG). Among cells of the immune system, CD11c$^+$ and DEC-205$^+$ DCs primarily and preferentially express α-DG. By selection, strains and variants of LCMV generated as quasi-species that bind α-DG with high affinity replicate in the majority of CD11c$^+$ and DEC-205$^+$ (>75%) DCs, causing a generalized immunosuppression, and establish a persistent infection. In contrast, viral strains and variants that bind with low affinity to α-DG display minimal replication in CD11c$^+$ and DEC-205$^+$ DCs (<10%), rarely replicate in the white pulp, and generate a robust anti-LCMV CTL re-

N. Sevilla, S. Kunz, D. McGavern, M.B.A. Oldstone (✉)
The Scripps Research Institute, Division of Virology,
Department of Neuropharmacology, 10550 N. Torrey Pines Road, La Jolla, CA 92037, USA

sponse that clears the virus infection. Hence, receptor-virus interaction on DCs in vivo is an essential step in the initiation of virus-induced immunosuppression and viral persistence. Investigation into the mechanism of how virus-infected DCs cause immunosuppression reveals loss of MHC class II surface expression and costimulatory molecules on surface of such DCs. As a consequence DCs are unable to act as APCs, initiate immune responses, and have a defect in migration into the T cell area. These data indicate that LCMV infection influences DC maturation and migration, leading to decreased T cell stimulatory capacity of DCs, events essential for the initiation of immune responses. Because several other viruses known to cause immunosuppression (HIV, measles) interact with DCs, the observations noted here are likely a common selective mechanism by which viruses also are able to evade the host's immune system.

1
Introduction

Viruses use a number of strategies, including escape from immune recognition or induction of immunosuppression, to avoid immunological surveillance and thereby persist in their host (reviewed in Borrow and Shaw 1998; Tortorella et al. 2000). The downregulation of immune responses provides the infecting pathogens with the opportunity to maximize their chances of survival, replication, and transmission. The generalized immunosuppression caused by viral infection is often associated with secondary infections with unrelated viral and/or bacterial pathogens and, as such, represents a serious clinical problem. Understanding the mechanisms involved in the induction of immunosuppression is therefore a crucial step for controlling and eliminating infections caused by several viruses including important human pathogens such as hepatitis B and C and human immunodeficiency virus (HIV). To better understand how viruses may cause immunosuppression, we have studied this phenomenon in an animal model system in which the noncytopathic lymphocytic choriomeningitis virus (LCMV) infects its natural host, the mouse, and causes transient to long-lasting immunosuppression (Oldstone 1989). In some ways, LCMV infection resembles HIV and HIV-triggered AIDS in humans (reviewed in Hazenberg et al. 2000; Lieberman et al. 2001; McMichael and Rowland-Jones 2001). In this review, we focus on one of the main strategies used by LCMV to subvert normal defense mechanisms and cause a generalized immunosuppression: infection and interference with the functions of dendritic cells (DCs).

Initiation of an adaptive immune response to infectious agents is mediated by a class of sentinel phagocytic leukocytes termed DCs, the primary function of which is to capture, process, and present antigens to unprimed T cells (Ibrahim et al. 1995; Steinman 1991). DCs are present in nonlymphoid organs such as the liver and skin in an immature state. During the course of an infection, DCs take up and process antigen, mature, and migrate to the local draining lymph node, where they efficiently activate both T and B cells (Banchereau and Steinman 1998). A reduced antigen-processing capacity and an increased cell surface expression of MHC class II and costimulatory molecules such as B7-1 and B7-2 occurs during the migration of DCs to lymph organs, thus making the DCs the most potent antigen-presenting cells (APCs) in the immune repertoire (Banchereau and Steinman 1998). Activated DCs are essential for the subsequent activation of both naïve and activated T cells. In addition, with the release of specific cytokines and chemokines, activated DCs either directly or indirectly play a crucial role in the recruitment and/or activation of other effector cells, including NK and NKT cells, macrophages, and B cells (Rescigno et al. 1999). In this review, we describe the infection of DCs by LCMV in vivo and the consequence of this infection on DC functionality.

2
LCMV and Immunosuppressive Variants

LCMV is the prototypic member of the family Arenaviridae, which includes important human pathogens such as Lassa fever and Machupo, Sabia, and Junin viruses (Buckley et al. 1970; Peters et al. 1996; Southern et al., 1987). Its genome consists of two negative-strand RNA segments, large (L) and short (S) fragments (Riviere et al. 1985; Singh et al. 1987; Southern 1996). Each RNA segment has an ambisense coding strategy, encoding two proteins in opposite orientations separated by an intergenic region. The L RNA encodes the viral RNA-dependent RNA polymerase (L) (Singh et al. 1987) and a small RING finger protein (Z) (Salvato and Shimomaye 1989). The S RNA encodes the three major structural proteins: the nucleoprotein (NP) and the precursor polypeptide, GP-C (Wright et al. 1990), which is processed into two products, the peripheral glycoprotein GP1 and the transmembrane glycoprotein GP2

(Buchmeier and Oldstone 1979). These mature forms of the glycoprotein form 5- to 10-nm-long surface projections (spikes) that are speculated to be composed of tetramers of GP2 that serve as platforms for tetramers of GP1 (Burns and Buchmeier 1993). The GP1 portion of the glycoprotein mediates the virus-receptor interaction based on the fact that neutralizing antibodies are formed exclusively against GP1 (Borrow and Oldstone 1992; Parekh and Buchmeier 1986). The main cellular receptor for LCMV has been identified as α-dystroglycan (α-DG) (Cao et al. 1998), which is a highly versatile cellular receptor for proteins of the extracellular matrix (ECM) (Ervasti and Campbell 1993; Ibraghimov-Beskrovnaya et al. 1992; Winder 2001). Analysis of the interaction of LCMV with α-DG is discussed later in this chapter.

LCMV, on congenital, neonatal, or in utero infection, can cause a persistent infection in its natural host, the mouse (reviewed in Borrow and Oldstone 1997). During this persistent infection, we (Ahmed and Oldstone 1988; Ahmed et al. 1984; Dockter et al. 1996; Evans et al. 1994; Sevilla et al. 2000) and others (Ahmed et al. 1991; Matloubian et al. 1993; Villarete et al. 1994) have described the emergence of LCMV variants. Viral variants were found with different biological properties: One variant predominated in the central nervous system (CNS) in neurons (Dockter et al. 1996; Evans et al. 1994; Rodriguez et al. 1983; Villarete et al. 1994) and another predominated in cells of the immune system such as lymphocytes and macrophages (Ahmed et al. 1991; Borrow et al. 1995; Dockter et al. 1996; Matloubian et al. 1993; Sevilla et al. 2000; Villarete et al. 1994). The parental virus, Armstrong 53b (ARM), and the vast majority of viruses and variants isolated from the CNS, when inoculated intravenously into adult mice, caused an acute infection characterized by profound $CD8^+$-T cell expansion and IFN-γ production that efficiently terminated the viral infection by 7–10 days after infection. The prototype variant that has tropism for cells of the immune system is called LCMV clone 13 (Cl 13) (Ahmed et al. 1991). Inoculation of Cl 13 intravenously into adult immunocompetent mice aborted cytotoxic T lymphocyte responses (CTL) and led to viral persistence (Matloubian et al. 1993; Salvato et al. 1991; Tishon et al. 1988). In contrast to the persistent infection initiated early in life that is due to thymic deletion of antiviral immune cells (Borrow and Oldstone 1997), persistent infection of adult mice with Cl 13 is due to exhaustion of virus-specific T cells and is associated with a generalized immunosuppression (Borrow et al. 1995; Tishon et al. 1993).

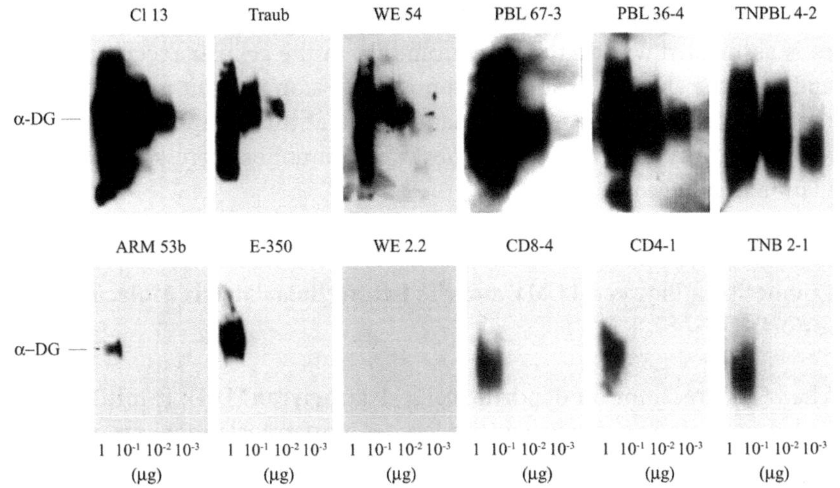

Fig. 1. LCMV binding affinity to α-DG. Virus overlay protein binding assay (VOPBA) was done with purified α-DG and LCMV variants and strains as described by Sevilla et al. (2000). Decreasing amounts of purified α-DG were loaded on an SDS-PAGE, electrophoresed, and transferred onto nitrocellulose. The nitrocellulose was incubated with 10^7 PFUs of each viral isolate, and viral proteins were detected with monoclonal antibodies specific for LCMV GP. The *top panel* shows the VOPBA of immunosuppressive LCMV strains (Traub and WE 54) and variants (Cl 13, PBL 67-3, PBL 36-4, and TNPBL 4-2); the *bottom panel* shows the VOPBA of nonimmunosuppressive LCMV strains (ARM 53b and E-350) and variants (WE 2.2, CD8-4, CD4-1, and TNB 2-1)

Sequence comparison between ARM and Cl 13 showed five nucleotide changes in the entire genome, and only two of them caused amino acid (aa) substitutions (Salvato et al. 1991; Salvato and Shimomaye 1989). One of these aa substitutions is in position 260 of GP-1, with ARM having phenylalanine (F) and Cl 13 leucine (L); the other aa change was in position 1079 of the L protein from lysine (K) (ARM) to glutamine (Q) (Cl 13). Recent studies of numerous LCMV strains and viral variants generated in vivo demonstrated a correlation between a F(260)→L/I mutation in the viral GP-1, the component that binds the cellular receptor, and the immunosuppressive phenotype (Sevilla et al. 2000; Smelt et al. 2001). Furthermore, all LCMV variants and strains with a L or I in position 260 of GP-1 showed high affinity binding to α-DG, with binding of at least 1 log and often 2–3 logs greater than that observed with viruses with F at position 260 (Fig. 1; Sevilla et al. 2000). These findings indicated that an aliphatic aa-like L or I in position 260 of GP-1 of LCMV viruses is associated with high-affinity binding to the cellular receptor α-DG and with the ability to cause immunosuppression in adult mice, whereas those viruses containing an aromatic (F) aa at this position bound with a lower affinity to α-DG and did not cause immunosuppression on inoculation of adult mice.

3 Competition Between LCMV and the Extracellular Matrix Molecules for α-DG

The LCMV receptor on dendritic cells, dystroglycan (DG), is initially encoded as a single protein and undergoes post-translational cleavage to form α-DG, a peripheral protein, and β-DG, a membrane protein. DG is expressed in most developing and adult tissues, typically in cell types that adjoin basement membranes (Durbeej et al. 1998). At those sites, DG plays a critical role in cell-mediated assembly of basement membranes (Ervasti and Campbell 1991, 1993; Gee et al. 1994; Henry and Campbell 1998; Talts et al. 1999; Williamson et al. 1997). α-DG undergoes high-affinity interactions with the ECM proteins laminin-1, laminin-2, agrin, and perlecan. In muscle tissue, DG is found associated with two additional types of membrane proteins, the sarcoglycans and sarcospan (Henry and Campbell 1999). α-DG is noncovalently associated with the membrane-spanning β-DG, which binds to the cytoskeletal proteins dystrophin and utrophin, as well as signal transduction molecules such as grb2 and the focal adhesion kinase FAK (Cavaldesi et al. 1999; Jung et al. 1995; Yang et al. 1995).

The complex binding pattern of α-DG results in the formation of tissue-specific α-DG complexes in the cell membrane. Because specific DG-associated proteins may enhance or block virus binding, their coexpression with DG may modulate the susceptibility of a specific cell for virus infection and represent an important determinant for viral tissue tropism in vivo. Given their expression in lymphoid organs, high binding affinity, and large size, ECM molecules may interfere with virus binding to α-DG and, therefore, provide a barrier for virus infection of specific DG^+ cell types such as $CD11c^+$ DCs. In vitro studies demonstrated that high-affinity binding LCMV variants like Cl 13 compete successfully with laminin-1 and laminin-2 for α-DG binding. In contrast, the low-affinity binding variant ARM was blocked from its receptor binding site by laminin (Kunz et al. 2001). Because recent binding studies suggest overlapping binding sites on α-DG for the ECM components laminin-1, laminin-2, perlecan, and agrin (Talts et al. 1999), these molecules likely provide a barrier for infection of cells by low-affinity binding LCMV variants.

The expression of α-DG-associated ECM in lymphoid tissues such as the spleen may be an important determinant for the differential anatomic distribution of LCMV variants with different receptor binding affinities and the ability to cause distinct pathologies. Immunohistochemical analysis of the expression of laminin in spleen tissue (Fig. 2) revealed particularly strong staining in the marginal zone at the periphery of the white pulp (Kunz et al. 2001; van den Berg et al. 1993), where large numbers of CD11c$^+$ DC are found (Kunz et al. 2001; Steinman et al. 1997). The overlap between laminin expression and infection with Cl 13 suggests that this virus efficiently infects cells surrounded by laminin. By contrast, the lack of association between laminin expression and the distribution of ARM suggests that laminin interferes with ARM infection. High receptor binding affinity likely enables variants like Cl 13 to displace α-DG-associated ECM (Fig. 3) molecules from the surface of CD11c$^+$ DCs in the marginal zone, resulting in infection of these cells. Infected DCs subsequently migrate into the white pulp and disseminate the virus into the T cell area (Lewicki et al., unpublished observation). In contrast to Cl 13 and the high-affinity binders, low-affinity binders like ARM may be unable to displace α-DG-associated ECM components (Fig. 3) and thus unable to effectively infect CD11c$^+$ DC. This may explain why LCMV variants like ARM are excluded from the white pulp and are found predominantly in the red pulp. The more efficient infection of cells in the red pulp exhibited by low-affinity binding variants may reflect their ability to use alternative cellular receptors for infection. The biochemical nature of this alternate receptor is currently unknown.

The scenario in the spleen may reflect the situation in other lymphoid organs, where immunohistochemical studies revealed expression of laminin and other ECM components that bind to α-DG. In lymph nodes, laminin is found in vessel walls, in a layer underlying the subcapsular sinus, and in an extensive array of ECM fibers (Jaspars et al. 1995; Kramer et al. 1988; van den Berg et al. 1993). In bone marrow, widespread laminin expression was detected in vascular basement membranes and intersinusoidal interstitial tissue (Gu et al. 1999; Nilsson et al. 1999). The laminin-containing ECM structures in lymphoid organs are thought to form

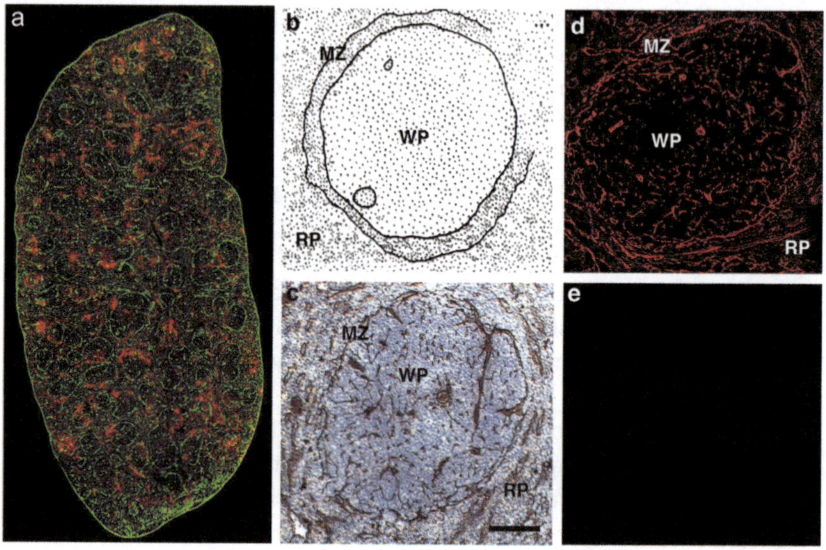

Fig. 2a–e. Expression pattern of laminin and CD11c in the spleen. Detection of laminin and CD11c in whole spleen sections (**a**): Laminin was detected with a rabbit anti-mouse laminin-1 antibody and a FITC-labeled secondary antibody (*green*). CD11c was detected with a hamster anti CD11c antibody and a rhodamine-X-conjugated secondary antibody (*red*). Overlapping fluorescence appears in yellow as a rhodamine-X/FITC double-filter device was used. Localization of laminin in the marginal zone of the white pulp (**b**): Schematic representation of spleen histology. In **c**, laminin was stained with rabbit anti-laminin-1 antibody with a HRP-conjugated secondary antibody and DAB as a chromophore. Sections were counterstained with hematoxylin and eosin. Laminin staining of a single white pulp area (**d**, **e**): Laminin was detected with a rabbit anti-laminin-1 antibody and a Texas-red conjugated secondary antibody (**d**); secondary antibody only is shown in **e**. *MZ*, marginal zone; *RP*, red pulp; *WP*, white pulp. Bars: **a**, 200 μm; **b–e**, 50 μm

a scaffold for localization and/or migration of immune cells (Kramer et al. 1988). However, despite a large number of studies describing laminin expression in lymphoid organs, the detailed distribution of laminin isoforms in these tissues is poorly understood. The identification of specific laminin isoforms that may be expressed in a specific temporal and spatial pattern in resting lymphoid tissue and during acute infection processes is clearly of great importance for an understanding of the molecular basis of LCMV-mediated immunosuppression. Furthermore, a similar in vivo selection of LCMV variants with high affinity for α-DG is also observed in nonlymphoid organs such as kidney, liver, and lung (Evans

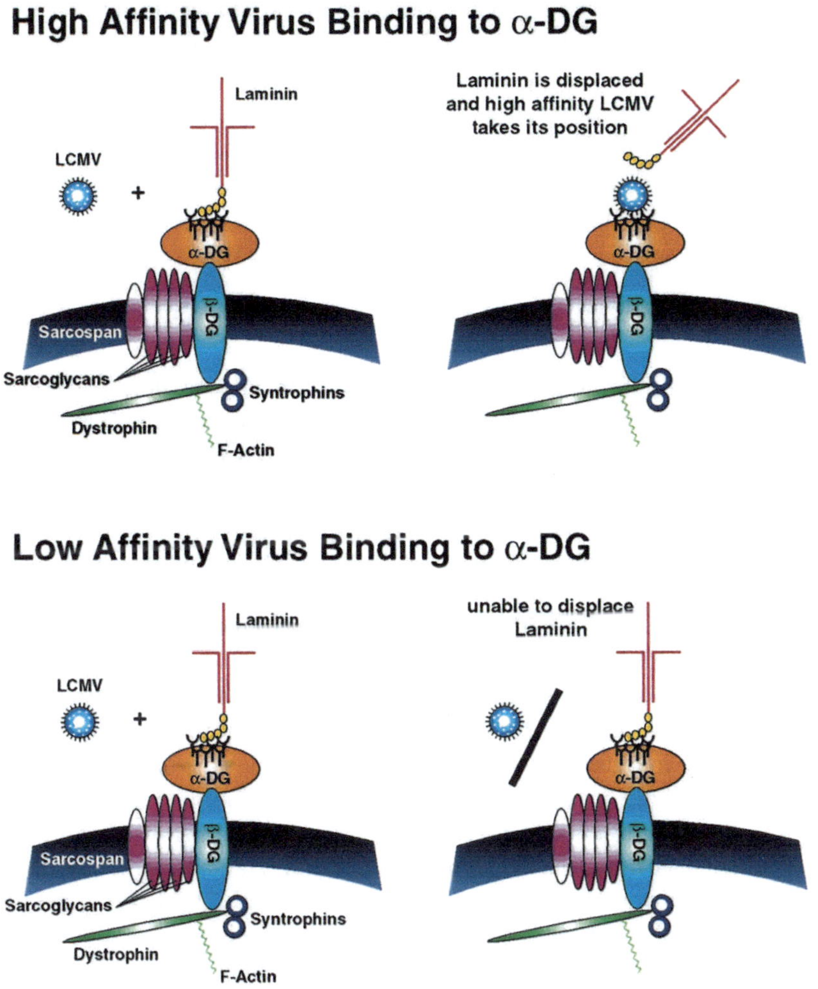

Fig. 3. Targeting DCs by LCMV: role of the ECM molecules in the interaction of LCMV with α-DG. α-DG undergoes high-affinity interactions with ECM such as laminin. These ECM components may interfere with virus binding to the cell and, therefore, provide a barrier for virus infection of specific cells such as CD11c^{+} DCs. High-affinity binding LCMV variants such as Cl 13 (*upper panel*) may compete successfully with laminin for α-DG, whereas low-affinity binders such as ARM (*lower panel*) are unable to displace laminin. The lack of association between laminin expression (Fig. 2) and ARM distribution (Fig. 5) suggests that laminin may interfere with ARM infection

et al. 1994), suggesting that the selection of variants with high receptor binding affinity is a more general phenomenon and may represent a selective advantage during the natural infectious process. Because α-DG binding laminin isoforms are widely expressed in these organs (Miner et al. 1997), it is likely that competition between virus and ECM components for α-DG may also play a role in the in vivo selection of viral variants in these tissues.

4 Replication of LCMV in DCs

The defect underlying the generalized immunosuppression in mice infected with immunosuppressive LCMV variants maps primarily to antigen presentation rather than T or B cell function per se (Althage et al. 1992; Odermatt et al. 1991) and would be predicted by the observation that these immunosuppressive viruses infect lymphocytes inefficiently both in vivo and in vitro (Borrow et al. 1991; Matloubian et al. 1993; Sevilla et al. 2000). The analysis of the anatomic distribution of immunosuppressive and nonimmunosuppressive variants and strains in the spleens of adult immunocompetent mice 3 days after infection showed a different tropism for both groups of viruses (Borrow et al. 1995; Sevilla et al. 2000; Smelt et al. 2001). Cl 13-like viruses localized exclusively in cells of the marginal zone and white pulp of spleen, whereas ARM-like viruses localized primarily in cells within the red pulp (Fig. 4). By days 3–5 after infection, mice infected with ARM or Cl 13 showed some loss of follicular structure as evidenced by laminin stains (Fig. 4). By day 7, all follicular organization was absent and $CD11c^+$ DCs had distributed throughout the entirely of the spleen. Interestingly, at day 10, $CD11c^+$ DCs in ARM-infected mice had returned to small nests suggestive of splenic reorganization, whereas Cl 13-infected mice maintained disorganized DCs and also showed elevated laminin staining (Fig. 4). However, by day 30 mice infected with either strain showed a complete restoration of spleen architecture, yet CD11c+ staining appeared to be less intense in Cl 13-infected spleens (Fig. 4).

This distinct distribution of virus in spleen indicated that different subsets of cells were infected by immunosuppressive and nonimmunosuppressive viruses. The identification of cells in spleen infected by LCMV revealed two subsets mainly infected by immunosuppressive viruses, CD11c- and DEC-205 positive DCs (Fig. 5; Sevilla et al. 2000).

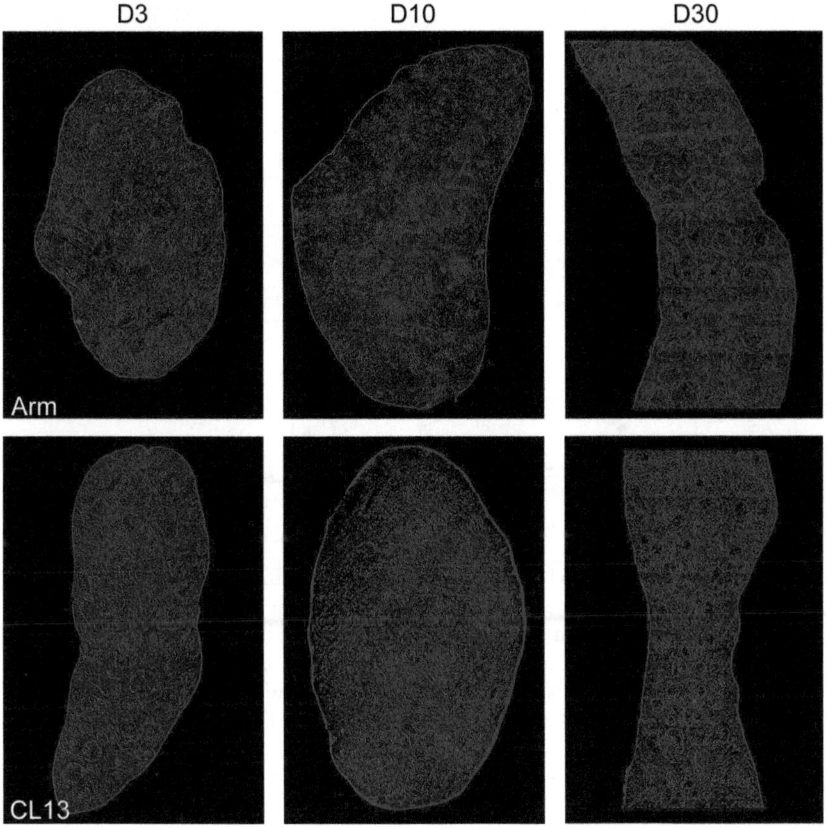

Fig. 4. Comparison of splenic architecture in Cl 13- and ARM-infected mice. The distribution of laminin and CD11c in whole spleen section from ARM- or Cl 13-infected mice at days 3, 10, and 30 after infection. Laminin was detected with a rabbit anti-mouse laminin-1 antibody and a FITC-labeled secondary antibody (*blue*). CD11c was detected with a hamster anti-CD11c antibody and a rhodamine-X-conjugated secondary antibody (*red*). Overlapping fluorescence appears in *pink*

$CD11c^+$ cells localized in the central white pulp, T cell areas (interdigitating cells), and "nests" of cells that interrupt the marginal zone and extend into the T cell but not the B cell areas. The latter have been termed "marginal zone DC." The DEC-205 antigen is expressed on a subset of DCs in spleen, and labeling appears to be confined largely to interdigitating DCs, localized into the T cell area (Steinman et al. 1997). In contrast, nonimmunosuppressive viruses infected mainly F4/80 positive

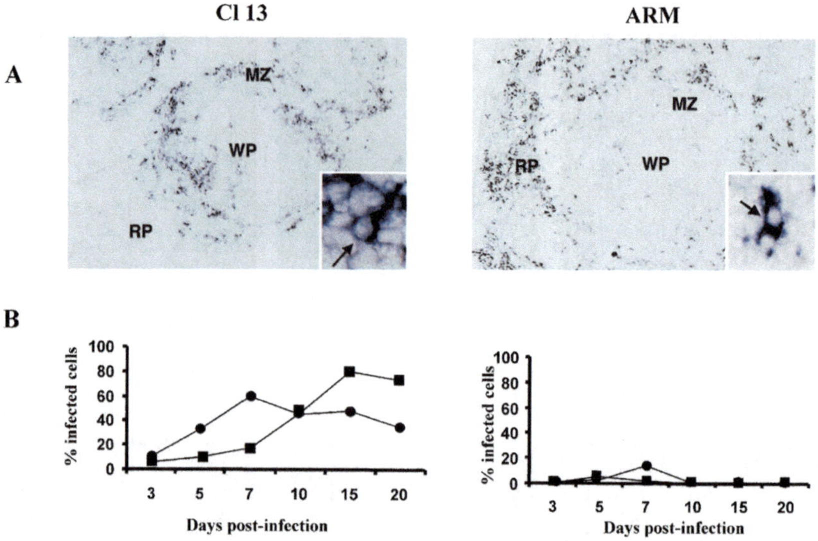

Fig. 5a, b. Infection of DCs by LCMV. **a** The comparison of the anatomic localization of viral nucleic acids in spleen between Cl 13 and ARM at day 3 after infection shows ARM nucleic acid localized predominantly in the red pulp (*RP*) and Cl 13 nucleic acid localized with the white pulp (*WP*) or its marginal zone (*MZ*). A higher-power view is shown of both panels. The *arrows* indicate viral nucleic detected in individual cells. **b** Replication of Cl 13 and ARM in DCs in vivo over a 20-day observation period. Four adult BALB/c ByJ mice were infected with ARM or Cl 13. CD11c$^+$ and DEC-205$^+$ cells obtained from spleens made into single splenocyte suspensions were labeled with specific antibodies for these cell subsets. (■) CD11c$^+$ DCs; (●) DEC-205$^+$ DCs

macrophages and minimal number of DCs (Fig. 5). The in vivo replication kinetics over the first 20 days after infection showed that immunosuppressive viruses infected more than 80% of DCs (Fig. 5), indicating that DCs are the primary cell infected in vivo by immunosuppressive LCMV variants (Sevilla et al. 2000). The fact that LCMV immunosuppressive and nonimmunosuppressive variants bound with different affinities to the viral receptor α-DG suggested a crucial role for virus-receptor interaction in the infection of DCs and the initiation of immunosuppression. Only those viruses binding with a high affinity to α-DG have a preferential tropism for DCs leading to a generalized suppression of the host immune system, including the anti-LCMV CTL response required for clearing the infection. Moreover, a recent study in which we evaluated the selective pressures involved in the emergence of immunosuppressive variants during a persistent infection indicated that only variants with high-affinity binding to the receptor were selected from the viral quasi-species to cause immunosuppression (Sevilla et al. 2000). All immunosuppressive variants had L or I in aa 260 of GP-1, instead of F found in nonimmunosuppressive viruses, implying the importance of this aa position in the selection of the immunosuppressive variants. This study suggested that those viral variants with increased binding receptor affinity might be selected within the replicating quasi-species to bind to $CD11c^{+}$ and $DEC\text{-}205^{+}$ cells. The selective pressure is likely based on the interaction of the virus with its cellular receptor, thereby offering a marked advantage to the virus in initiating the dysfunction of the host APCs, aborting the necessary host immune responses required to terminate the viral infection. In addition, the major cell population expressing α-DG in spleen is $CD11c^{+}$ cells (Sevilla et al. 2000), which further confirms the role of virus-receptor interaction in the pathogenesis of LCMV.

5
Impairment of LCMV-Infected DCs

The observation that LCMV is able to infect more than 80% of DCs in vivo suggested the possibility that LCMV may functionally impair DCs. Indeed, DCs isolated from Cl 13-infected mice showed a functional deficit in antigen presentation as evidenced by the inability to stimulate the proliferation of naïve T cells in a primary mixed lymphocyte reaction (MLR; Borrow et al. 1995; Sevilla et al., manuscript in preparation). The prolonged interaction between naïve T cells and APCs establishes that APCs are highly adhesive and costimulatory. In fact, adhesion molecules allow formation of stable APC-T cell conjugates that are essential to deliver the sustained TCR triggering (Sykulev et al. 1996). Additionally, costimulation increases the possibility that this scheduled threshold is reached by lowering the time required for commitment. Consequently, mature DCs that express high levels of adhesion and costimulatory molecules can form stable clusters (Austyn et al. 1988). Because DCs from Cl 13-infected mice were impaired in MLR, it was important to map out any phenotypic changes that could potentially contribute to this dysfunction. Analysis of in vivo-isolated DCs resulted in the identification of two phenotypic populations. In ARM-infected mice a $CD11c^{+high}$ MHC class II^{high} population of DCs were isolated, whereas $CD11c^{+high}$

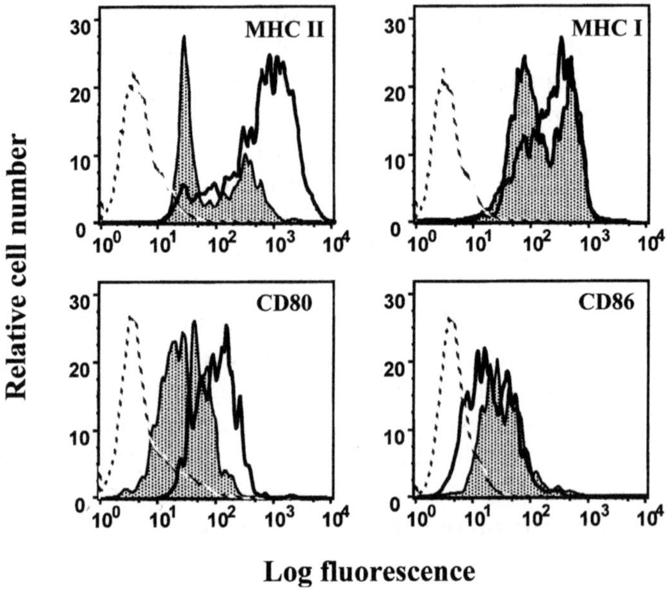

Fig. 6. LCMV modulates expression of functionally important surface molecules on mature DCs. At day 15 after infection, spleens from ARM- or Cl 13-infected mice were collected and individual splenocyte suspensions were prepared from each mouse. Splenocytes were stained with specific monoclonal antibodies for CD11c, MHC II, MHC I, CD80, and CD86. The figure shows FACS histograms of surface expression of MHC I, MHC II, CD80, and CD86 on $CD11c^+$ cells. The *gray filled-in curves* represent expression of the corresponding marker molecules on DCs from Cl 13-infected mice, whereas the *black curves* show surface expression on DCs from ARM-infected mice. *Broken lines* depict staining of DCs with an irrelevant antibody (isotype control). In the *X*-axis, the fluorescence intensity (log scale) is given, whereas the *Y*-axis depicts the relative cell number

MHC class II^{low} DCs were found in Cl 13-infected mice (Fig. 6). In addition, phenotypic analysis of infected DCs showed severely reduced expression of CD80 but normal expression of CD86 during the course of the persistent infection (Fig. 6; Sevilla et al., manuscript in preparation). The engagement of CD80 with CD28 on T cells can direct naïve T cells to differentiate along a T_H1 pathway whereas the ligation of CD28 with CD86 directs the differentiation of T_H precursors to a T_H2 pathway (Kuchroo et al. 1995; Webb and Feldmann 1995). The consequences of downregulation of CD80 on DCs could impair the development of a T_H1 immune response that mediates the elimination of LCMV. Moreover,

CD80 acts as a triggering signal for NK cell-mediated cytolysis (Chambers et al. 1996). The downregulation of CD80 molecules on LCMV-infected DCs could also render DCs less susceptible to NK cell-mediated lysis. In comparison to the viral effect on MHC class II expression, we found a less pronounced albeit significant downregulation of MHC class I molecules by LCMV infected DCs (Fig. 6). This downregulation could blunt the generation of antiviral $CD8^+$ T lymphocytes by infected DC, thereby weakening the host defense.

Virus infection also inhibited production of cytokines such as IL-12. IL-12 is a critical cytokine produced by DCs that directs naïve T cells to differentiate toward a T_H1 phenotype (Macatonia et al. 1995). However, LCMV infection does not depend on the presence of IL-12 for the induction of protective type 1 T cell responses (Oxenius et al. 1999). A T_H2 immunodeviation, as suggested with measles virus (reviewed in Patterson et al. 2001), has not been observed in the LCMV system. The expression of IL-4, as the more potent polarizing signal for T_H2 differentiation (Ohshima and Delespesse 1997), was not found to be increased in LCMV Cl 13 infected mice (Sevilla et al., manuscript in preparation). On the basis of these findings, the infection of DCs allows LCMV to evade T cell attack by disrupting antigen presentation and, hence, the initiation of an efficient antiviral immune response. However, the role of other mechanisms is currently under investigation. Elucidation of these mechanisms will help to further understand LCMV infection and other immunosuppressive viral infections.

6
Concluding Remarks

This chapter has considered the more recent studies regarding the interaction of LCMV with DCs. This interaction, as in many other viral systems, leads to immunosuppression in the host. In fact, infection plus impairment of DC function provides a selective advantage to the virus. The aforementioned studies clearly demonstrate that it is beneficial for the virus to directly interact with DCs in vivo. Variants and strains of LCMV showing high-affinity binding with the cellular receptor α-DG infect DCs and cause immunosuppression in mice. This high receptor binding affinity may allow immunosuppressive variants to displace α-DG-associated ECM molecules and infect DCs. The infection of DCs by LCMV leads to function impairment. Phenotypic analysis of infected DCs revealed that MHC class II and costimulatory molecules were expressed at reduced levels. Although further experiments are required to elucidate the specific mechanisms involved in the impairment of infected DC function, the model system described in this review should allow dissection of the molecules involved with virus-DC interactions.

References

Ahmed R, Hahn C. S, Somasundaram T, Villarete L, Matloubian M, and Strauss J. H. (1991). Molecular basis of organ-specific selection of viral variants during chronic infection. J Virol 65 4242–4247

Ahmed R, and Oldstone M. B. (1988). Organ-specific selection of viral variants during chronic infection. J Exp Med 167 1719–1724

Ahmed R, Salmi A, Butler L. D, Chiller J. M, and Oldstone M. B. (1984). Selection of genetic variants of lymphocytic choriomeningitis virus in spleens of persistently infected mice. Role in suppression of cytotoxic T lymphocyte response and viral persistence. J Exp Med 160 521–540

Althage A, Odermatt B, Moskophidis D, Kundig T, Hoffman-Rohrer U, Hengartner H, and Zinkernagel R. M. (1992). Immunosuppression by lymphocytic choriomeningitis virus infection: competent effector T and B cells but impaired antigen presentation. Eur J Immunol 22 1803–1812

Austyn J. M, Weinstein D. E, and Steinman R. M. (1988). Clustering with dendritic cells precedes and is essential for T-cell proliferation in a mitogenesis model. Immunology 63 691–696

Banchereau J, and Steinman R. M. (1998). Dendritic cells and the control of immunity. Nature 392 245–252

Borrow P, Evans C. F, and Oldstone M. B. (1995). Virus-induced immunosuppression: immune system-mediated destruction of virus-infected dendritic cells results in generalized immune suppression. J Virol 69 1059–1070

Borrow P, and Oldstone M. B. (1992). Characterization of lymphocytic choriomeningitis virus-binding protein(s): a candidate cellular receptor for the virus. J Virol 66 7270–7281

Borrow P, and Oldstone M. B. (1997). Lymphocytic choriomeningitis virus. Viral pathogenesis N Nathanson et al Eds Lippincott-Raven Publishers Philadelphia 593–627

Borrow P, and Shaw G. M. (1998). Cytotoxic T-lymphocyte escape viral variants: how important are they in viral evasion of immune clearance in vivo? Immunol Rev 164 37–51

Borrow P, Tishon A, and Oldstone M. B. (1991). Infection of lymphocytes by a virus that aborts cytotoxic T lymphocyte activity and establishes persistent infection. J Exp Med 174 203–212

Buchmeier M. J, and Oldstone M. B. (1979). Protein structure of lymphocytic choriomeningitis virus: evidence for a cell-associated precursor of the virion glycopeptides. Virology 99 111–120

Buckley S. M, Casals J, and Downs W. G. (1970). Isolation and antigenic characterization of Lassa virus. Nature 227 174

Cao W, Henry M. D, Borrow P, Yamada H, Elder J. H, Ravkov E. V, Nichol S. T, Compans R. W, Campbell K. P, and Oldstone M. B. (1998). Identification of alpha-dystroglycan as a receptor for lymphocytic choriomeningitis virus and Lassa fever virus. Science 282 2079–2081

Cavaldesi M, Macchia G, Barca S, Defilippi P, Tarone G, and Petrucci T. C. (1999). Association of the dystroglycan complex isolated from bovine brain synaptosomes with proteins involved in signal transduction. J Neurochem 72 1648–1655

Chambers B. J, Salcedo M, and Ljunggren H. G. (1996). Triggering of natural killer cells by the costimulatory molecule CD80 (B7-1). Immunity 5 311–317

Dockter J, Evans C. F, Tishon A, and Oldstone M. B. (1996). Competitive selection in vivo by a cell for one variant over another: implications for RNA virus quasispecies in vivo. J Virol 70 1799–1803

Durbeej M, Henry M. D, Ferletta M, Campbell K. P, and Ekblom P. (1998). Distribution of dystroglycan in normal adult mouse tissues. J Histochem Cytochem 46 449–457

Ervasti J. M, and Campbell K. P. (1991). Membrane organization of the dystrophin-glycoprotein complex. Cell 66 1121–1131

Ervasti J. M, and Campbell K. P. (1993). Dystrophin and the membrane skeleton. Curr Opin Cell Biol 5 82–87

Evans C. F, Borrow P, de la Torre J. C, and Oldstone M. B. (1994). Virus-induced immunosuppression: kinetic analysis of the selection of a mutation associated with viral persistence. J Virol 68 7367–7373

Gee S. H, Montanaro F, Lindenbaum M. H, and Carbonetto S. (1994). Dystroglycan-alpha a dystrophin-associated glycoprotein is a functional agrin receptor. Cell 77 675–686

Hazenberg M. D, Hamann D, Schuitemaker H, and Miedema F. (2000). T cell depletion in HIV-1 infection: how CD4+ T cells go out of stock. Nat Immunol 1 285–289

Henry M. D, and Campbell K. P. (1998). A role for dystroglycan in basement membrane assembly. Cell 95 859–870

Henry M. D, and Campbell K. P. (1999). Dystroglycan inside and out. Curr Opin Cell Biol 11 602–607

Ibraghimov-Beskrovnaya O, Ervasti J. M, Leveille C. J, Slaughter C. A, Sernett S. W, and Campbell K. P. (1992). Primary structure of dystrophin-associated glycoproteins linking dystrophin to the extracellular matrix. Nature 355 696–702

Ibrahim M. A, Chain B. M, and Katz D. R. (1995). The injured cell: the role of the dendritic cell system as a sentinel receptor pathway. Immunol Today 16 181–186

Jaspars L. H, Bloemena E, Bonnet P, Van der Valk P, and Meijer C. J. (1995). Distribution of extracellular matrix components and their receptors in human lymphoid tissue and B-cell non-Hodgkin lymphomas. Histopathology 26 113–121

Jung D, Yang B, Meyer J, Chamberlain J. S, and Campbell K. P. (1995). Identification and characterization of the dystrophin anchoring site on beta-dystroglycan. J Biol Chem 270 27305–27310

Kramer R. H, Rosen S. D, and McDonald K. A. (1988). Basement-membrane components associated with the extracellular matrix of the lymph node. Cell Tissue Res 252 367–375

Kuchroo V. K, Das M. P, Brown J. A, Ranger A. M, Zamvil S. S, Sobel R. A, Weiner H. L, Nabavi N, and Glimcher L. H. (1995). B7–1 and B7–2 costimulatory molecules activate differentially the Th1/Th2 developmental pathways: application to autoimmune disease therapy. Cell 80 707–718

Kunz S, Sevilla N, McGavern D. B, Campbell K. P, and Oldstone M. B. (2001). Molecular analysis of the interaction of LCMV with its cellular receptor [alpha]-dystroglycan. J Cell Biol 155 301–310

Lieberman J, Shankar P, Manjunath N, and Andersson J. (2001). Dressed to kill? A review of why antiviral CD8 T lymphocytes fail to prevent progressive immunodeficiency in HIV-1 infection. Blood 98 1667–1677

Macatonia S. E, Hosken N. A, Litton M, Vieira P, Hsieh C. S, Culpepper J. A, Wysocka M, Trinchieri G, Murphy K. M, and O'Garra A. (1995). Dendritic cells produce IL-12 and direct the development of Th1 cells from naive CD4+ T cells. J Immunol 154 5071–5079

Matloubian M, Kolhekar S. R, Somasundaram T, and Ahmed R. (1993). Molecular determinants of macrophage tropism and viral persistence: importance of single amino acid changes in the polymerase and glycoprotein of lymphocytic choriomeningitis virus. J Virol 67 7340–7349

McMichael A. J, and Rowland-Jones S. L. (2001). Cellular immune responses to HIV. Nature 410 980–987

Odermatt B, Eppler M, Leist T. P, Hengartner H, and Zinkernagel R. M. (1991). Virus-triggered acquired immunodeficiency by cytotoxic T-cell-dependent destruction of antigen-presenting cells and lymph follicle structure. Proc Natl Acad Sci USA 88 8252–8256

Ohshima Y, and Delespesse G. (1997). T cell-derived IL-4 and dendritic cell-derived IL-12 regulate the lymphokine-producing phenotype of alloantigen-primed naive human CD4 T cells. J Immunol 158 629–636

Oldstone M. B. (1989). Viral persistence. Cell 56 517–520

Oxenius A, Karrer U, Zinkernagel R. M, and Hengartner H. (1999). IL-12 is not required for induction of type 1 cytokine responses in viral infections. J Immunol 162 965–973

Parekh B. S, and Buchmeier M. J. (1986). Proteins of lymphocytic choriomeningitis virus: antigenic topography of the viral glycoproteins. Virology 153 168–178

Patterson J. B, Manchester M, and Oldstone M. B. (2001). Disease model: dissecting the pathogenesis of the measles virus. Trends Mol Med 7 85–88

Peters C. J, Buchmeier M. J, Rollin P. E, and Kisiazek T. G. (1996). Arenavirus. DM Knipe BN Fields and PM Howlley (ed) Fields Virology 3rd edition Lippincott-Raven Publishers Philadelphia 2 1521–1551

Rescigno M, Granucci F, and Ricciardi-Castagnoli P. (1999). Dendritic cells at the end of the millennium. Immunol Cell Biol 77 404–410

Riviere Y, Ahmed R, Southern P. J, Buchmeier M. J, Dutko F. J, and Oldstone M. B. (1985). The S RNA segment of lymphocytic choriomeningitis virus codes for the nucleoprotein and glycoproteins 1 and 2. J Virol 53 966–968

Rodriguez M, Buchmeier M. J, Oldstone M. B, and Lampert P. W. (1983). Ultrastructural localization of viral antigens in the CNS of mice persistently infected with lymphocytic choriomeningitis virus (LCMV). Am J Pathol 110 95–100

Salvato M, Borrow P, Shimomaye E, and Oldstone M. B. (1991). Molecular basis of viral persistence: a single amino acid change in the glycoprotein of lymphocytic choriomeningitis virus is associated with suppression of the antiviral cytotoxic T-lymphocyte response and establishment of persistence. J Virol 65 1863–1869

Salvato M. S, and Shimomaye E. M. (1989). The completed sequence of lymphocytic choriomeningitis virus reveals a unique RNA structure and a gene for a zinc finger protein. Virology 173 1–10

Sevilla N, Kunz S, Holz A, Lewicki H, Homann D, Yamada H, Campbell K. P, de La Torre J. C, and Oldstone M. B. (2000). Immunosuppression and resultant viral persistence by specific viral targeting of dendritic cells. J Exp Med 192 1249–1260

Singh M. K, Fuller-Pace F. V, Buchmeier M. J, and Southern P. J. (1987). Analysis of the genomic L RNA segment from lymphocytic choriomeningitis virus. Virology 161 448–456

Smelt S. C, Borrow P, Kunz S, Cao W, Tishon A, Lewicki H, Campbell K. P, and Oldstone M. B. (2001). Differences in affinity of binding of lymphocytic choriomeningitis virus strains to the cellular receptor alpha-dystroglycan correlate with viral tropism and disease kinetics. J Virol 75 448–457

Southern P. J. (1996). Arenaviridae: the viruses and their replication. DM Knipe BN Fields and PM Howlley (ed) Fields Virology 3rd edition Lippincott-Raven Publishers Philadelphia 2 1505–1551

Southern P. J, Singh M. K, Riviere Y, Jacoby D. R, Buchmeier M. J, and Oldstone M. B. (1987). Molecular characterization of the genomic S RNA segment from lymphocytic choriomeningitis virus. Virology 157 145–155

Steinman R. M. (1991). The dendritic cell system and its role in immunogenicity. Annu Rev Immunol 9 271–296

Steinman R. M, Pack M, and Inaba K. (1997). Dendritic cells in the T-cell areas of lymphoid organs. Immunol Rev 156 25–37

Sykulev Y, Joo M, Vturina I, Tsomides T. J, and Eisen H. N. (1996). Evidence that a single peptide-MHC complex on a target cell can elicit a cytolytic T cell response. Immunity 4 565–571

Talts J. F, Andac Z, Gohring W, Brancaccio A, and Timpl R. (1999). Binding of the G domains of laminin alpha1 and alpha2 chains and perlecan to heparin sulfatides alpha-dystroglycan and several extracellular matrix proteins. EMBO J 18 863–870

Tishon A, Borrow P, Evans C, and Oldstone M. B. (1993). Virus-induced immunosuppression. 1. Age at infection relates to a selective or generalized defect. Virology 195 397–405

Tishon A, Southern P. J, and Oldstone M. B. (1988). Virus-lymphocyte interactions. II. Expression of viral sequences during the course of persistent lymphocytic choriomeningitis virus infection and their localization to the L3T4 lymphocyte subset. J Immunol 140 1280–1284

Tortorella D, Gewurz B. E, Furman M. H, Schust D. J, and Ploegh H. L. (2000). Viral subversion of the immune system. Annu Rev Immunol 18 861–926 van den Berg T. K, van der Ende M, Dopp E. A, Kraal G, and Dijkstra C. D. (1993). Localization of beta 1 integrins and their extracellular ligands in human lymphoid tissues. Am J Pathol 143 1098–1110

Villarete L, Somasundaram T, and Ahmed R. (1994). Tissue-mediated selection of viral variants: correlation between glycoprotein mutation and growth in neuronal cells. J Virol 68 7490–7496

Webb L. M, and Feldmann M. (1995). Critical role of CD28/B7 costimulation in the development of human Th2 cytokine-producing cells. Blood 86 3479–3486

Williamson R. A, Henry M. D, Daniels K. J, Hrstka R. F, Lee J. C, Sunada Y, Ibraghimov-Beskrovnaya O, and Campbell K. P. (1997). Dystroglycan is essential for early embryonic development: disruption of Reichert's membrane in Dag1-null mice. Hum Mol Genet 6 831–841

Winder S. J. (2001). The complexities of dystroglycan. Trends Biochem Sci 26 118–124

Wright K. E, Spiro R. C, Burns J. W, and Buchmeier M. J. (1990). Post-translational processing of the glycoproteins of lymphocytic choriomeningitis virus. Virology 177 175–183

Yang B, Jung D, Motto D, Meyer J, Koretzky G, and Campbell K. P. (1995). SH3 domain-mediated interaction of dystroglycan and Grb2. J Biol Chem 270 11711–11714

Chapter 7

The Interaction Between Dendritic Cells and Herpes Simplex Virus-1

D. Kobelt, M. Lechmann, A. Steinkasserer

Abstract. Dendritic cells (DCs) are the most potent antigen-presenting cells, because they are also able to induce native T cells. Thus they are crucial in the induction of antiviral immune responses. Several viral immune escape mechanisms have been described; here we concentrate on the interaction between DCs and herpes simplex virus type 1 (HSV-1). DCs can be infected by HSV-1; however, only immature DCs generate infectious viral particles, whereas mature DCs do not support virus production and only immediate-early and early viral transcripts are generated. To induce potent immune responses DCs must mature. Interestingly, HSV-1 interferes with this maturation process, thus inhibiting antivi-

D. Kobelt, M. Lechmann, A. Steinkasserer (✉)
Department of Dermatology, University of Erlangen,
Hartmannstrasse 14, 91052 Erlangen, Germany

ral T cell stimulation. Furthermore, HSV-1 strongly interferes with DC-mediated T cell proliferation. A striking finding was the complete degradation of CD83, the best-known marker for mature DC, after HSV-1 infection in lysosomal compartments. This CD83 degradation coincided with a clearly reduced T cell stimulation representing an additional new escape strategy. The functional role and the importance of CD83 are discussed in detail.

1 Introduction

1.1 Herpes Simplex Virus Type 1

The Herpesviridae family consists of three subfamilies, the α-, β- and γ-herpesviruses, differing in their pathogenicity and tropisms. Human pathogen members are the herpes simplex viruses HSV-1, -2 and –6 to -8 as well as the (human) cytomegalovirus (CMV), Epstein-Barr virus (EBV) and the varicella zoster virus (VZV).

HSV-1 belongs to the family of large DNA viruses replicating in the nuclei of infected mammalian cells, able to persistently infect host cells. The virus is able to enter the body at epithelial and mucosal surfaces (mouth, face and eyes). The infection of sensory neurons innervating the mucosa is caused by the lytic infection of epithelial cells and fibroblasts followed by the rapid establishment of latent infection in the cell bodies. Reactivation as well as initial infection can result in serious human disease including encephalitis, keratitis (Whitley 1996; Geiss et al. 2000) and genital herpes, although most cases of genital herpes are caused by HSV-2 (Slomka et al. 1998).

The Herpes-1 genome consists of 72 genes and has a length of 152 kb. The virion is able to pack large genetic insertions (Hammerschmidt 2000), offering the possibility of using genetically modified viruses for immunotherapy (Jenne et al. 2001). However, it is of utmost importance to identify viral vectors which do not interfere with dendritic cell (DC) functions (see chapter by Humrich and Jenne, this volume).

Although the indispensable *cis*- and *trans*-acting factors involved in herpesviral DNA-replication are well characterised (for a review, see Hammerschmidt 2000), the genes responsible for immunoevasive strategies, for cytotoxicity as well as those responsible for the modulation of

DC-mediated T cell stimulation have yet to be identified. The removal of immediate-early genes (IE genes) could possibly reduce cytotoxicity. However, one must keep in mind that deletion of particular genes may complicate the production of infectious virions. For example, mutations that inactivate the vhs polypeptide cause a 5- to 10-fold reduction in the production of progeny virus in cell culture system (Everly and Read 1999). On the other hand, Howard and colleagues have shown that HSV-1 viruses with inactivating mutations show lower toxicity as well as a higher gene delivery efficiency (Howard et al. 1998). Nevertheless, more research is needed to identify and subsequently eliminate gene products which interfere with DC biology.

1.2 Function of Immediate-Early Proteins

Recent studies have concentrated on investigating the function of the immediate-early proteins (IE proteins) to analyse their cytotoxic effect and to determine how the virus switches between the lytic infection in the epithelia and periods of latency when the genome is residing in neurons in a mainly quiescent state. Initial synthesis of IE proteins is essential for the lytic infection.

The IE-gene product ICP4 interacts with TFIID through TAF250 thereby influencing the basal transcription apparatus (Carrozza and DeLuca 1996). ICP27 shows two different functions by acting at both the transcriptional and post-transcriptional levels (Phelan and Clements 1998). ICP22 acts by enhancing the levels of late gene expression in selected cell types (Sears et al. 1985).

Several studies have been performed using viruses with lesions in IE genes. The results show that in the absence of ICP0 the latent state is favoured, strengthening the hypothesis that ICP0 is involved in the control of the balance between lytic and latent infection (Preston and Nicholl 1997; Samaniego et al. 1998). This hypothesis has been further supported by the findings that inactivation of ICP0 and absence of VP16 lead to a profound defect in the ability of HSV-1 to proceed to lytic infection (Mossman and Smiley 1999).

In addition, Roger Everett was able to demonstrate that ICP0 induces a striking accumulation of conjugated ubiquitin at nuclear substructures known as ND10, centromeres and other locations where ICP0 accumulates and that the RING finger region of ICP0 was essential for this activity. These data strongly support the hypothesis that ICP0 functions through the ubiquitin-proteasome pathway to induce the targeted destruction of specific cellular proteins (Everett 2000). This is further supported by the findings that alpha-herpesvirus proteins related to HSV-1 ICP0 induce the formation of co-localising, conjugated ubiquitin (Parkinson and Everett 2001).

1.3
Course of Infection

The infection of susceptible cells is initiated by the attachment of virions to cell surface heparan sulphate proteoglycans via glycoproteins C and/ or B (gC/gB) (Herold et al. 1994). This process is followed by the binding of glycoprotein D (gD) to specific receptors: herpesvirus entry mediator (Hve-A) (Montgomery et al. 1996), which is expressed in lymphocytes, Hve-B (Warner et al. 1998), which is also found in lymphocytes, showing a more prominent expression in epithelial cells, some neuronal cells and cells of fibroblastic origin. The latter receptor mediates the entry of some mutants of HSV-1 that do not enter via Hve-A (Spear et al. 2000). Hve-C (Geraghty et al. 1998; Cocchi et al. 1998) expressed on epithelial and neuronal cells allows the infection of cells located on mucosal surfaces followed by subsequent spreading of HSV-1 to the nervous system (Spear et al. 2000). The mechanisms of entry of HSV-1 into human cells are still under investigation, and recent publications show the existence of splice variant isoforms of Hve-B (also designated as nectin2/PRR2α and PPR2δ) (Lopez et al. 2000).

An important step during infection is the inhibition of host cell protein synthesis that can be divided into an early and an intermediate phase. An early shutoff involves the disruption of pre-existing polysomes and the degradation of host mRNAs (reviewed in Lu et al. 2001). In a recent work it was shown that the virion host shutoff (vhs) protein encoded by HSV gene UL41 acts as an RNAse employing more than one mode of substrate recognition (Lu et al. 2001). The vhs polypeptide enters the cell as a component of infecting virions, destabilising all three temporal classes of viral messages and most cellular messages studied to date, with the exception of rRNAs (Strelow and Leib 1995; Oroskar and Read 1989). The activity of vhs is regulated by VP16; the direct interaction between these two proteins has been shown (Mossmann 2000; Lam 1996; Smibert 1994). Also, HSV-1 encoded ICP27 plays a role in controlling the rate of mRNA turnover by inhibiting the splicing of pre-mRNA (Everly and Read 1999; Hardwicke et al. 1994; Hardy et al. 1994).

1.4
Lytic Infection and Latency

Herpesviruses can infect proliferating cells as well as resting cells because they encode the genes necessary for DNA replication, enabling them to replicate their viral DNA autonomously in almost any cell they infect (Challberg 1996).

The neurotrophic virus HSV-1 is able to infect axonal nerve terminals before retrograde transport to neuronal cell bodies where latency is established. The HSV genome remains extrachromosomal, and the genes encoding the latency-associated transcripts (LATs) are transcribed (Palmer et al. 2000, Stevens et al. 1987).

It must be mentioned that the HSV-1 virions contain linear DNA. ON infection of a susceptible cell the termini of the DNA are fused, causing the circularisation of the herpesviral genome (Hammerschmidt and Mankertz 1991; Hammerschmidt 2000).

Expression of IE genes initiates the cascade of lytic gene expression. The induction of the lytic pathway is triggered by a protein called VP16 (also known as Vmw65 or alpha-TIF) that is carried into the infected cell as part of the tegument. The four IE genes encoded by HSV-1, ICP0, ICP4, ICP22 and ICP27 contain multiple OCT/TAATGARAT-type binding motifs (Aprhys et al. 1989; Mullen et al. 1995) that normally mediate strong positive responses to VP16, although all four promoters show a high basal-level activity (Mullen et al. 1995). IE genes have the function to transactivate a second set of viral genes – the early genes. These genes are directly involved in viral DNA replication and nucleotide biosynthesis.

Intermediate Products: Viral DNA replication produces largely concatemeric DNA molecules which must be cleaved into linear virion DNA units of discrete length. Late gene expression: Structural gene products such as viral capsid proteins, surface membrane glycoproteins and other non-structural components of the virion (reviewed by Hammerschmidt 2000).

HSV-1 can spread in infected persons via infectious virions or by virus-induced syncytium formation (Novotny et al. 1996). Nevertheless, the functions of many genes encoded by HSV-1 still must be analysed. For example, the expression of U_L43 is still under investigation. Studies by Carter et al. (1996) and Ward et al. (1996) report that a sense message as well as an anti-sense message are expressed at different times after infection.

2
Infection of Dendritic Cells by HSV-1

DCs ells express receptors necessary for the entry of Alpha-herpesviruses Hve-A and Hve-B (Salio et al. 1999). Over 90% of mature monocyte-derived DCs can be infected with HSV-1 at a MOI of 1 (Kruse et al. 2000a). This infection leads to the generation of DC clusters. Interestingly, only immediate and early viral transcripts could be detected, whereas no late transcripts were generated. In addition, no infectious viral particles were generated in infected DC. The phenotypic examination by FACS of HSV-1 infected DC revealed a dramatic down-modulation of CD83, the best-known cell surface marker for mature DCs (Zhou et al. 1995), within the first 10 h after infection. The functional implications of the CD83 down-modulation are discussed below. Other molecules including CD25, CD40, CD80, CD86, CD95 and MHC molecules were, however, not influenced at these early time points.

The infection of immature DCs has also been recently reported (Mikloska et al. 2001). In contrast to mature DCs, immature DCs allow for viral replication and generation of infectious particles. This led to a largely reduced survival of DCs and induction of cell death. In addition, these authors reported the down-modulation of several surface molecules including CD1a, CD40, CD54, CD80 and CD86. However, because HSV-1 infection induces a high degree of cytopathicity in immature DC (20%–45% of DC die within the first 24–48 h) the observed down-modulation could be due to cell death rather than a specific viral mechanism. Nevertheless, because only mature DC induce strong antiviral immune responses, elimination of immature DC, and thereby inhibition of DC maturation, reveals an immunoevasive strategy of HSV-1.

3 Interference with DC Maturation

Immature DCs reside in the periphery and are specialised in antigen uptake. However, only fully mature DCs are able to induce potent immune responses, whereas immature DCs might be more specialised in the induction of tolerance (Steinman 1991; Banchereau et al. 2000; Jonuleit et al. 2001). Thus inhibition of DC maturation would possibly be a very effective viral strategy to overcome the host immune response. Maturation can be induced by various viral and bacterial molecules or via inflammatory stimuli. Thus the induction of cell death and down-modulation of several crucial molecules on immature HSV-1 DCs, as described above, will definitely inhibit DC maturation and subsequently antiviral T cell stimulation.

In addition, the specific inhibition of DC maturation by HSV-1 has also been described (Salio et al. 1999). Using a recombinant disabled infectious single-cycle (DISC) HSV-1 to infect DCs, the authors demonstrated that infected DCs were unable to up-regulate co-stimulatory molecules which are essential for T cell stimulation. Even the addition of LPS to the cell cultures did not induce DC maturation. Furthermore, these infected DCs did not produce any cytokines [such as tumour necrosis factor (TNF)-α, interleukin (IL)-6, IL-10 and IL-12]. A vital aspect in DC maturation is the up-regulation of CCR7, a crucial chemokine receptor for DC migration. However, infected DCs were completely unresponsive to chemokines required for migration to secondary lymphoid organs. This again demonstrates the development of a new and effective immune escape strategy.

4 Interference with DC-Mediated T Cell Stimulation

Only mature DCs are able to induce strong T cell responses. Apart from cytokine production and migratory properties, a very powerful tool to investigate the functional properties of DC is the allogeneic T cell proliferation. In this respect, several groups reported that HSV-1 strongly interferes with DC-mediated T cell proliferation. Salio et al. described a HSV-1 mediated block in DC maturation. Furthermore, when they checked their capacity to stimulate proliferation and polarisation of allogeneic naive T cells, only an extremely low T cell stimulatory capacity was registered (Salio et al. 1999). Furthermore, DCs which were infected after induction of maturation were also poor T cell stimulators. Thus HSV-1 infection leads to a down-modulation of DC-mediated T cell responses. Mikloska et al. reported the down-modulation of several surface molecules and a high degree of cytopathicity in immature DCs after infection with HSV-1 (Mikloska et al. 2001). Although the authors did not investigate the functional implication on the T cell level, one may expect that such a strong modulation of DCs would also impair their functional properties, i.e. their capacity to stimulate T cells.

Investigations performed in our laboratories with mature DCs clearly showed that HSV-1 infections led to a complete degradation of CD83 (Kruse et al. 2000). Already 10 h after infection CD83 was completely downmodulated from the cell surface, whereas other known costimulatory molecules, including CD80 and CD86, were not affected. Because no reduction of CD83 at the mRNA level could be detected after 10 h, CD83 down-modulation must be caused by post-transcriptional events. To further investigate this phenomenon and to investigate where CD83 might be degraded, Vero cells were microinjected with a CD83-encoding plasmid and the subcellular localisation was analysed with CD83- and cathepsin D-specific antibodies. With the use of these tools it became clear that CD83 was degraded in lysosomal compartments as CD83 and cathepsin D co-localised within the same lysosomal compartments. Thus HSV-1 induces the degradation of CD83.

Interestingly, CD83 degradation clearly coincided with a reduced T cell stimulatory capacity of the infected DC. This was the first hint that CD83 might be of functional importance for DC biology.

Another group reported that human cytomegalovirus (HCMV) in late-stage infected fibroblasts led to the down-modulation of CD83 and inhibition of DC maturation, which cross-presented these infected cells (Arrode et al. 2002). These DCs showed a defect in inducing HCMV-specific CTLs. The authors point out that HCMV-infected fibroblasts secrete the soluble transforming growth factor-β 1 (TGF-β 1) and speculate that this might be a possible mediator for the reduced CD83 expression and inhibited stimulatory function of DCs.

Together, these data indicate that modulation of CD83 expression is yet another new viral mechanism to escape host-specific antiviral immune responses.

5 Functional Role of CD83

To further investigate the functional role of CD83 several mechanistic studies were performed.

5.1 Interference with CD83 mRNA Transport Modulates DC Function

It was observed previously that CD83 expression is closely mirrored by the expression of the eukaryotic initiation factor 5A (eIF-5A). eIF-5A was originally described as an initiation factor (Park et al. 1993a). Later studies, however, revealed that eIF-5A is not an initiator of protein translation (Kang et al. 1994) but is involved in a specific nuclear export pathway (Bevec et al. 1996; Elfgang et al. 1999). It has been shown that eIF-5A interacts with the general nuclear export receptor CRM1 at the nucleoplasmic site of the nuclear pore complex and is translocated from the nucleus to the cytoplasm in mammalian cells (Rosorius et al. 1999).

Interestingly, the eIF-5A expression level is very low in immature DCs but, similarly to CD83, is upregulated during DC maturation. An additional unique feature of eIF-5A is its post-translational amino acid hyposine modification, which is essential for its biological function (Park et al. 1993a,b). One of the two enzymes responsible for catalysing this modification can be inhibited by the low-molecular-weight inhibitor GC7 (N^1-guanyl-1,7-diaminoheptane). Incubation of maturing DCs with GC7 inhibited the CD83mRNA export from the nucleus into the cytoplasm, and thus CD83 protein expression was also inhibited. Interestingly, this correlated with a clearly reduced T cell stimulatory capacity of these DCs (Kruse et al. 2000b).

These results, i.e. viral infection and interference with CD83 mRNA transport, showed a clear correlation between CD83 down-modulation and inhibition of the stimulatory function of the DC. However, this does not prove that CD83 modulation alone is responsible for the inhibition of DC biology, because it could not be excluded that other molecules also contributed to these effects.

5.2 The Soluble Extracellular CD83 Domain Suppresses DC-Mediated T Cell Proliferation

To prove that CD83 indeed has functional implications for DC biology the extracellular domain of human CD83 (hCD83ext) was expressed in *Escherichia coli* (Lechmann et al. 2002a,b) and used in functional mixed leucocyte reaction (MLR) analyses (Lechmann et al. 2001). Interestingly, hCD83ext completely abrogated the DC-mediated primary allostimulation of T cells in a dose-dependent manner. This provided the first direct proof of the principle that CD83 is indeed necessary for DC function. In addition, with an influenza matrix peptide M1-specific cytotoxic lymphocyte clone, it could be demonstrated that hCD83ext also inhibits the interaction between DC and peptide-specific cytotoxic T cells (Lechmann et al. 2001).

These inhibitory effects were clearly not due to cytotoxicity on either the DC or T cell level. Furthermore, the proliferation arrest of T cells which were inhibited by hCD83ext during the primary stimulation could be overcome by a secondary stimulation, either with IL-2 or with untreated allogeneic DC. These data imply that the hCD83ext-mediated inhibition represents a novel mechanism which is induced by the inhibition of the B7-CD28 costimulatory pathway (Boussiotis et al. 1994) and is not connected to T cell anergy.

Additional functional data regarding CD83 were obtained with a soluble human CD83-Ig fusion protein (hCD83-Ig). Again, the allogenic MLR could be inhibited in a dose-dependent manner (Lechmann et al. 2001). Because hCD83ext, which was expressed in *E. coli*, had the same inhibitory effect as the eukaryotically expressed CD83-Ig fusion protein, it is clear that CD83-glycosylation is not a prerequisite for the immunosuppressive function of soluble CD83 molecules.

Human and murine CD83 show a high degree of amino acid homology, and therefore one could speculate that they have similar functions. Indeed, it could be demonstrated that human soluble CD83 also inhibits murine allo-responses in a dose-dependent manner (Lechmann et al. 2001). In addition, recombinant soluble CD83 molecules from the mouse, either expressed in *E. coli* or with a baculovirus expression system, also inhibited DC-mediated allogeneic T cell proliferation in the murine system in a dose-dependent manner (Steinkasserer et al., unpublished data).

Another group confirmed these observations in the mouse system and showed clearly that soluble CD83 molecules are also inhibitory in vivo. Immunogenic P815 tumour cells were implanted into recipient mice, and subsequently mCD83-Ig fusion protein was injected intraperitoneally. The effect was that tumours in mice treated with mCD83-Ig were twice as large as those in control mice receiving no mCD83-Ig. Additional analysis revealed that splenocytes, harvested from mCD83-Ig-treated mice 15 days after onset of the experiment, showed a significantly lower lysis of P815 cells than splenocytes from the control group (Scholler et al. 2002).

Together, these results clearly show that soluble CD83 is immunosuppressive in vitro and in vivo and that this function is conserved between the different species. These findings gain further biological importance when viewed together with the fact that a soluble form of CD83 was detected in normal human sera (Hock et al. 2001). It was demonstrated that a soluble form of CD83 is released from DCs and B lymphocytes and that the concentration level of soluble CD83 is increased during DC maturation. Inhibition of the de novo protein synthesis did not prevent the release of CD83, and thus the authors suggest that proteolytic shading of the ectodomain of membrane-bound CD83 represents the mechanism for CD83 release (Hock et al. 2001). Together, these observations propose a role for soluble CD83 in the regulation of immune responses in vivo.

5.3 The Soluble Extracellular CD83 Domain Interferes with DC Maturation

An additional interesting finding was that immature DC incubated together with soluble hCD83ext could no longer be fully matured. Even the presence of the potent maturation cocktail, composed of IL-1β, TNF-α and prostaglandin E_2 (PGE_2) failed to mature hCD83ext-treated DCs. hCD83ext treatment lead to the specific down-modulation of CD80 cell surface expression, a crucial costimulatory molecule, and of CD83 itself. The number of $CD14^+$ cells was increased by hCD83ext treatment, correlating with an immature phenotype (Lechmann et al. 2001). These data suggest that soluble forms of CD83 might induce a maturation block in DCs and could represent a mechanism to control and down-modulate immune response in vivo.

In addition, mature DCs which were incubated with hCD83ext showed only a minimal influence on CD80 expression, whereas CD83 expression was clearly reduced (Lechmann et al. 2001). This indicates a regulatory mechanism also on mature DCs and would explain the inhibitory effect of soluble CD83 in mixed lymphocyte reactions. The reduction of CD83 on mature DCs by soluble CD83 correlates with the downmodulation of CD83 after HSV-1 infection, both leading to inhibition of DC-mediated T cell stimulation.

The mechanism of this inhibition is unclear. The soluble molecule may induce inhibition by delivering a negative signal after internalisation. Another possible explanation could be that soluble CD83 binds to the receptor on DC, thereby inhibiting the induction of signal transduction normally induced by the interaction between membrane-bound CD83 and its ligand. While soluble CD83 binds to this receptor, the receptor is blocked and the activation signal can no longer be transmitted. However, these hypotheses must be proven in further experimental settings.

5.4 CD83 Influences $CD4^+$ T Cell Development in the Thymus

Very recently, Fujimoto et al. reported the generation of CD83-deficient ($CD83^{-/-}$) mice. Interestingly these, mice showed an impaired, i.e. 68% reduced $CD4^+$ single-positive thymocyte development (Fujimoto et al. 2002), although the numbers of $CD4^-$ $CD8^-$ double-negative thymocytes which enter the thymus and the numbers of the later differentiation stage $CD4^+$ $CD8^+$ double-positive thymocytes were normal.

The specific block in $CD4^+$ single-positive thymocyte development also resulted in a selective 75%–90% reduction in peripheral T cells. The number of $CD8^+$ single-positive thymocytes was not increased, whereas the $CD8^+$ T cell numbers were increased in peripheral blood of $CD83^{-/-}$ mice, although their numbers in peripheral tissues remained normal. Peripheral $B220^+$ B cells, bone marrow leucocytes, peripheral monocytes, neutrophils and $CD3^-CD16^+$ NK cells developed normally. However, low MHC class II antigen expression by resting B cells from $CD83^{-/-}$ mice and reduced MHC class II and CD86 induction after activation also suggested additional requirements for CD83 expression in B cell activation.

Thus CD83 expressed on thymic epithelial cells and DC plays an important role in the development of the $CD4^+$ T cell lineage development.

Considering the different cell types which are able to express CD83, such as activated DC, activated B cell and activated $CD8^+$ T cells, CD83 might play a differentiated role in the communication among these key cells of the immune system. Moreover, soluble CD83 also seems to have an important regulatory effect on the immune response in vitro and in vivo.

Finally, with tools such as $CD83^{-/-}$ mice and soluble CD83 molecules one may be able to explore CD83 function(s) on a molecular level in more detail. In addition, characterisations of the three-dimensional structure of CD83 will also help to clarify and analyse its function. However, also from these studies, it became clear that targeting CD83 expression in vivo, as developed by HSV-1, is a very powerful strategy to escape antiviral immunity.

Acknowledgements. This work was supported by the Deutsche Forschungsgemeinschaft (Sonderforschungsbereich 466), by the Bundesministerium für Bildung und Forschung [Interdisziplinäres Zentrum für Klinische Forschung (IZKF) Erlangen] and by the ELAN programme of the University Hospital of Erlangen. We thank Heidi Joao for excellent proofreading and critical comments.

References

Arrode G, Boccaccio C, Abastado JP, Davriinche C: Cross-presentation of human cytomegalovirus pp65 (UL83) to $CD8^+$ T cells is regulated by virus-induced, soluble-mediator-dependent maturation of dendritic cells. J. Virol. 2002; 76:142

Aprhys C, O'Neill E, Kelly TJ, Hayward GS (1989) Overlapping octamer and TAATGARAT motifs in the VF65-response elements in herpes simplex virus immediate-early promoters represent independent binding sites for cellular nuclear factor III. J. Virol. 63: 2798–2812

Banchereau J, Briere F, Caux C, Davoust J, Lebecque S, Liu YJ, Pulendran B, Palucka K. (2000) Immunobiology of dendritic cells. Annu. Rev. Immunol. 18: 767–811

Bevec D, Jaksche H, Oft M, Wöhl T, Himmelspach M, Pacher A, Schebsta M, Koettnitz K, Dobrovnik M, Csonga R, Lottspeich F, Hauber J. (1996) Inhibition of HIV-1 replication in lymphocytes by mutants of the REV cofactor eIF-5A. Science. 271:1858

Boussiotis VA, Gribben JG, Freeman GJ, Naddler LM. (1994) Blockade of the CD28 co-stimulatory pathway: a means to induce tolerance. Curr. Opin. Immunol. 6:797

Carrozza MJ and DeLuca NA (1996) Interaction of the viral activator protein ICP4 with TFIID through TAF250. Mol. Cell. Biol. 295: 1103–1112

Carter KL, Ward PL, Roizman B (1996) Characterization of the products of the U_L43 gene of herpes simplex virus 1: Potential implications for regulation of gene expression by antisense transcription. J. Virol. 70: 7663–7668

Challberg M (1996) Herpesvirus DNA replication. DNA Replication in Eukaryotic Cells. DePamphilis ML (Ed), Cold Spring Harbor Laboratory Press, NY, USA: 721–750

Cocchi F, Menotti L, Mirandola P, Lopez M, Campadelli-Fiume G. (1998) The ectodomain of a novel member of the immunoglobulin subfamily related to the poliovirus receptor has the attributes of a bona fide receptor for herpes simplex virus types 1 and 2 in human cells. J Virol. 72(12):9992–10002

Elfgang C, Rosorius O, Hofer L, Jaksche H, Hauber J, Bevec D. (1999) Evidence for specific nucleocytoplasmic transport pathways used by leucin-rich nuclear export signals. Proc. Natl. Acad. Sci. USA 96:6229

Engelmayer J, Larsson M, Subklewe M et al. (1999) Vaccinia virus inhibits the maturation of human dendritic cells: a novel mechanism of immune evasion. J. Immunol. 163: 6762–6768

Everett RD (2000) ICP0 induces the accumulation of colocalizing conjugated ubiquitin. J. Virol. 74: 9994–10005

Everly DN and Read GS (1999) Site-directed mutagenesis of the virion host shutoff gene (UL41) of Herpes Simplex virus (HSV): Analysis of Functional Differences between HSV Type 1 (HSV-1) and HSV-2 alleles. J. Virol. 73: 9117–9129

Fujimoto Y, Tu L, Miller AS, Bock C, Fujimoto M, Doyle C, Steeber DA, Tedder TF. (2002) CD83 Expression influences $CD4^+$ T cell development in the thymus. Cell 108:755

Geiss BJ, Smith TJ, Leib DA, Morrison LA (2000) Disruption of virion host shutoff activity improves the immunogenicity and protective capacity of a replication-incompetent Herpes Simplex Virus Type 1 vaccine strain. J.Virol. 74: 11137–11144

Geraghty RJ, Krummenacher C, Cohen GH, Eisenberg RJ, Spear PG (1998) Entry of alphaherpesviruses mediated by poliovirus receptor-related protein 1 and poliovirus receptor. Science 280: 1618–1620

Hammerschmidt W and Mankertz J (1991) Herpesviral DNA replication: Between the known and the unknown. Seminars Virology 2: 257–269

Hammerschmidt W (2000) Herpesvirus vectors come of age. Curr. Opin.Mol. Ther. 2(5): 532–539

Hardwicke MA, Sandri-Goldin RM (1994) The herpes simplex virus regulatory protein ICP27 contributes to the decrease in cellular mRNA levels during infection. J. Virol. 68: 4797–4810

Hardy WR, Sandri-Goldin RM (1994) Herpes simplex virus inhibits host cell splicing, and the regulatory protein ICP27 is required for this effect. J. Virol. 68: 7790–7799

Herold BC, Visalli RJ, Susmarski N, Brandt CR, Spear PG. (1994) Glycoprotein C-independent binding of herpes simplex virus to cells requires cell surface heparan sulphate and glycoprotein B. J Gen Virol. 75 (Pt 6):1211–22

Hock BD, Kato M, McKenzie JL, Hart DNJ. (2001) A soluble form of CD83 is released from activated dendritic cells and B lymphocytes, and is detectable in normal human sera. Int. Immunol. 13:959

Howard MK, Kershaw T, Gibb B, Storey N, MacLean AR, Zeng BY, Tel BC, Jenner P, Brown SM, Woolf CJ, Anderson PN, Coffin RS, Latchman DS (1998) High efficient gene transfer to the central nervous system of rodents and primates using herpes virus vectors lacking functional ICP27 and ICP34.5. Gene Ther. 5(8): 1137–1147

Jenne L, Schuler G, Steinkasserer A. (2001) Viral vectors for dendritic cell-based immunotherapy. Trends Immunol. 22(2):102–107

Jonuleit H, Schmitt E, Steinbrink K, Enk AH. (2001) Dendritic cells as a tool to induce anergic and regulatory T cells. Trends Immunol. 22(7):394–400

Kang HA, Hershey JWB. (1994) Effect of initiation factor eIF-5A depletion on protein synthesis and proliferation of *Saccharomyces cervisiae*. J. Biol. Chem. 269:3934

Kruse M, Rosorius M, Krätzer F, Bevec D, Kuhnt C, Steinkasserer A, Schuler G, Hauber J (2000a) Inhibition of CD83 cell surface expression during dendritic cell maturation by interference with nuclear export of CD83 mRNA. J. Ex. Med. 191(9): 1581–1590

Kruse M, Rosorius O, Krätzer F, Bevec D, Kuhnt C, Steinkasserer A, Sxhuler G, Hauber J. (2000b) Inhibition of CD83 cell surface expression during dendritic cell maturation by interference with nuclear export of CD83 mRNA. J. Exp. Med. 191:1581

Lam Q, Smibert CA, Koop KE, Lavery C, Capone JP, Weinheimer SP, Smiley JR (1996) Herpes simplex virus VP16 rescues viral mRNA from destruction by the virion host shutoff function. EMBO J. 15: 2575–2581

Lechmann M, Krooshoop DJEB, Dudziak D, Kremmer E, Kuhnt C, Figdor CG, Schuler G, Steinkasserer A. (2001) The extracellular domain of CD83 inhibits dendritic cell-mediated T cell stimulation and binds to a ligand on dendritic cells. J. Exp. Med. 194:1813

Lechmann M, Kremmer E, Sticht H, Steinkasserer A. (2002a) Overexpression, purification, and biochemical characterization of the extracellular human CD83 domain and generation of monoclonal antibodies. Prot. Exp. Purif. 2002; 24:445

Lechmann M, Berchtold S, Hauber J, Steinkasserer A. (2002b) CD83 on dendritic cells: more than just a marker for maturation. Trends Immunol. 23(6):273–275

Lopez M, Cocchi F, Menotti L, Avitabile E, Dubreuil P, Campadelli-Fiume G (2000) Nectin2α (PRR2α or HveB) and Nectin2δ are low-efficiency mediators for entry of Herpes Simplex virus mutants carrying the Leu25Pro Substitution in Glycoprotein D. J. Virol. 74: 1267–1274

Lu P, Safran HA, Smiley JR (2001) The vhs1 mutant form of Herpes Simplex Virus virion host shutoff protein retains significant internal ribosome entry site-directed RNA cleavage activity. J. Virol. 75: 1072–1076

Mikloska Z, Bosnjak L, Cunningham AL (2001) Immature Monocyte-derived dendritic cells are productively infected with Herpes Simplex Virus Type 1. J. Virol. 75: 5958–5964

Montgomery RI, Warner MS, Lum BJ, Spear PG (1996) Herpes simplex virus-1 entry into cells mediated by a novel member of the TNF/NGF receptor family. Cell 87: 427–436

Mossman KL, Sherburne R, Lavery C, Duncan J, Smiley JR (2000) Evidence that Herpes Simplex Virus VP16 is required for viral egress downstream of the initial envelopment event. J. Virol. 74: 6287–6299

Mossman KL and Smiley JR (1999) Truncation of the C-terminal acidic transcriptional activation domain of herpes simplex VP16 renders expression of the immediate-early genes almost entirely dependent on ICP0. J. Virol. 73: 9726–9733

Mullen MA, Gerstberger S, Ciufo DM, Mosca JD, Hayward GS (1995) Evaluation of colocalization interactions between the IE110, IE175, and IE63 transactivator proteins of Herpes Simplex virus within subcellular punctate structures. J. Virol. 69: 476–491

Novotny MJ, Parish ML, Spear PG (1996) Variability of Herpes Simplex Virus 1 gL and anti-gl antibodies that inhibit cell fusion but not viral infectivity. Virology 221: 1–13

Oroskar AA and Read GS (1989) Control of mRNA stability by the virion host shutoff function of herpes simplex virus. J. Virol. 63: 1897–1906

Palmer JA, Branston RH, Lilley CE, Robinson MJ, Groutsi F, Smith J, Latchman DS, Coffin RS (2000) Development and optimization of Herpes Simplex Virus vectors for multiple long-term gene delivery to the peripheral nervous system. J. Virol. 74: 5604–5618

Park MH, Wolff EC, Folk JE. (1993a) Hypusine: its post-translational formation in eukaryotic initiation factor 5A and its potential role in cellular regulation. Biofactors 1993; 4:95

Park MH, Wolff EC, Folk JE. (1993b) Is hyposine essential for eukaryotic cell proliferation? Trends Biochem. Sci. 18:475

Parkinson J and Everett RD (2001) Alphaherpesvirus proteins related to Herpes Simplex Virus Type 1 ICP0 induce the formation of colocalizing, conjugated Ubiquitin. J. Virol. 75: 5357–5362

Phelan A and Clements JB (1998) Post-transcriptional regulation in herpes simplex virus. Semin. Virol. 8: 309–318

Preston CM and Nicholl MJ (1997) Repression of gene expression upon infection of cells with herpes simplex virus type 1 mutants impaired for immediate-early protein synthesis. J. Virol. 71: 7807–7813

Rosorius O, Reichart B, Krätzer F, Heger P, Dabauvalle MC, Hauber J. (1999) Nuclear pore localization and nucleocytoplasmic transport of eIF-5A: evidence for direct interaction with the export receptor CRM1. J. Cell. Sci. 112:2369

Salio M, Cella M, Suter M, Lanzavecchia A (1999) Inhibition of dendritic cell maturation by herpes simplex virus. Eur. J. Immunol. 29: 3245–3253

Scholler N, Hayden-Ledbetter M, Dahlin A, Hellström I, Hellström K-E, Ledbetter JA. (2002) CD83 regulates the development of cellular immunity. J. Immunol. 168:2599

Sears AE, Halliburton IW, Meignier B, Silver S, Roizman B (1985) Herpes simplex virus 1 mutant deleted in the α22gene: growth and gene expression in permissive and restrictive cells and establishment of latency in mice. J. Virol. 55: 338–346

Slomka MJ, Emery L, Munday PE, Moulsdale M, Brown DW (1998) A comparison of PCR with virus isolation and direct antigen detection for diagnosis and typing of genital herpes. J. Med. Virol. 55: 177–183

Smibert CA, Popova B, Xiao P, Capone JP, Smiley JR (1994) Herpes simplex virus VP16 forms a complex with the virion host shutoff protein vhs. J. Virol. 68: 2339–2346

Stevens JG, Wagner EK, Devi-Rao GB, Cook ML, Feldman LT (1987) RNA complementary to a herpesvirus alpha gene mRNA is prominent in latently infected neurons. Science 235:1056–1059

Spear PG, Eisenberg RJ, Cohen GH (2000) Three classes of cell surface receptors for alphaherpesvirus entry. Virology 275: 1–8

Steinman RM (1991) The dendritic cell system and its role in immunogenicity. Annu. Rev. Immunol. 9: 271–296

Strelow LI and Leib DA (1995) Role of the Virion Host Shutoff (vhs) of Herpes Simplex Virus Type 1 in Latency and Pathogenesis. J. Virol. 69: 6779–6786

Ward PL, Barker E, Roizman B (1996) A novel herpes simplex virus 1 gene, U_L43.5, maps antisense to the U_L43 gene and encodes a protein which colocalizes in nuclear structures with capsid proteins. J. Virol. 70: 2684–2690

Warner MS, Geraghty RJ, Martinez WM, Montgomery RI, Whitbeck JC, Xu R, Eisenberg RJ, Cohen GH, Spear PG (1998) A cell surface protein with herpesvirus entry activity (HveB) confers susceptibility to infection by mutants of herpes simplex virus type 1, herpes simplex virus type 2, and pseudorabies virus. Virology 246: 179–189

Whitley RJ and Gnann JW Jr. (1992) Acyclovir: a decade later. N. Engl. J. Med. 327: 782–789. [Errata, (1993) 328: 671, (1997) 337:1703]

Whitley RJ (1996) Herpes simplex viruses. B. Fields and D. Knipe, ed. Virology, 3rd ed. Lippincott-Raven Publishers, Philadelphia, PA

Zhou LJ and Tedder TF (1995) Human blood dendritic cells selectively express CD83, a member of the immunoglobulin superfamily. J. Immunol. 154: 3821–3835

Chapter 8

Dendritic Cell-Based Immunotherapy

T. G. Berger, E. S. Schultz

T.G. Berger, E.S. Schultz (✉)
Department of Dermatology, University of Erlangen,
Hartmannstrasse 14, 91052 Erlangen, Germany

Abstract. Dendritic cell (DC)-based vaccinations represent a promising approach for the immunotherapy of cancer and infectious diseases as DCs play an essential role in initiating cellular immune responses. A number of clinical trials using ex vivo-generated DCs have been performed so far and only minor toxicity has been reported. Both the induction of antigen-specific T cells and clinical responses have been observed in vaccinated cancer patients. Nevertheless, DC-based immunotherapy is still in its infancy and there are many issues to be addressed such as antigen loading procedures, DC source and maturational state, migration properties, route, frequency, and dosage of DC vaccination. The increasing knowledge of DC biology should be used to improve the efficacy of this new therapy.

1 Introduction

There is consensus that tumors can be recognized by the immune system. Melanoma is one of the best-defined model tumors. Spontaneous antitumor immune responses have been observed in patients, including regressions of primary tumors. On the other hand, local tumor recurrence and systemic spread is seen in many patients even years after excision of the primary tumor. One may, therefore, speculate as to the existence of a continuous battle between the immune system and occult tumor cells. In this scenario, tumor progression could be a consequence of a compromised immune system or could be due to tumor escape mechanisms. Tumor cells can downregulate or completely lose expression of tumor antigens and/or MHC molecules, thus avoiding recognition by tumor-specific T cells. Furthermore, T cells may become nonfunctional as a result of changes in T-cell receptor signal transduction or as a consequence of the influence of an altered cytokine milieu at the tumor site leading to inhibited functions of antigen-presenting cells. To overcome the multitude of defense mechanisms developed by the tumor, immunotherapy of cancer must aim at the induction of a strong and broad antitumor immune response, which should combine innate and adaptive immunity. Dendritic cell (DC)-based immunotherapy represents one of the most promising approaches, as DCs regulate several components of the immune system. DCs can activate different cell types that mediate immune resistance to tumors. $CD8^+$ cytolytic T cells (CTLs) directly kill tumor cells expressing the appropriate MHC-peptide complexes, whereas NK and NKT cells eliminate targets that downregulate MHC class I expression to escape a CTL attack. $CD4^+$ helper T cells assist in inducing and maintaining CTL responses. Furthermore, $CD4^+$ T cells can recruit inflammatory cells with tumoricidal activity, such as macrophages and granulocytes at the tumor site. Inhibition of tumoral angiogenesis by secretion of IFN-γ and direct recognition of MHC class II-expressing tumor cells are additional effector functions of $CD4^+$ helper T cells. These different types of lymphocytes can now be activated directly in patients with ex vivo-generated DCs.

Important observations have been made in studies with healthy volunteers who have been immunized with DCs loaded with model antigens, such as KLH and influenza virus matrix peptide. A single injection of mature antigen-loaded DCs rapidly induced antigen-specific $CD4^+$ and $CD8^+$ T cell responses in vivo. With immature DCs, however, T cell immunity can be dampened by the induction of IL-10-producing regulatory T cells or the inhibition of preexisting effector T cell functions. A number of phase I and phase II clinical studies using ex vivo-generated DCs have been performed, with only minor toxicity being observed. Induction of antigen-specific T cells has been detected in fresh blood samples, and clinical responses have been observed in some patients. Nevertheless, DC-based immunotherapy is still in its infancy and the increasing knowledge of DC biology should be used to improve the efficacy of this new therapy. Some relevant issues include antigen loading and DC maturation procedures, the migratory properties of the injected DCs, and their longevity after injection (Fig. 1). The optimal vaccination scheme including the route, dose, and frequency of DC injections as well as the source of tumor antigens still has to be defined.

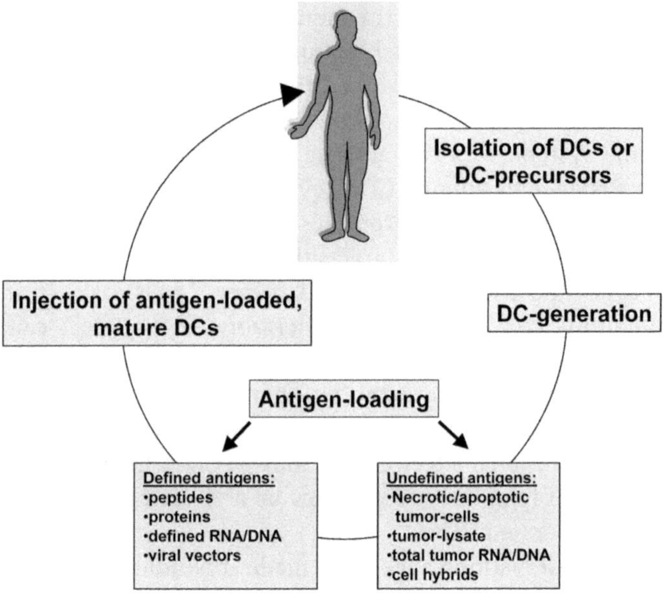

Fig. 1. Pinciples of DC-based immunotherapy

2 Antigen Presentation and Induction of Cellular Immune Responses

2.1 $CD8^+$ T Cells

Classically, endogenous cellular antigens are presented by the MHC class I presentation pathway. Cytosolic proteins are subject to proteasomal proteolysis, and the resulting peptides are shuttled to the endoplasmic reticulum (ER) via TAP transporters. In the ER lumen peptides are loaded on empty MHC class I molecules and the MHC-peptide complexes are transported to the cell surface via the *trans*-Golgi network. In addition to this classic pathway, DCs are able to present peptides derived from exogenous antigens to $CD8^+$ T cells, a phenomenon referred to as "cross-presentation." Numerous mechanisms seem to be involved in these "exogenous class I presentation pathways." In some cases, specific receptors are known, such as the Fcγ receptor binding immune complexes (Regnault et al. 1999) and a glycolipid binding the B subunit of

Shiga toxin (Haicheur et al. 2000). Other mechanisms include the uptake of dead or dying cells and the delivery of peptides chaperoned by heat-shock proteins, such as gp96 and hsp70 (Blachere et al. 1997; Arnold et al. 1995). In vitro liposomes (Ignatius et al. 2000a), exosomes (Zitvogel et al. 1998), hepatitis B virus (Arnold et al. 1997; You et al. 2000), and the HIV-1 tat protein (Kim et al. 1997) have been shown to efficiently target the MHC class I presentation pathway.

2.2 $CD4^+$ T Cells

DCs efficiently take up exogenous antigens and present them to $CD4^+$ T cells as peptides bound in the groove of MHC class II molecules. Immature DCs are characterized by a high ability of endocytosis and expression of relatively low levels of surface MHC and costimulatory molecules. Thus they are very efficient in antigen uptake but less efficient in T cell stimulation. Most immature DCs possess three mechanisms to take up antigen: macropinocytosis, phagocytosis, and clathrin-mediated endocytosis (reviewed in Mellman et al. 2001). Macropinocytosis is a process in which large vesicles containing extracellular fluid are formed. The enormous associated influx and efflux of fluid seems to be mediated by specific aquaporins (de Baey and Lanzavecchia 2000). Phagocytosis of organisms and dead or dying cells is mediated via many different receptors, including Fc receptors and integrins (Albert et al. 1998; Savill and Fadok 2000; Inaba et al. 1998). The multilectin receptors DEC 205 and the macrophage mannose receptor are involved in the adsorptive endocytosis via clathrin-coated vesicles (Sallusto et al. 1995; Mahnke 2000; Jiang et al. 1995). Most DCs in peripheral tissues in situ are of the immature phenotype, the prototype being Langerhans cells in the epidermis. On maturation DCs downregulate their endocytic/phagocytic activity and upregulate the expression of MHC, adhesion, and costimulatory molecules, making them extremely effective in T cell stimulation (reviewed in Banchereau et al. 1998). The HLA class II-peptide complexes remain on the cell surface for several days, allowing interaction with $CD4^+$ T cells (Inaba et al. 1998; Cella et al. 1997). After recognition of the MHC-peptide complex $CD4^+$ T cells can differentiate into either Th1 or Th2 cells. The production of IL-12 by mature DCs is critical for the induction of IFN-γ-producing TH1 cells, which are thought to be important for antitumor immunity.

2.3
NK Cells

Given the fact that tumors frequently escape antigen-specific T cells by downregulating the expression of MHC class I molecules, the additional activation of NK cells by DC-based immunotherapy could be valuable for the induction of antitumor immunity. In mice, DCs directly activate NK cells, which elicit antitumor effects (Fernandez et al. 1999). Recently, it has been shown that human DCs stimulate resting NK cells, a process that mainly involves the NKp30 natural cytotoxicity receptor (Ferlazzo et al. 2002). Together, there is growing evidence that DCs may induce both adaptive and innate antitumor immune responses.

2.4
NKT Cells

NKT cells represent a unique subpopulation of T cells building a link between innate and adaptive immunity. NKT cells can be activated by glycolipid and phospholipid antigens presented on CD1d molecules and by IL-12. They have been shown to produce large amounts of Th1 and Th2 cytokines on stimulation and to kill tumor targets by a perforin-dependent mechanism (reviewed in Smyth et al. 2002). A synthetic glycolipid, α-galactosylceramide (α-GalCer), binds to CD1d and efficiently activates NKT cells (Kawano et al. 1997). DCs pulsed with α-GalCer efficiently induce antitumor activity in mice (Toura et al. 1999). Therefore, vaccination with α-GalCer-pulsed DCs may be a potential way to induce NKT cells with antitumor activity in patients.

3
Source of Antigen

3.1
Peptides

The identification of tumor-associated antigens was followed by the determination of immunodominant peptides that are recognized on tumor cells by T cells. A broad array of tumor-specific peptides presented by different HLA class I molecules and recognized by $CD8^+$ CTLs has been identified, and recently several $CD4^+$ helper T cell epitopes have been added. These defined tumor peptides can be readily synthesized at GMP quality and used to load onto ex vivo-generated DCs. The sequence of the natural occurring peptides can be modified by substitution of single amino acids to improve their binding to a given HLA molecule or the affinity of the HLA-peptide complex for the T cell receptor. Vaccination with peptide-pulsed DCs has been shown to induce both peptide-specific $CD8^+$ and $CD4^+$ T cells in healthy volunteers and even in advanced cancer patients (Dhodapkar et al. 1999, 2000; Schuler-Thurner et al. 2002). Although straightforward and technically easy, the peptide-based approach has some major drawbacks. First, the choice of peptides is restricted to the HLA typing of the patient, at least for HLA class I peptides, which are less promiscuous binders than HLA class II peptides. Second, vaccination with peptide-pulsed DCs should only induce a T cell response directed against a limited number of tumor antigens, which may not be sufficient to effectively combat the tumor. In this scenario, the tumor might escape the immune response directed against a small array of peptides and emergence of antigen-loss tumor cell variants may occur. Third, repetitive vaccinations using relatively high peptide concentrations may favor the induction of low-affinity T cells that are not able to recognize the tumor cells.

3.2 Exosomes

Exosomes may present an attractive source to load DCs with antigen. These are small membrane vesicles resulting from fusion of the plasma membrane with multivesicular endosomes. They contain adhesion and costimulatory molecules, MHC products, and heat shock proteins and are secreted by various cell types including dendritic and tumor cells (Wolfers et al. 2001; Zitvogel et al. 1999). Exosomes derived from peptide-pulsed DCs induce antitumor immune responses in mice (Zitvogel et al. 1998). Tumor-specific CTLs can be activated with DCs loaded with exosomes derived from tumor cells (Wolfers et al. 2001).

3.3 Dead or Dying Tumor Cells

To optimize the antitumor effects of DC-based immunotherapy it is tempting to allow the DCs to present the whole antigenic spectrum of a given tumor. This strategy should lead to the induction of an antitumor T cell response directed against a broad array of tumor antigens including antigens derived from tumor-specific mutations potentially relevant for oncogenesis (Dubey et al. 1997; Mandruzzato et al. 1997; Wolfel et al. 1995). Thus the probability of tumor escape via loss of antigen should be reduced. One intriguing way is to let the DCs phagocytose whole tumor cells or their fragments, resulting in cross-presentation of tumor antigens on both MHC class I and II molecules. This would allow the simultaneous induction of tumor-specific CTLs and $CD4^+$ helper T cells, which play a critical role in antitumor immunity (reviewed in Toes et al. 1999).

Several concerns have been raised regarding this approach. First, it is often difficult to obtain sufficient quantities of autologous tumor material from patients. The use of allogeneic tumor cell lines may present an alternative to overcome this problem and even amplify the immune response by activation of alloreactive T cells. Second, immunizing with DCs loaded with whole tumor cell preparations bears the potential risk of inducing autoimmunity (Ludewig et al. 1999, 2000), as the DCs will not only present tumor-specific antigens but also numerous self-peptides. However, it has been suggested that immature DCs induce tolerance to self-antigens derived from apoptotic cells (Steinman et al. 2000). Thus most patients should be tolerant to many self-peptides that DCs can present to T cells after having phagocytosed apoptotic cells. Third, reliable monitoring of the immune response may be difficult in the latter case compared with the use of defined tumor antigens. Fourth, T cells may be generated that recognize peptides only processed by the immunoproteasome expressed by mature DCs (Macagno et al. 1997) and not by the constitutive proteasome expressed by tumor cells (Morel et al. 2000; Schultz et al. 2002). Therefore, the tumor cells would not be recognized by these T cells unless expression of the immunoproteasome by tumor cells could be induced by IFN-γ production, e.g., by $CD4^+$ helper T cells at the tumor site. Finally, the optimal preparation of antigen, e.g., tumor cell lysates, necrotic or apoptotic material, still has to be defined.

3.4 Recombinant Viruses

DCs can be readily infected with recombinant viruses containing the cDNA coding for a given tumor antigen. With the use of adenoviral or influenza viral vectors transduction rates of more than 90% can be achieved (reviewed in Jenne et al. 2001b). Infected immature DCs efficiently express the transgene, can be matured with standard stimuli, and activate antigen-specific T cells (Zhong et al. 1999; Cella et al. 1999; Strobel et al. 2000b; Dietz et al. 1998). Both viral vectors exhibit little or no cytopathic effects on the infected DCs. Poxvirus vectors such as avipox and vaccinia are also very suitable for the transduction of DCs; however, infection is followed by a significant decrease in viability of immature DCs, which undergo apoptosis (Subklewe et al. 1999). Furthermore, infected immature DCs show a block in maturation, impairing their T cell stimulatory properties (Sevilla et al. 2000; Jenne et al. 2000; Engelmayer et al. 1999). Nevertheless, efficient presentation of the transgene and induction of antigen-specific T cells has been demonstrated in vitro (Subklewe et al. 1999; Engelmayer et al. 2001). This may be due to cross-presentation of antigens derived from the uptake of dying infected DCs by noninfected DCs (Munz et al. 2000; Ignatius et al. 2000b). Several other viral vectors suitable for the transduction of DCs have been described, such as lentiviruses (Dyall et al. 2001) and retroviruses (Henderson et al. 1996).

One major concern in using viral vectors for immunotherapy is the induction of antiviral cellular and humoral immune responses in patients, which may impair the desired induction of antitumor immunity. Although the virus should be hidden from potentially neutralizing antibodies when using ex vivo-infected DCs, a strong cellular immune response against viral antigens may lead to the destruction of the infected DCs themselves. Clinical studies are required to assess the therapeutic potential of virus-infected DCs and to compare this approach with the use of viruses alone.

3.5
DNA Transfection

An elegant approach to circumvent the disadvantages associated with the use of viral vectors is to directly transfect DCs with plasmid DNA coding for full-length tumor antigens. Transfected DCs present the relevant antigens to human T cells in vitro (Smith et al. 2001). Plasmids can be readily constructed to not only encode a tumor antigen but also other sequences that lead to better antigen processing and T cell stimulation (Pardoll 1998; Seder and Hill 2000). One major problem remains the difficulty of transfecting DCs with any efficiency. This obstacle might be overcome by implementation of a newly described cationic CL22 peptide carrier (Irvine et al. 2000).

3.6 RNA Transfection

Alternatively, DCs can be transfected with RNA. If tumor material is available, whole tumor RNA can be used directly or after amplification from a few tumor cells to transfect the DCs by simply adding RNA (Strobel et al. 2000a) or by use of lipofection (Nair et al. 2000) or electroporation (Van Tendeloo et al. 2001). Vaccination with mRNA-loaded DCs has been shown to induce protective and therapeutic antitumor responses in mice (Ashley et al. 1997; Koido et al. 2000). RNA transfection represents a promising approach to engineer DCs to present the whole and unique antigenic spectrum of a patient's tumor.

3.7 Cell Hybrids

Another method to allow DCs the presentation of many different tumor antigens is to fuse tumor cells and DCs with high electric voltage or polyethylene glycol. Vaccination with the resulting cell hybrids has been shown to induce regressions of established carcinomas, lymphomas, and melanomas in mice (Koido et al. 2000; Gong et al. 2000). A similar approach fusing autologous tumor and allogeneic dendritic cells has been used to vaccinate patients with advanced renal cell cancer (Kugler et al. 2000).

3.8 In Vivo Targeting of DCs

Considering the complexity of in vitro DC generation, large-scale vaccination of many patients remains a difficult task. Thus it is necessary to develop vaccines that can provide potent immune responses with minimal quantities of vaccine. This may be achieved by targeting resident DCs in vivo. Modern vaccination strategies using, for example, naked DNA (Tang et al. 1992) may prove advantageous in comparison to traditional approaches such as Edward Jenner's first "vaccination" with attenuated virus in 1796, as so-called "DC-targeted vaccines" may address different DC subsets and offer the possibility of controlling the type and potency of immune responses (Takashima and Morita 1999). DC poietins (e.g., GM-CSF, FLT3-L) may further augment vaccination efficacy by increasing the number of resident DCs, which can be activated in vivo by adjuvants such as IFN-α (Le Bon et al. 2000) or CpG oligonucleotides (Jakob et al. 1999).

4
DC Source and Subsets

An increasing number of circulating DC subsets have been identified in humans. $CD34^+$ hematopoietic stem cells give rise to $CD11c^+$ $CD1a^+$ immature DCs, which migrate to the epidermis and differentiate into Langerhans cells, and $CD11c^+$ $CD1a^-$ immature DCs, which migrate to the dermis to become interstitial DCs. In addition, preDC1 (monocytes) and preDC2 (plasmacytoid cells) develop from $CD34^+$ progenitor cells. PreDC1 differentiate into immature myeloid DCs (DC1s) after in vitro culture with GM-CSF and IL-4, whereas preDC2 differentiate into immature DC2s in response to IL-3 or after viral stimulation (recently reviewed in Liu 2001). Both cell types exhibit different functional properties. In short, DC1s are considered as promoters of Th1 responses, whereas DC2s induce either Th1 or Th2 responses depending on the stimulus (Blom et al. 2000; Maraskovsky et al. 2000). In addition, circulating DCs have been identified and isolated via the novel BDCA-surface markers (Dzionek et al. 2000). Because of the scarcity of circulatory DCs, isolation of sufficient cell numbers for multiple vaccinations is a major obstacle. For example, a complete leukapheresis is needed to prepare a single DC vaccination with DCs obtained directly from fresh blood (Peshwa 1998). To date, the majority of DC-based vaccination trials have been performed with monocyte-derived DCs. These cells resemble interstitial DCs and are relatively easy to generate in large numbers and purity. In advanced cancer patients approximately 200 Mio or more immature DCs can be obtained from a single leukapheresis after culture in the presence of GM-CSF and IL-4 (Thurner et al. 1999b). Maturation can be induced by different stimuli, and the resulting mature DCs can be cryopreserved "ready for use," which further facilitates their application (Feuerstein et al. 2000). Nevertheless, the generation of DCs still remains too laborious to treat large numbers of patients and, therefore, major effort is currently being applied to increase the yield of functional DCs and simplify the generation methods at the same time. One approach is the modification of the widely used plastic adherence to obtain DC precursors. A completely closed, automated system would greatly enhance the applicability of DC therapy. Alternatively, $CD14^+$ DC precursors can be enriched by paramagnetic microbeads coated with anti-CD14-antibodies (Miltenyi et al. 1990). Although this approach is well established in the laboratory setting, it awaits detailed evaluation for clinical use.

$CD34^+$ hematopoietic stem cells from blood, bone marrow, or cordblood are another source for DC generation. These DC preparations consists of LC-like and interstitial DC-like cells (Caux et al. 1996, 1997). In a clinical study in which patients with advanced cancer have been vaccinated with $CD34^+$ progenitor-derived DCs, immunological and clinical responses could be observed (Banchereau et al. 2001a). Given the sparse numbers of $CD34^+$ cells in adult blood, patients must be pretreated with GM-CSF or G-CSF before leukapheresis. Moreover, the DC preparation from $CD34^+$ cells requires a more complex cytokine supplementation compared with the relatively easy moDC generation. There is some evidence from in vitro studies that CD34-derived DCs may be more immunogenic than moDC (Ferlazzo et al. 1999); however, this finding awaits further confirmation.

5 Maturational State

Mature DCs are more immunogenic than immature DCs in mice (Schuurhuis et al. 2000: Labeur et al. 1999; Inaba et al. 2000), and there is good evidence that this also applies to humans. Mature DCs express a higher number of costimulatory molecules and more MHC-peptide complexes with a longer half-life (Kukutsch et al. 2000; Kampgen et al. 1991). In direct comparison in melanoma patients, intranodally injected peptide-pulsed mature DCs led to a potent T cell response whereas immature DCs failed to do so (Jonuleit et al. 2001). Recent studies have shown that immature DCs can even silence the immune system. Repetitive stimulation of naive $CD4^+$ T cells with immature DCs results in IL-10-producing regulatory T cells (Jonuleit et al. 2000). In healthy volunteers an antigen-specific $CD8^+$ T cell response to the influenza virus matrix peptide was dampened by vaccination with immature DCs (Dhodapkar et al. 2001).

One concern regarding the use of fully mature DCs is that their Th1 polarizing potential is limited to a short time period. Already 24 h after LPS stimulation cytokine production (e.g., IL-12 p70) dramatically drops. These "exhausted" DCs promote Th2 rather than Th1 responses in vitro (Langenkamp et al. 2000). There are several lines of evidence that Th1 responses are advantageous in antitumor immunity. IFN-γ-producing Th1 cells are more tumor protective in mouse models (Nishimura et al. 1999), may better home to inflamed tissues (Austrup et al. 1997), and help to sustain a $CD8^+$ T cell response via CD40-CD40L interaction with DCs (Ridge 1998; Bennett 1998; Schoenberger et al. 1998;). Th1 cells may exert direct cytotoxicity (Hahn et al. 1995; Schultz et al. 2000; Takahashi 1995; Thomas 1998) and antiangiogenic effects via IFN-γ production (Coughlin et al. 1998; Qin and Blankenstein 2000).

In vivo studies of healthy volunteers (Dhodapkar et al. 1999) and advanced melanoma patients with fully mature, potentially "exhausted" DCs have, nevertheless, demonstrated that both antigen-specific $CD8^+$ T cells (Schuler-Thurner et al. 2000; Thurner et al. 1999a) and IFN-γ-producing Th1 T cells (Schuler-Thurner et al. 2002) can be rapidly induced.

In summary, we strongly recommend the use of mature DCs for cancer immunotherapy. Mature DCs exhibit a stable phenotype and are more immunogeneic, easier to cryopreserve (Feuerstein et al. 2000), and even resistant to CTL-mediated lysis (Medema et al. 2001).

6 Maturation Stimuli

There is an ongoing debate about the optimal maturation stimulus of DCs used for vaccination approaches. DC maturation can be achieved by the addition of monocyte-conditioned medium (MCM), which is obtained from monocytes bound to immunoglobulin-coated plastic surfaces (Bender et al. 1996). After identification of the major components of MCM, a cocktail of proinflammatory cytokines and prostaglandins ("MCM-mimic") was introduced (Jonuleit et al. 1997) and subsequently applied in many clinical trials. MCM-mimic elicits reliable DC maturation and is more practical than MCM, which does not only vary in quality from donor to donor but is time consuming to produce and requires monocytes that are thereafter not available for DC generation. Other groups have used TNF-α alone or various combinations of the cytokines IL-1β, IL-6, and TNF-α with or without PGE_2. In our experience TNF-α

alone does not yield fully mature DCs and especially PGE_2 is necessary for stable matured DCs. Criticism of the use of PGE_2 resulting from in vitro studies showing a Th2- rather than Th1-polarizing potential of PGE_2-treated cells (Kalinski 1997) could not be confirmed in vivo. We and others have used PGE_2-treated DCs in clinical trials and demonstrated potent Th1 responses to the control antigen KLH as well as to HLA class II-restricted tumor antigens (Schuler-Thurner et al. 2002; Dhodapkar 1999). Recent observations that DCs generated in the presence of IL-4 may have an impaired arachidonic acid metabolism that can be restored by the addition of external prostaglandins underscores the necessity for their supplementation in the MCM-mimic (Thurnher et al. 2001). Many other substances have been shown to mature DCs. Generally, they can be categorized into internal and external "danger signals." Internal danger signals can be proinflammatory cytokines, prostaglandins, interferons, and CD40L, mimicking the DC-T cell interaction. External danger signals are numerous. Via different receptors, importantly Toll-like receptors (Kaisho and Akira 2000) but also yet unidentified structures, DCs can be activated by foreign material such as LPS, monophosphoryl lipid A, dsRNA (Verdijk et al. 1999; Alexopoulou et al. 2001), bacterial DNA, and synthetic CPG-oligonucleotides (Sparwasser et al. 1998). Some of these stimuli exert special DC properties, e.g., enhanced cytokine production and, concomitantly, polarization of T helper cell responses (de Jong et al. 2002). However, depending on the desired application, different maturation stimuli and combinations thereof may be optimal. Only a direct comparison in vivo would answer this interesting question.

7 Migration

To induce strong T cell responses via DC-based immunotherapy it is crucial that a significant number of antigen-bearing DCs reach the draining lymph node and remain viable to efficiently activate T cells. Consequently, migration properties and longevity of the injected DCs play an important role. DC migration is regulated by their response to chemokines, which is influenced by their maturation status. For instance, immature DCs express receptors for inflammatory chemokines, such as CCR1, CCR2, CCR5, CCR6, and CXCR1, which guide them to sites of inflammation where antigen uptake and induction of maturation can take place

(Dieu et al. 1998; Sallusto et al. 1998b; Sozzani et al. 1998). On maturation DCs downregulate receptors for inflammatory chemokines and rapidly express receptors for constitutive chemokines such as CXCR4 and CCR7. The latter regulates their trafficking into the lymphatic vessels where the CCR7 ligand CCL21/6Ckine/SLC is produced and then to the T cell areas of the draining lymph node (LN) in response to another CCR7 ligand, CCL19/MIP3ß/ELC (Sallusto et al. 1998a; Sallusto et al. 1999; Kellermann et al. 1999). In mice, only a few DCs migrate to the draining LNs after s.c. injection (Josien et al. 2000). Treatment with viability-enhancing CD40L or TRANCE/RANKL before injection can increase the number and persistence of antigen-presenting DCs in the lymph node (Josien et al. 2000). However, the majority of the injected DCs do not reach the LNs. In humans, a small percentage (<1%) of DCs injected intradermally (i.d.) were detected in the draining LNs (Morse et al. 1999), whereas no migration could be observed when DCs were injected s.c. Further studies are clearly needed to optimize DC migration into lymphatics and their life span on reaching the lymph node.

8 Route, Dose, and Schedule of DC Vaccination

As in vaccines against infectious diseases, the route, dose, and frequency of DC vaccination may have great influence on vaccination efficacy. In the majority of studies DCs were administered into the skin, without doubt the easiest approach. DCs were also injected i.v., into lymphatic vessels or directly into regional LNs. The skin approach may prove advantageous for several reasons. In mice, more DCs were found in the lymph nodes after i.d. or s.c. vaccination compared with i.v. administration. (Eggert et al. 1999). In a clinical study, patients with prostate cancer were immunized with antigen-loaded DCs via i.d., intralymphatic (i.l.), or i.v. injection. All patients developed specific T cell responses, regardless of the route; however, IFN-γ production consistent with a Th1 induction was only detected after i.d. and i.l. injection (Fong et al. 2001b). In melanoma patients i.v. injection of peptide-pulsed DCs was found to dampen the immune response detected in the peripheral blood (Thurner et al. 1999b). Nevertheless, only a minority of DCs injected into the skin reach the draining LN (Josien et al. 2000; Morse et al. 1999). An alternative application, which can be relatively easily and safely performed, is the injection of DCs directly into the regional LNs (Nestle et al. 1998). A possible drawback may be the destruction of the lymph node architecture.

The number of injected DCs may be equally important. High DC-to-T cell ratios polarize helper responses toward Th1-type in vitro and give rise to higher-affinity T cells (Tanaka et al. 2000). In particular, when DCs are pulsed with different peptides and injected separately into the skin, the number of DCs finally reaching the draining LNs may simply be too low to effectively induce a T cell response. However, in previous studies the number of injected DCs varied from 4 to 40 Mio cells per vaccination without striking differences being observed (Banchereau et al. 2001b).

The schedule of DC vaccinations must be determined. Is it sufficient to vaccinate every 2–4 weeks as previously published, or does one cause activation-induced cell death by vaccinating too frequently? Repetitive vaccinations can also cause killing of antigen-loaded DCs by antigen-specific CTLs that may diminish the immune response (Hermans et al. 2000). Should one stop vaccinating after tumor regression, or is it a potentially lifelong therapy? How often should one vaccinate patients with a high risk of recurrence but without current tumor manifestations in the adjuvant/prophylactic setting? All these questions must be taken into account in the planning phase of DC-based vaccination trials.

9
Clinical Studies in Cancer Patients

The initial clinical DC vaccination studies were performed primarily as phase I/II trials to demonstrate feasibility, immunogenicity, and lack of toxicity. Some studies revealed T cell responses to the applied antigens and also clinical improvement even in advanced cancer patients. The heterogeneity of treated patients, vaccination schemes, DC sources, and generation protocols as well as the lack of standardized methods to determine a successful induction of immunity make it difficult to draw firm conclusions about the efficacy of DC-based immunotherapy. Nevertheless, many of these studies provided valuable information and a brief summary of some examples is given below.

9.1 Melanoma

The first human tumor-associated antigen was identified in melanoma (van der Bruggen et al. 1991), and currently this is the best-studied malignancy in the context of DC-therapy. The pioneering work with peptide-loaded, monocyte-derived DCs (moDCs) has proven the concept of DC vaccination to elicit tumor-specific CTL responses even in advanced cancer patients (Schuler-Thurner et al. 2000; Thurner et al. 1999a). Although these patients were heavily pretreated, even clinical responses could be observed in some patients. In another clinical study melanoma patients were vaccinated with immature, peptide- or tumor lysate-loaded DCs. Several clinical responses were observed, but a detailed immunological read-out was lacking (Nestle et al. 1998). CD34-derived DCs have also been used in melanoma vaccination trials. Whereas in one study only moderate immune and minor clinical responses were achieved (Mackensen et al. 2000), strong immune responses as well as tumor regressions were observed in a recent study (Banchereau et al. 2001a). Recently, with mature, peptide-pulsed moDCs a potent induction of tumor-specific $CD4^+$ T cell responses could be detected in melanoma patients accompanied with a stabilization of disease in 8 of the 16 patients evaluated (Schuler-Thurner et al. 2002). Together, DC-based immunotherapy in melanoma patients is particularly promising, as melanoma is an immunogeneic tumor for which a multitude of tumor antigens have been described. Apart from minor side effects (fever, swelling of lymphnodes, cutaneous reactions at the vaccination site) or the development of vitiligo in a few patients, no major toxicity was reported. Further efforts are clearly required to enhance the immune response, e.g., by simultaneous activation of different immune cells (T cells, NK and NKT cells).

9.2 Solid Tumors

In contrast to melanoma, many solid tumors are considered less immunogeneic and only few TAAs have been identified for these tumors. Clinical studies have been performed with DCs loaded with peptides as well as with crude tumor extracts or tumor-derived RNA.

CEA is a self-antigen overexpressed in colorectal, lung, and breast cancer. In a phase I study vaccinating patients with advanced CEA-expressing malignancies with peptide-pulsed DCs minor responses and stabilization of disease were observed (Nair et al. 1999). Other overexpressed self-antigens are Her2/neu, which is expressed in up to 30% of human breast and ovarian carcinoma, and MUC1, which is found in an aberrant form in these tumors. Patients with advanced breast or ovarian cancer were vaccinated with DCs pulsed with Her2/neu- and MUC1-derived peptides after high-dose chemotherapy. T cell responses could be detected in 50% of patients (Brossart et al. 2000).

Renal cancer is relatively resistant to chemotherapy; however, spontaneous remissions have been observed as well as responses to unspecific immunotherapy with IL-2 (Rieser et al. 1999; Holtl 1998). For specific immunotherapy, DCs have been pulsed with tumor lysate (Thurnher et al. 1998) or, in a recent study, have been fused with autologous tumor cells. Although 9 of 17 patients exhibited striking tumor regressions in this study, the stability of DC tumor hybrids and the induction of specific immune responses were not demonstrated (Kugler et al. 2000). More preclinical data are needed to prove the effectiveness of this approach.

Prostate cancer is another promising candidate for DC-based immunotherapy, because serum markers to monitor the tumor burden and defined antigens such as prostate acid phosphatase, prostate-specific antigen, and prostate-specific membrane antigen are known. Several HLA-A*0201-restricted peptides are available and have been used in a number of clinical trials with moDCs (Murphy et al. 1996, 1999a,b, 2000; Tjoa 1998). Others have pulsed blood-derived DCs with a GM-CSF/PAP fusion protein and reported a significant decrease in serum PSA-levels, reflecting the reduction of tumor burden in some patients (Small et al. 2000; Burch 2000). Recently, 21 patients with metastatic prostate cancer were vaccinated with DCs pulsed with the xenoantigen mouse PAP (Fong et al. 2001a). All patients developed T cell responses to mouse PAP and 11 also to the human homolog. Of 21 patients, 6 showed a stabilization of disease. Vaccination with xenoantigens may break tolerance to self-antigens and increase the clinical efficacy of DC-based immunizations.

9.3 Virus-Associated Malignancies

The targets for immune intervention strategies in these malignancies are different viral antigens. Because they are foreign to the immune system, the challenge is to circumvent viral immune escape mechanisms rather than induce autoimmunity. Examples of virus-associated malignancies include hepatocellular carcinoma (hepatitis B or C virus), Burkitt lymphoma and nasopharyngeal carcinoma (Epstein-Barr virus), squamous cell cancer of the skin and cervical carcinoma (human papillomavirus). Studies to vaccinate against HPV-associated cervical cancer have been performed with non-DC vaccines. Some immunological effects have been reported that, however, were only detectable after intensive in vitro restimulation of T cells (zur Hausen 1996; Da Silva et al. 2001).

9.4 Other Malignancies

Because of the immune-privileged status of the central nervous system (CNS) it is questionable as to whether DC-based immunotherapy can be applied to patients with these tumors. In fact, in a recent study, patients suffering from malignant glioblastoma or astroglioma were vaccinated with immature moDCs pulsed with peptides eluted from autologous tumor cells (Yu et al. 2001). Two patients showed $CD8^+$ infiltrates within the tumor, but no tumor regressions occurred. There is also evidence that DCs pulsed with crude tumor lysate can induce immunity and clinical responses. A study in pediatric cancer patients showed a substantial tumor regression in a child with fibrosarcoma after DC vaccination. Tumor-specific $CD8^+$ lymphocytes and some clinical effects were also demonstrated in two recent studies using tumor lysate-pulsed DCs in gynecological malignancies (Santin et al. 2002; Hernando et al. 2002). However, it is difficult to monitor the successful antigen-loading and T cell response with lysates from tumors for which defined antigens have not yet been described.

9.5 Hematological Malignancies

Myeloma and B cell lymphoma respond initially to standard chemotherapy, but are ultimately incurable. For both tumors, idiotypic proteins have been used as specific antigens. B cell lymphoma was the first human malignancy to be targeted by a DC-based vaccination (Hsu et al. 1996). Isolated blood DCs were pulsed with KLH and the patient-specific idiotype-protein (Id), which is expressed on the surface or secreted by the tumor cells. This early DC trial with four patients showed that administration of DCs was safe and immunogeneic with priming to the neo-antigen KLH. In a subsequent study, 35 patients with B cell lymphoma were treated with Id-pulsed DCs. T cell and humoral anti-Id responses as well as durable tumor regressions or stabilization were reported for the majority of patients (Timmerman et al. 2002). In myeloma, idiotype- and Id-KLH-pulsed DCs were used to vaccinate patients after high-dose chemotherapy and stem cell transplantation (Reichardt et al. 1999; Liso et al. 2000; Lim et al. 1999). Although the majority of these patients developed potent KLH-responses, only minor immune responses against the tumor antigens or clinical remissions could be observed.

10 DC Vaccination in Infectious Diseases

The prevention of infectious diseases through vaccination represents one of medicine's greatest triumphs. However, many infectious agents such as HIV, Dengue virus, *Mycobacterium tuberculosis*, *Malaria plasmodiae*, and numerous others escape the immune system and standard vaccination approaches fail. The underlying immune evasion mechanisms may be attributed to a great extent to interference of the microbial pathogen with dendritic cells and their function, e.g., inhibition of DC maturation (Urban et al. 1999; Rescigno and Borrow 2001). Augmentation of the immune responses by ex vivo-generated DCs is therefore an attractive consideration. However, successful DC vaccination trials have not been reported so far. The feasibility of this approach is nevertheless supported by in vitro studies with human cells as well as in vivo models in the mouse system. Induction of antiparasitic CTLs has been demonstrated in vitro with DCs loaded with helminthic proteins (Jenne et al. 2001a), and protective immunity against bacteria has been achieved by

DC vaccination in mice (Worgall et al. 2001). Protection against pulmonary infection with *Pseudomonas aeruginosa* after immunization with *P. aeruginosa*-pulsed dendritic cells has also been demonstrated (Mbow et al. 1997). One of the best-studied animal models in this regard is leishmaniasis (Moll et al. 2001). Lessons learned from this model may have important implications for future DC-based vaccination strategies in infectious diseases.

11 Quality Control

"Quality control" refers to the vaccine itself, methods to monitor the immune response, as well as the study design. We suggest monitoring the quality of the DC vaccine by reliable, reproducible, and relatively simple tests to maintain their feasibility. For each of the tests, thresholds must be defined for the conditions under which the vaccine fulfils the release criteria. Mature DCs are superior by far to immature DCs in the induction of immunity. Therefore, the maturational state must be determined before vaccination. DCs should be clearly positive for CD83, CD80, CD86, and DC-lamp. The majority of DCs are expected to be negative for CD14 and MCSF-R. The cells should be stable in a prolonged in vitro culture even without cytokines and should exhibit a potent T cell stimulatory capacity (e.g., in an allogeneic MLR). The cells should also exhibit the typical DC morphology (nonadherent cells with large, mobile veils). Of course, the vaccine must be hygienically perfect.

Study designs must be modified for DC-based and other immunotherapies because traditional criteria for the evaluation of anticancer drugs (e.g., toxicity) do not apply here. The same is true for the evaluation of the clinical response. Tumor stabilization and prevention of recurrence may be valid clinical end points instead of regression of established metastasis (Korn et al. 2001). The reliable interpretation of the clinical outcome requires larger patient groups. A currently ongoing multicenter trial in Germany is the first to compare DC vaccination versus standard chemotherapy in 200 melanoma patients.

12 Immune Monitoring

DC-based immunotherapy provides an opportunity to learn more about antitumor immune responses and their impact on the clinical outcome of cancer patients. A detailed analysis of the immune response in vaccinated patients will be crucial to optimize future vaccination schedules. Control antigens, such as viral T cell epitopes or neoantigens such as KLH should be included in the vaccine to obtain information on the quality of the injected DCs and the competence of the patient's immune system.

The ELISPOT assay is a sensitive, fast method to quantify antigen-specific T cells on the basis of their cytokine secretion on stimulation. Alternatively, effector cytokine production can be measured by flow cytometry or quantitative real-time PCR. An elegant method for the quantification of antigen-specific $CD8^+$ T cells involves the use of tetrameric HLA class I-peptide complexes to directly label the corresponding T cell receptor (Jager et al. 2000; Lee 1999; Altman 1996). Combined with FACS sorting, the tetramer technology allows a phenotypical and functional analysis of the relevant T cell population. The recent development of HLA class II (Nepom et al., 2002) and CD1d tetramers (Benlagha et al. 2000; Matsuda et al. 2000) makes the additional monitoring of $CD4^+$ and NKT cells feasible. To better understand the correlation between immunological and clinical response. immune monitoring should not be limited to the peripheral blood but should also include the tumor site and secondary lymphoid organs (Andersen et al. 2001).

13 Conclusion

DC-based immunotherapy represents a promising way to fight cancer as DCs play a key role in inducing antitumor immunity. Early clinical studies have demonstrated that vaccination with DCs can induce immunological and clinical responses in patients with advanced cancer. Additionally, DC vaccination may be a prophylactic and therapeutic option for many infectious diseases that are otherwise difficult to treat or incurable. There are still many open questions concerning the optimal vaccination strategy, but combining the increasing knowledge in DC biology and new techniques for immune monitoring will help us to improve the efficacy of this new therapeutic concept.

Acknowledgements. This work was supported in part by grants from the Deutsche Forschungsgemeinschaft (SCHU 1264/2-1) and the Bundesministerium für Bildung und Forschung (BMBF; FKZ 01GE9911/7), the ELAN program, and the Johannes and Frieda Marohn Stiftung of the University of Erlangen-Nuremberg, Germany.

References

Albert ML, Pearce SF, Francisco LM, Sauter B, Roy P, Silverstein RL, Bhardwaj N: Immature dendritic cells phagocytose apoptotic cells via alphavbeta5 and CD36, and cross-present antigens to cytotoxic T lymphocytes. J.Exp.Med. 1998, 188:1359–1368.

Alexopoulou L, Holt AC, Medzhitov R, Flavell RA: Recognition of double-stranded RNA and activation of NF-kappaB by Toll-like receptor 3. Nature 2001, 413:732–738.

Altman JD, Moss PH, Goulder PR, Barouch DH, McHeyzer-Williams MG, Bell JI, McMichael AJ, Davis MM: Phenotypic analysis of antigen-specific T lymphocytes [published erratum appears in Science 1998 Jun 19;280(5371):1821]. Science 1996, 274:94–96.

Andersen MH, Pedersen LO, Capeller B, Brocker EB, Becker JC, Thor SP: Spontaneous cytotoxic T-cell responses against survivin-derived MHC class I-restricted T-cell epitopes in situ as well as ex vivo in cancer patients. Cancer Res. 2001, 61:5964–5968.

Arnold D, Faath S, Rammensee H, Schild H: Cross-priming of minor histocompatibility antigen-specific cytotoxic T cells upon immunization with the heat shock protein gp96. J.Exp.Med. 1995, 182:885–889.

Arnold D, Wahl C, Faath S, Rammensee HG, Schild H: Influences of transporter associated with antigen processing (TAP) on the repertoire of peptides associated with the endoplasmic reticulum-resident stress protein gp96. J.Exp.Med. 1997, 186:461–466.

Ashley DM, Faiola B, Nair S, Hale LP, Bigner DD, Gilboa E: Bone marrow-generated dendritic cells pulsed with tumor extracts or tumor RNA induce antitumor immunity against central nervous system tumors. J Exp.Med. 1997, 186:1177–1182.

Austrup F, Vestweber D, Borges E, Lohning M, Brauer R, Herz U, Renz H, Hallmann R, Scheffold A, Radbruch A, Hamann A: P- and E-selectin mediate recruitment of T-helper-1 but not T-helper-2 cells into inflamed tissues. Nature 1997, 385:81–83.

Banchereau J, Palucka AK, Dhodapkar M, Burkeholder S, Taquet N, Rolland A, Taquet S, Coquery S, Wittkowski KM, Bhardwaj N, Pineiro L, Steinman R, Fay J: Immune and clinical responses in patients with metastatic melanoma to CD34(+) progenitor-derived dendritic cell vaccine. Cancer Res. 2001a, 61: 6451–6458.

Banchereau J, Schuler-Thurner B, Palucka AK, Schuler G: Dendritic cells as vectors for therapy. Cell 2001b, 106:271–274.

Banchereau J, Steinman RM: Dendritic cells and the control of immunity. Nature 1998, 392:245–252.

Bender A, Sapp M, Schuler G, Steinman RM, Bhardwaj N: Improved methods for the generation of dendritic cells from nonproliferating progenitors in human blood. J.Immunol.Methods 1996, 196:121–135.

Benlagha K, Weiss A, Beavis A, Teyton L, Bendelac A: In vivo identification of glycolipid antigen-specific T cells using fluorescent CD1d tetramers. J Exp.Med. 2000, 191:1895–1903.

Bennett SR: Help for cytotoxic-T-cell responses is mediated by CD40 signalling. Nature. 1998, 393:478–480.

Blachere NE, Li Z, Chandawarkar RY, Suto R, Jaikaria NS, Basu S, Udono H, Srivastava PK: Heat shock protein-peptide complexes, reconstituted in vitro, elicit peptide-specific cytotoxic T lymphocyte response and tumor immunity. J.Exp.Med. 1997, 186:1315–1322.

Blom B, Ho S, Antonenko S, Liu YJ: Generation of interferon alpha-producing predendritic cell (Pre-DC)2 from human CD34(+) hematopoietic stem cells. J Exp.Med. 2000, 192:1785–1796.

Brossart P, Wirths S, Stuhler G, Reichardt VL, Kanz L, Brugger W: Induction of cytotoxic T-lymphocyte responses in vivo after vaccinations with peptide-pulsed dendritic cells. Blood 2000, 96:3102–3108.

Burch PA, Breen JK, Buckner JC, Gastineau DA, Kaur JA, Laus RL, Padley DJ, Peshwa MV, Pitot HC, Richardson RL, Smits BJ, Sopapan P, Strang G, Valone FH, Vuk-Pavlovic S: Priming tissue-specific cellular immunity in a phase I trial of autologous dendritic cells for prostate cancer. Clin.Cancer Res. 2000, 6:2175–2182.

Caux C, Massacrier C, Vanbervliet B, Dubois B, Durand I, Cella M, Lanzavecchia A, Banchereau J: CD34+ hematopoietic progenitors from human cord blood differentiate along two independent dendritic cell pathways in response to granulocyte-macrophage colony-stimulating factor plus tumor necrosis factor alpha: II. Functional analysis. Blood 1997, 90:1458–1470.

Caux C, Vanbervliet B, Massacrier C, Dezutter-Dambuyant C, Saint-Vis B, Jacquet C, Yoneda K, Imamura S, Schmitt D, Banchereau J: CD34+ hematopoietic progenitors from human cord blood differentiate along two independent dendritic cell pathways in response to GM-CSF+TNF alpha. J Exp.Med. 1996, 184: 695–706.

Cella M, Engering A, Pinet V, Pieters J, Lanzavecchia A: Inflammatory stimuli induce accumulation of MHC class II complexes on dendritic cells. Nature 1997, 388:782–787.

Cella M, Salio M, Sakakibara Y, Langen H, Julkunen I, Lanzavecchia A: Maturation, activation, and protection of dendritic cells induced by double-stranded RNA. J.Exp.Med. 1999, 189:821–829.

Coughlin CM, Salhany KE, Gee MS, LaTemple DC, Kotenko S, Ma X, Gri G, Wysocka M, Kim JE, Liu L, Liao F, Farber JM, Pestka S, Trinchieri G, Lee WM: Tumor cell responses to IFNgamma affect tumorigenicity and response to IL-12 therapy and antiangiogenesis. Immunity. 1998, 9:25–34.

Da Silva DM, Eiben GL, Fausch SC, Wakabayashi MT, Rudolf MP, Velders MP, Kast WM: Cervical cancer vaccines: emerging concepts and developments. J Cell Physiol 2001, 186:169–182.
de Baey A, Lanzavecchia A: The role of aquaporins in dendritic cell macropinocytosis. J.Exp.Med. 2000, 191:743–748.
de Jong EC, Vieira PL, Kalinski P, Schuitemaker JH, Tanaka Y, Wierenga EA, Yazdanbakhsh M, Kapsenberg ML: Microbial compounds selectively induce Th1 cell-promoting or Th2 cell-promoting dendritic cells in vitro with diverse th cell-polarizing signals. J Immunol. 2002, 168:1704–1709.
Dhodapkar MV, Krasovsky J, Steinman RM, Bhardwaj N: Mature dendritic cells boost functionally superior CD8(+) T-cell in humans without foreign helper epitopes. J Clin.Invest 2000, 105:R9-R14.
Dhodapkar MV, Steinman RM, Krasovsky J, Munz C, Bhardwaj N: Antigen-specific inhibition of effector T cell function in humans after injection of immature dendritic cells. J.Exp.Med. 2001, 193:233–238.
Dhodapkar MV, Steinman RM, Sapp M, Desai H, Fossella C, Krasovsky J, Donahoe SM, Dunbar PR, Cerundolo V, Nixon DF, Bhardwaj N: Rapid generation of broad T-cell immunity in humans after a single injection of mature dendritic cells. J.Clin.Invest. 1999, 104:173–180.
Dietz AB, Vuk-Pavlovic S: High efficiency adenovirus-mediated gene transfer to human dendritic cells. Blood 1998, 91:392–398.
Dieu MC, Vanbervliet B, Vicari A, Bridon JM, Oldham E, Ait-Yahia S, Briere F, Zlotnik A, Lebecque S, Caux C: Selective recruitment of immature and mature dendritic cells by distinct chemokines expressed in different anatomic sites. J Exp.Med. 1998, 188:373–386.
Dubey P, Hendrickson RC, Meredith SC, Siegel CT, Shabanowitz J, Skipper JC, Engelhard VH, Hunt DF, Schreiber H: The immunodominant antigen of an ultraviolet-induced regressor tumor is generated by a somatic point mutation in the DEAD box helicase p68. J Exp.Med. 1997, 185:695–705.
Dyall J, Latouche JB, Schnell S, Sadelain M: Lentivirus-transduced human monocyte-derived dendritic cells efficiently stimulate antigen-specific cytotoxic T lymphocytes. Blood 2001, 97:114–121.
Dzionek A, Fuchs A, Schmidt P, Cremer S, Zysk M, Miltenyi S, Buck DW, Schmitz J: BDCA-2, BDCA-3, and BDCA-4: three markers for distinct subsets of dendritic cells in human peripheral blood. J Immunol. 2000, 165:6037–6046.
Eggert AA, Schreurs MW, Boerman OC, Oyen WJ, de Boer AJ, Punt CJ, Figdor CG, Adema GJ: Biodistribution and vaccine efficiency of murine dendritic cells are dependent on the route of administration. Cancer Res. 1999, 59:3340–3345.
Engelmayer J, Larsson M, Lee A, Lee M, Cox WI, Steinman RM, Bhardwaj N: Mature dendritic cells infected with canarypox virus elicit strong anti-human immunodeficiency virus CD8+ and CD4+ T-cell responses from chronically infected individuals. J Virol. 2001, 75:2142–2153.
Engelmayer J, Larsson M, Subklewe M, Chahroudi A, Cox WI, Steinman RM, Bhardwaj N: Vaccinia virus inhibits the maturation of human dendritic cells: a novel mechanism of immune evasion. J.Immunol. 1999, 163:6762–6768.

Ferlazzo G, Tsang ML, Moretta L, Melioli G, Steinman RM, Munz C: Human dendritic cells activate resting natural killer (NK) cells and are recognized via the NKp30 receptor by activated NK cells. J Exp.Med. 2002, 195: 343–351.

Ferlazzo G, Wesa A, Wei WZ, Galy A: Dendritic cells generated either from CD34+ progenitor cells or from monocytes differ in their ability to activate antigen-specific CD8+ T cells. J Immunol. 1999, 163:3597–3604.

Fernandez NC, Lozier A, Flament C, Ricciardi-Castagnoli P, Bellet D, Suter M, Perricaudet M, Tursz T, Maraskovsky E, Zitvogel L: Dendritic cells directly trigger NK cell functions: cross-talk relevant in innate anti-tumor immune responses in vivo. Nat.Med. 1999, 5:405–411.

Feuerstein B, Berger TG, Maczek C, Roder C, Schreiner D, Hirsch U, Haendle I, Leisgang W, Glaser A, Kuss O, Diepgen TL, Schuler G, Schuler-Thurner B: A method for the production of cryopreserved aliquots of antigen-preloaded, mature dendritic cells ready for clinical use. J Immunol Methods 2000, 245:15–29.

Fong L, Brockstedt D, Benike C, Breen JK, Strang G, Ruegg CL, Engleman EG: Dendritic cell-based xenoantigen vaccination for prostate cancer immunotherapy. J Immunol. 2001a, 167:7150–7156.

Fong L, Brockstedt D, Benike C, Wu L, Engleman EG: Dendritic cells injected via different routes induce immunity in cancer patients. J.Immunol. 2001b, 166:4254–4259.

Gong J, Nikrui N, Chen D, Koido S, Wu Z, Tanaka Y, Cannistra S, Avigan D, Kufe D: Fusions of human ovarian carcinoma cells with autologous or allogeneic dendritic cells induce antitumor immunity. J Immunol. 2000, 165: 1705–1711.

Hahn S, Gehri R, Erb P: Mechanism and biological significance of CD4-mediated cytotoxicity. Immunol.Rev. 1995, 146:57–79.

Haicheur N, Bismuth E, Bosset S, Adotevi O, Warnier G, Lacabanne V, Regnault A, Desaymard C, Amigorena S, Ricciardi-Castagnoli P, Goud B, Fridman WH, Johannes L, Tartour E: The B subunit of Shiga toxin fused to a tumor antigen elicits CTL and targets dendritic cells to allow MHC class I-restricted presentation of peptides derived from exogenous antigens. J.Immunol. 2000, 165: 3301–3308.

Henderson RA, Nimgaonkar MT, Watkins SC, Robbins PD, Ball ED, Finn OJ: Human dendritic cells genetically engineered to express high levels of the human epithelial tumor antigen mucin (MUC-1). Cancer Res. 1996, 56:3763–3770.

Hermans IF, Ritchie DS, Yang J, Roberts JM, Ronchese F: CD8+ T cell-dependent elimination of dendritic cells in vivo limits the induction of antitumor immunity. J Immunol. 2000, 164:3095–3101.

Hernando J, Park TW, Kubler K, Offergeld R, Schlebusch H, Bauknecht T: Vaccination with autologous tumour antigen-pulsed dendritic cells in advanced gynaecological malignancies: clinical and immunological evaluation of a phase I trial. Cancer Immunol.Immunother. 2002, 51:45–52.

Holtl L: CD83+ blood dendritic cells as a vaccine for immunotherapy of metastatic renal-cell cancer. Lancet. 1998, 352:1358.

Hsu FJ, Benike C, Fagnoni F, Liles TM, Czerwinski D, Taidi B, Engleman EG, Levy R: Vaccination of patients with B-cell lymphoma using autologous antigen-pulsed dendritic cells. Nat.Med. 1996, 2:52–58.

Ignatius R, Mahnke K, Rivera M, Hong K, Isdell F, Steinman RM, Pope M, Stamatatos L: Presentation of proteins encapsulated in sterically stabilized liposomes by dendritic cells initiates CD8(+) T-cell responses in vivo. Blood 2000a, 96:3505–3513.

Ignatius R, Marovich M, Mehlhop E, Villamide L, Mahnke K, Cox WI, Isdell F, Frankel SS, Mascola JR, Steinman RM, Pope M: Canarypox virus-induced maturation of dendritic cells is mediated by apoptotic cell death and tumor necrosis factor alpha secretion. J Virol. 2000b, 74:11329–11338.

Inaba K, Turley S, Iyoda T, Yamaide F, Shimoyama S, Sousa C, Germain RN, Mellman I, Steinman RM: The formation of immunogenic major histocompatibility complex class II-peptide ligands in lysosomal compartments of dendritic cells is regulated by inflammatory stimuli. J Exp.Med. 2000, 191:927–936.

Inaba K, Turley S, Yamaide F, Iyoda T, Mahnke K, Inaba M, Pack M, Subklewe M, Sauter B, Sheff D, Albert M, Bhardwaj N, Mellman I, Steinman RM: Efficient presentation of phagocytosed cellular fragments on the major histocompatibility complex class II products of dendritic cells. J.Exp.Med. 1998, 188:2163–2173.

Irvine AS, Trinder PK, Laughton DL, Ketteringham H, McDermott RH, Reid SC, Haines AM, Amir A, Husain R, Doshi R, Young LS, Mountain A: Efficient nonviral transfection of dendritic cells and their use for in vivo immunization. Nat.Biotechnol. 2000, 18:1273–1278.

Jager E, Nagata Y, Gnjatic S, Wada H, Stockert E, Karbach J, Dunbar PR, Lee SY, Jungbluth A, Jager D, Arand M, Ritter G, Cerundolo V, Dupont B, Chen YT, Old LJ, Knuth A: Monitoring CD8 T cell responses to NY-ESO-1: correlation of humoral and cellular immune responses. Proc.Natl.Acad.Sci.U.S.A 2000, 97:4760–4765.

Jakob T, Walker PS, Krieg AM, von Stebut E, Udey MC, Vogel JC: Bacterial DNA and CpG-containing oligodeoxynucleotides activate cutaneous dendritic cells and induce IL-12 production: implications for the augmentation of Thl responses. Int.Arch.Allergy Immunol. 1999, 118:457–461.

Jenne L, Arrighi JF, Sauter B, Kern P: Dendritic cells pulsed with unfractionated helminthic proteins to generate antiparasitic cytotoxic T lymphocyte. Parasite Immunol. 2001a, 23:195–201.

Jenne L, Hauser C, Arrighi JF, Saurat JH, Hugin A: Poxvirus as a vector to transduce human dendritic cells for immunotherapy: abortive infection but reduced APC function. Gene Therapy 2000, 7:1575–1583.

Jenne L, Schuler G, Steinkasserer A: Viral vectors for dendritic cell-based immunotherapy. Trends Immunol 2001b, 22:102–107.

Jiang W, Swiggard WJ, Heufler C, Peng M, Mirza A, Steinman RM, Nussenzweig MC: The receptor DEC-205 expressed by dendritic cells and thymic epithelial cells is involved in antigen processing. Nature 1995, 375:151–155.

Jonuleit H, Giesecke-Tuettenberg A, Tuting T, Thurner-Schuler B, Stuge TB, Paragnik L, Kandemir A, Lee PP, Schuler G, Knop J, Enk AH: A comparison of two types of dendritic cell as adjuvants for the induction of melanoma-specific T-cell responses in humans following intranodal injection. Int.J.Cancer 2001, 93:243–251.

Jonuleit H, Kuhn U, Muller G, Steinbrink K, Paragnik L, Schmitt E, Knop J, Enk AH: Pro-inflammatory cytokines and prostaglandins induce maturation of potent immunostimulatory dendritic cells under fetal calf serum-free conditions. Eur.J.Immunol. 1997, 27:3135–3142.

Jonuleit H, Schmitt E, Schuler G, Knop J, Enk AH: Induction of interleukin 10-producing, nonproliferating CD4(+) T cells with regulatory properties by repetitive stimulation with allogeneic immature human dendritic cells. J.Exp.Med. 2000, 192:1213–1222.

Josien R, Li HL, Ingulli E, Sarma S, Wong BR, Vologodskaia M, Steinman RM, Choi Y: TRANCE, a tumor necrosis factor family member, enhances the longevity and adjuvant properties of dendritic cells in vivo. J Exp.Med. 2000, 191:495–502.

Kaisho T, Akira S: Critical roles of Toll-like receptors in host defense. Crit Rev.Immunol. 2000, 20:393–405.

Kalinski P, Hilkens CM, Snijders A, Snijdewint FG, Kapsenberg ML: IL-12-deficient dendritic cells, generated in the presence of prostaglandin E2, promote type 2 cytokine production in maturing human naive T helper cells. J Immunol. 1997, 159:28–35.

Kampgen E, Koch N, Koch F, Stoger P, Heufler C, Schuler G, Romani N: Class II major histocompatibility complex molecules of murine dendritic cells: synthesis, sialylation of invariant chain, and antigen processing capacity are down-regulated upon culture. Proc.Natl.Acad.Sci.U.S.A 1991, 88:3014–3018.

Kawano T, Cui J, Koezuka Y, Toura I, Kaneko Y, Motoki K, Ueno H, Nakagawa R, Sato H, Kondo E, Koseki H, Taniguchi M: CD1d-restricted and TCR-mediated activation of valpha14 NKT cells by glycosylceramides. Science 1997, 278:1626–1629.

Kellermann SA, Hudak S, Oldham ER, Liu YJ, McEvoy LM: The CC chemokine receptor-7 ligands 6Ckine and macrophage inflammatory protein-3 beta are potent chemoattractants for in v. J Immunol. 1999, 162:3859–3864.

Kim DT, Mitchell DJ, Brockstedt DG, Fong L, Nolan GP, Fathman CG, Engleman EG, Rothbard JB: Introduction of soluble proteins into the MHC class I pathway by conjugation to an HIV tat peptide. J.Immunol. 1997, 159:1666–1668.

Koido S, Kashiwaba M, Chen D, Gendler S, Kufe D, Gong J: Induction of antitumor immunity by vaccination of dendritic cells transfected with MUC1 RNA. J Immunol. 2000, 165:5713–5719.

Korn EL, Arbuck SG, Pluda JM, Simon R, Kaplan RS, Christian MC: Clinical trial designs for cytostatic agents: are new approaches needed? J Clin.Oncol. 2001, 19:265–272.

Kugler A, Stuhler G, Walden P, Zoller G, Zobywalski A, Brossart P, Trefzer U, Ullrich S, Muller CA, Becker V, Gross AJ, Hemmerlein B, Kanz L, Muller GA, Ringert RH: Regression of human metastatic renal cell carcinoma after vaccination with tumor cell-dendritic cell hybrids. Nat.Med. 2000, 6:332–336.

Kukutsch NA, Rossner S, Austyn JM, Schuler G, Lutz MB: Formation and kinetics of MHC class I-ovalbumin peptide complexes on immature and mature murine dendritic cells. J Invest Dermatol. 2000, 115:449–453.

Labeur MS, Roters B, Pers B, Mehling A, Luger TA, Schwarz T, Grabbe S: Generation of tumor immunity by bone marrow-derived dendritic cells correlates with dendritic cell maturation stage. J Immunol. 1999, 162:168–175.

Langenkamp A, Messi M, Lanzavecchia A, Sallusto F: Kinetics of dendritic cell activation: impact on priming of TH1, TH2 and nonpolarized T cells. Nat.Immunol. 2000, 1:311–316.

Le Bon A, Schiavoni G, D'Agostino G, Gresser I, Belardelli F, Tough DF: Type i interferons potently enhance humoral immunity and can promote isotype switching by stimulating dendritic cells in vivo. Immunity. 2001, 14:461–470.

Lee PP, Yee C, Savage PA, Fong L, Brockstedt D, Weber JS, Johnson D, Swetter S, Thompson J, Greenberg PD, Roederer M, Davis MM: Characterization of circulating T cells specific for tumor-associated antigens in melanoma patients. Nat.Med. 1999, 5:677–685.

Lim SH, Bailey-Wood R: Idiotypic protein-pulsed dendritic cell vaccination in multiple myeloma. Int.J Cancer 1999, 83:215–222.

Liso A, Stockerl-Goldstein KE, Auffermann-Gretzinger S, Benike CJ, Reichardt V, van Beckhoven A, Rajapaksa R, Engleman EG, Blume KG, Levy R: Idiotype vaccination using dendritic cells after autologous peripheral blood progenitor cell transplantation for multiple myeloma. Biol.Blood Marrow Transplant. 2000, 6:621–627.

Liu YJ: Dendritic cell subsets and lineages, and their functions in innate and adaptive immunity. Cell 2001, 106:259–262.

Ludewig B, Ochsenbein AF, Odermatt B, Paulin D, Hengartner H, Zinkernagel RM: Immunotherapy with dendritic cells directed against tumor antigens shared with normal host cells results in severe autoimmune disease. J Exp.Med. 2000, 191:795–804.

Ludewig B, Odermatt B, Ochsenbein AF, Zinkernagel RM, Hengartner H: Role of dendritic cells in the induction and maintenance of autoimmune diseases. Immunol.Rev. 1999, 169:45–54.

Macagno A, Gilliet M, Sallusto F, Lanzavecchia A, Nestle FO, Groettrup M: Dendritic cells up-regulate immunoproteasomes and the proteasome regulator PA28 during maturation. Eur.J Immunol. 1999, 29:4037–4042.

Mackensen A, Herbst B, Chen JL, Kohler G, Noppen C, Herr W, Spagnoli GC, Cerundolo V, Lindemann A: Phase I study in melanoma patients of a vaccine with peptide-pulsed dendritic cells generated in vitro from CD34(+) hematopoietic progenitor cells. Int.J.Cancer 2000.May.;86.(3.):385.-392. 86:385–392.

Mahnke K, Guo M, Lee S, Sepulveda H, Swain SL, Nussenzweig M, Steinman RM: The dendritic cell receptor for endocytosis, DEC-205, can recycle and enhance antigen presentation via major histocompatibility complex class II-positive lysosomal compartments. J Cell Biol. 2000, 151:673–684.

Mandruzzato S, Brasseur F, Andry G, Boon T, van der BP: A CASP-8 mutation recognized by cytolytic T lymphocytes on a human head and neck carcinoma. J Exp.Med. 1997, 186:785–793.

Maraskovsky E, Daro E, Roux E, Teepe M, Maliszewski CR, Hoek J, Caron D, Lebsack ME, McKenna HJ: In vivo generation of human dendritic cell subsets by Flt3 ligand. Blood 2000, 96:878–884.

Matsuda JL, Naidenko OV, Gapin L, Nakayama T, Taniguchi M, Wang CR, Koezuka Y, Kronenberg M: Tracking the response of natural killer T cells to a glycolipid antigen using CD1d tetramers. J Exp.Med. 2000, 192:741–754.

Mbow ML, Zeidner N, Panella N, Titus RG, Piesman J: *Borrelia burgdorferi*-pulsed dendritic cells induce a protective immune response against tick-transmitted spirochetes. Infect.Immun. 1997, 65:3386–3390.

Medema JP, Schuurhuis DH, Rea D, van Tongeren J, de Jong J, Bres SA, Laban S, Toes RE, Toebes M, Schumacher TN, Bladergroen BA, Ossendorp F, Kummer JA, Melief CJ, Offringa R: Expression of the serpin serine protease inhibitor 6 protects dendritic cells from cytotoxic T lymphocyte-induced apoptosis: differential modulation by T helper type 1 and type 2 cells. J Exp.Med. 2001, 194:657–667.

Mellman I, Steinman RM: Dendritic cells: specialized and regulated antigen processing machines. Cell 2001, 106:255–258.

Miltenyi S, Muller W, Weichel W, Radbruch A: High gradient magnetic cell separation with MACS. Cytometry 1990, 11:231–238.

Moll H, Berberich C: Dendritic cell-based vaccination strategies: induction of protective immunity against leishmaniasis. Immunobiology 2001, 204:659–666.

Morel S, Levy F, Burlet-Schiltz O, Brasseur F, Probst-Kepper M, Peitrequin AL, Monsarrat B, Van Velthoven R, Cerottini JC, Boon T, Gairin JE, Van den Eynde BJ: Processing of some antigens by the standard proteasome but not by the immunoproteasome results in poor presentation by dendritic cells. Immunity. 2000, 12:107–117.

Morse MA, Coleman RE, Akabani G, Niehaus N, Coleman D, Lyerly HK: Migration of human dendritic cells after injection in patients with metastatic malignancies. Cancer Res. 1999, 59:56–58.

Munz C, Bickham KL, Subklewe M, Tsang ML, Chahroudi A, Kurilla MG, Zhang D, O'Donnell M, Steinman RM: Human CD4(+) T lymphocytes consistently respond to the latent Epstein- Barr virus nuclear antigen EBNA1. J Exp.Med. 2000, 191:1649–1660.

Murphy G, Tjoa B, Ragde H, Kenny G, Boynton A: Phase I clinical trial: T-cell therapy for prostate cancer using autologous dendritic cells pulsed with HLA-A0201-specific peptides from prostate-specific membrane antigen. Prostate 1996, 29:371–380.

Murphy GP, Tjoa BA, Simmons SJ, Jarisch J, Bowes VA, Ragde H, Rogers M, Elgamal A, Kenny GM, Cobb OE, Ireton RC, Troychak MJ, Salgaller ML, Boynton AL: Infusion of dendritic cells pulsed with HLA-A2-specific prostate-specific membrane antigen peptides: a phase II prostate cancer vaccine trial involving patients with hormone-refractory metastatic disease. Prostate 1999a, 38:73–78.

Murphy GP, Tjoa BA, Simmons SJ, Ragde H, Rogers M, Elgamal A, Kenny GM, Troychak MJ, Salgaller ML, Boynton AL: Phase II prostate cancer vaccine trial: report of a study involving 37 patients with disease recurrence following primary treatment. Prostate 1999b, 39:54–59.

Murphy GP, Tjoa BA, Simmons SJ, Rogers MK, Kenny GM, Jarisch J: Higher-dose and less frequent dendritic cell infusions with PSMA peptides in hormone-refractory metastatic prostate cancer patients. Prostate 2000, 43:59–62.

Nair SK, Heiser A, Boczkowski D, Majumdar A, Naoe M, Lebkowski JS, Vieweg J, Gilboa E: Induction of cytotoxic T cell responses and tumor immunity against unrelated tumors using telomerase reverse transcriptase RNA transfected dendritic cells. Nat.Med. 2000, 6:1011–1017.

Nair SK, Hull S, Coleman D, Gilboa E, Lyerly HK, Morse MA: Induction of carcinoembryonic antigen (CEA)-specific cytotoxic T- lymphocyte responses in vitro using autologous dendritic cells loaded with CEA peptide or CEA RNA in patients with metastatic malignancies expressing CEA. Int.J Cancer 1999, 82:121–124.

Nepom GT, Buckner JH, Novak EJ, Reichstetter S, Reijonen H, Gebe J, Wang R, Swanson E, Kwok WW: HLA class II tetramers: tools for direct analysis of antigen-specific CD4+ T cells. Arthritis Rheum. 2002, 46:5–12.

Nestle FO, Alijagic S, Gilliet M, Sun Y, Grabbe S, Dummer R, Burg G, Schadendorf D: Vaccination of melanoma patients with peptide- or tumor lysate-pulsed dendritic cells. Nat.Med. 1998, 4:328–332.

Nishimura T, Iwakabe K, Sekimoto M, Ohmi Y, Yahata T, Nakui M, Sato T, Habu S, Tashiro H, Sato M, Ohta A: Distinct role of antigen-specific T helper type 1 (Th1) and Th2 cells in tumor eradication in vivo. J Exp.Med. 1999, 190:617–627.

Pardoll DM: Cancer vaccines. Nat.Med. 1998, 4:525–531.

Peshwa MV: Induction of prostate tumor-specific CD8+ cytotoxic T-lymphocytes in vitro using antigen-presenting cells pulsed with prostatic acid phosphatase peptide. Prostate. 1998, 36:129–138.

Qin Z, Blankenstein T: CD4+ T cell-mediated tumor rejection involves inhibition of angiogenesis that is dependent on IFN gamma receptor expression by nonhematopoietic cells. Immunity. 2000, 12:677–686.

Regnault A, Lankar D, Lacabanne V, Rodriguez A, Thery C, Rescigno M, Saito T, Verbeek S, Bonnerot C, Ricciardi-Castagnoli P, Amigorena S: Fcgamma receptor-mediated induction of dendritic cell maturation and major histocompatibility complex class I-restricted antigen presentation after immune complex internalization. J.Exp.Med. 1999, 189:371–380.

Reichardt VL, Okada CY, Liso A, Benike CJ, Stockerl-Goldstein KE, Engleman EG, Blume KG, Levy R: Idiotype vaccination using dendritic cells after autologous peripheral blood stem cell transplantation for multiple myeloma – a feasibility study. Blood 1999, 93:2411–2419.

Rescigno M, Borrow P: The host-pathogen interaction: new themes from dendritic cell biology. Cell 2001, 106:267–270.

Ridge JP: A conditioned dendritic cell can be a temporal bridge between a CD4+ T-helper and a T-killer cell. Nature. 1998, 393:474–478.

Rieser C, Ramoner R, Holtl L, Rogatsch H, Papesh C, Stenzl A, Bartsch G, Thurnher M: Mature dendritic cells induce T-helper type-1-dominant immune responses in patients with metastatic renal cell carcinoma. Urol.Int. 1999, 63:151–159.

Sallusto F, Cella M, Danieli C, Lanzavecchia A: Dendritic cells use macropinocytosis and the mannose receptor to concentrate macromolecules in the major histocompatibility complex class II compartment: downregulation by cytokines and bacterial products. J.Exp.Med. 1995, 182:389–400.

Sallusto F, Lanzavecchia A, Mackay CR: Chemokines and chemokine receptors in T-cell priming and Th1/Th2-mediated responses. Immunol Today 1998a, 19:568–574.

Sallusto F, Palermo B, Lenig D, Miettinen M, Matikainen S, Julkunen I, Forster R, Burgstahler R, Lipp M, Lanzavecchia A: Distinct patterns and kinetics of chemokine production regulate dendritic cell function. Eur.J Immunol. 1999, 29:1617–1625.

Sallusto F, Schaerli P, Loetscher P, Schaniel C, Lenig D, Mackay CR, Qin S, Lanzavecchia A: Rapid and coordinated switch in chemokine receptor expression during dendritic cell maturation. Eur.J Immunol 1998b, 28:2760–2769.

Santin AD, Bellone S, Ravaggi A, Roman JJ, Pecorelli S, Parham GP, Cannon MJ: Induction of tumour-specific CD8(+) cytotoxic T lymphocytes by tumour lysate-pulsed autologous dendritic cells in patients with uterine serous papillary cancer. Br.J Cancer 2002, 86:151–157.

Savill J, Fadok V: Corpse clearance defines the meaning of cell death. Nature 2000, 407:784–788.

Schoenberger SP, Toes RE, van der Voort EI, Offringa R, Melief CJ: T-cell help for cytotoxic T lymphocytes is mediated by CD40-CD40L interactions [see comments]. Nature 1998, 393:480–483.

Schuler-Thurner B, Dieckmann D, Keikavoussi P, Bender A, Maczek C, Jonuleit H, Roder C, Haendle I, Leisgang W, Dunbar R, Cerundolo V, von Den DP, Knop J, Brocker EB, Enk A, Kampgen E, Schuler G: Mage-3 and influenza-matrix peptide-specific cytotoxic T cells are inducible in terminal stage HLA-A2.1+ melanoma patients by mature monocyte-derived dendritic cells. J.Immunol. 2000, 165:3492–3496.

Schuler-Thurner B, Schultz ES, Berger T, Weinlich G, Ebner S, Woerl P, Bender A, Feuerstein B Fritsch PO, Romani N, Schuler G: Rapid induction of tumor-specific type 1 helper T cells in metastatic melanoma patients by vaccination with mature, cryopreserved, peptide-loaded monocyte-derived dendritic cells. J Exp.Med. 2002.

Schultz ES, Chapiro J, Lurquin C, Claverol S, Burlet-Schiltz O, Warnier G, Russo V, Morel S, Levy F, Boon T, Van den Eynde BJ, van der BP: The production of a new MAGE-3 peptide presented to cytolytic T lymphocytes by HLA-B40 requires the immunoproteasome. J.Exp.Med. 2002, 195:391–399.

Schultz ES, Lethe B, Cambiaso CL, Van Snick J, Chaux P, Corthals J, Heirman C, Thielemans K, Boon T, van der BP: A MAGE-A3 peptide presented by HLA-DP4 is recognized on tumor cells by CD4+ cytolytic T lymphocytes. Cancer Res. 2000, 60:6272–6275.

Schuurhuis DH, Laban S, Toes RE, Ricciardi-Castagnoli P, Kleijmeer MJ, van der Voort EI, Rea D, Offringa R, Geuze HJ, Melief CJ, Ossendorp F: Immature dendritic cells acquire CD8(+) cytotoxic T lymphocyte priming capacity upon activation by T helper cell-independent or -dependent stimuli. J Exp.Med. 2000, 192:145–150.

Seder RA, Hill AV: Vaccines against intracellular infections requiring cellular immunity. Nature 2000, 406:793–798.

Sevilla N, Kunz S, Holz A, Lewicki H, Homann D, Yamada H, Campbell KP, de La Torre JC, Oldstone MB: Immunosuppression and resultant viral persistence by specific viral targeting of dendritic cells. J Exp.Med. 2000, 192:1249–1260.

Small EJ, Fratesi P, Reese DM, Strang G, Laus R, Peshwa MV, Valone FH: Immunotherapy of hormone-refractory prostate cancer with antigen-loaded dendritic cells. J Clin.Oncol. 2000, 18:3894–3903.

Smith SG, Patel PM, Porte J, Selby PJ, Jackson AM: Human dendritic cells genetically engineered to express a melanoma polyepitope DNA vaccine induce multiple cytotoxic T-cell responses. Clin.Cancer Res. 2001, 7:4253–4261.

Smyth MJ, Crowe NY, Hayakawa Y, Takeda K, Yagita H, Godfrey DI: NKT cells - conductors of tumor immunity? Curr.Opin.Immunol. 2002, 14:165–171.

Sozzani S, Allavena P, D'Amico G, Luini W, Bianchi G, Kataura M, Imai T, Yoshie O, Bonecchi R, Mantovani A: Differential regulation of chemokine receptors during dendritic cell maturation: a model for their trafficking properties. J Immunol. 1998, 161:1083–1086.

Sparwasser T, Koch ES, Vabulas RM, Heeg K, Lipford GB, Ellwart JW, Wagner H: Bacterial DNA and immunostimulatory CpG oligonucleotides trigger maturation and activation of murine dendritic cells. Eur.J Immunol. 1998, 28:2045–2054.

Steinman RM, Turley S, Mellman I, Inaba K: The induction of tolerance by dendritic cells that have captured apoptotic cells. J.Exp.Med. 2000, 191:411–416.

Strobel I, Berchtold S, Gotze A, Schulze U, Schuler G, Steinkasserer A: Human dendritic cells transfected with either RNA or DNA encoding influenza matrix protein M1 differ in their ability to stimulate cytotoxic T lymphocytes. Human Gene Therapy 2000a,In Press.

Strobel I, Krumbholz M, Menke A, Hoffmann E, Dunbar PR, Bender A, Hobom G, Steinkasserer A, Schuler G, Grassmann R: Efficient expression of the tumor-associated antigen MAGE-3 in human dendritic cells, using an avian influenza virus vector. Hum.Gene Ther. 2000b, 11:2207–2218.

Subklewe M, Chahroudi A, Schmaljohn A, Kurilla MG, Bhardwaj N, Steinman RM: Induction of Epstein-Barr virus-specific cytotoxic T-lymphocyte responses using dendritic cells pulsed with EBNA-3A peptides or UV-inactivated, recombinant EBNA-3A vaccinia virus. Blood 1999, 94:1372–1381.

Takahashi T: Reactivity of autologous CD4+ T lymphocytes against human melanoma. Evidence for a shared melanoma antigen presented by HLA-DR15. J.Immunol. 1995, 154:772–779.

Takashima A, Morita A: Dendritic cells in genetic immunization. J Leukoc.Biol. 1999, 66:350–356.

Tanaka H, Demeure CE, Rubio M, Delespesse G, Sarfati M: Human monocyte-derived dendritic cells induce naive T cell differentiation into T helper cell type 2 (Th2) or Th1/Th2 effectors. Role of stimulator/responder ratio. J Exp.Med. 2000, 192:405–412.

Tang DC, DeVit M, Johnston SA: Genetic immunization is a simple method for eliciting an immune response. Nature 1992, 356:152–154.

Thomas WD: CD4 T cells kill melanoma cells by mechanisms that are independent of Fas (CD95). Int.J.Cancer. 1998, 75:384–390.

Thurner B, Haendle I, der C, Dieckmann D, Keikavoussi P, Jonuleit H, Bender A, Maczek C, Schreiner D, von Den Driesch P, Br, Steinman RM, Enk A, mpgen E, Schuler G: Vaccination with Mage-3A1 peptide-pulsed mature, monocyte-derived dendritic cells expands specific Cytotoxic T Cells and induces regression of some metastases in advanced stage IV melanoma. J.Exp.Med. 1999a, 190: 1669–1678.

Thurner B, Roder C, Dieckmann D, Heuer M, Kruse M, Glaser A, Keikavoussi P, Kampgen E, Bender A, Schuler G: Generation of large numbers of fully mature and stable dendritic cells from leukapheresis products for clinical application. J.Immunol.Methods 1999b, 223:1–15.

Thurnher M, Rieser C, Holtl L, Papesh C, Ramoner R, Bartsch G: Dendritic cell-based immunotherapy of renal cell carcinoma. Urol.Int. 1998, 61:67–71.

Thurnher M, Zelle-Rieser C, Ramoner R, Bartsch G, Holtl L: The disabled dendritic cell. FASEB J 2001, 15:1054–1061.

Timmerman JM, Czerwinski DK, Davis TA, Hsu FJ, Benike C, Hao ZM, Taidi B, Rajapaksa R, Caspar CB, Okada CY, van Beckhoven A, Liles TM, Engleman EG, Levy R: Idiotype-pulsed dendritic cell vaccination for B-cell lymphoma: clinical and immune responses in 35 patients. Blood 2002, 99:1517–1526.

Tjoa BA: Evaluation of phase I/II clinical trials in prostate cancer with dendritic cells and PSMA peptides. Prostate. 1998, 36:39–44.

Toes RE, Ossendorp F, Offringa R, Melief CJ: CD4 T cells and their role in antitumor immune responses. J.Exp.Med. 1999, 189:753–756.

Toura I, Kawano T, Akutsu Y, Nakayama T, Ochiai T, Taniguchi M: Cutting edge: inhibition of experimental tumor metastasis by dendritic cells pulsed with alpha-galactosylceramide. J.Immunol. 1999, 163:2387–2391.

Urban BC, Ferguson DJ, Pain A, Willcox N, Plebanski M, Austyn JM, Roberts DJ: *Plasmodium falciparum*-infected erythrocytes modulate the maturation of dendritic cells. Nature 1999, 400:73–77.

van der Bruggen P, Traversari C, Chomez P, Lurquin C, De Plaen E, Van den EB, Knuth A, Boon T: A gene encoding an antigen recognized by cytolytic T lymphocytes on a human melanoma. Science 1991, 254:1643–1647.

Van Tendeloo VF, Ponsaerts P, Lardon F, Nijs G, Lenjou M, Van Broeckhoven C, Van Bockstaele DR, Berneman ZN: Highly efficient gene delivery by mRNA electroporation in human hematopoietic cells: superiority to lipofection and passive pulsing of mRNA and to electroporation of plasmid cDNA for tumor antigen loading of dendritic cells. Blood 2001, 98:49–56.

Verdijk RM, Mutis T, Esendam B, Kamp J, Melief CJ, Brand A, Goulmy E: Polyriboinosinic polyribocytidylic acid (poly(I:C)) induces stable maturation of functionally active human dendritic cells. J.Immunol. 1999, 163: 57–61.

Wolfel T, Hauer M, Schneider J, Serrano M, Wolfel C, Klehmann-Hieb E, De Plaen E, Hankeln T, Meyer zum Buschenfelde KH, Beach D: A p16INK4a-insensitive CDK4 mutant targeted by cytolytic T lymphocytes in a human melanoma. Science 1995, 269:1281–1284.

Wolfers J, Lozier A, Raposo G, Regnault A, Thery C, Masurier C, Flament C, Pouzieux S, Faure F, Tursz T, Angevin E, Amigorena S, Zitvogel L: Tumor-derived exosomes are a source of shared tumor rejection antigens for CTL cross-priming. Nat.Med. 2001, 7:297–303.

Worgall S, Kikuchi T, Singh R, Martushova K, Lande L, Crystal RG: Protection against pulmonary infection with *Pseudomonas aeruginosa* following immunization with *P. aeruginosa*-pulsed dendritic cells. Infect.Immun. 2001, 69:4521–4527.

You Z, Huang XF, Hester J, Rollins L, Rooney C, Chen SY: Induction of vigorous helper and cytotoxic T cell as well as B cell responses by dendritic cells expressing a modified antigen targeting receptor-mediated internalization pathway. J.Immunol. 2000, 165:4581–4591.

Yu JS, Wheeler CJ, Zeltzer PM, Ying H, Finger DN, Lee PK, Yong WH, Incardona F, Thompson RC, Riedinger MS, Zhang W, Prins RM, Black KL: Vaccination of malignant glioma patients with peptide-pulsed dendritic cells elicits systemic cytotoxicity and intracranial T-cell infiltration. Cancer Res. 2001, 61:842–847.

Zhong L, Granelli-Piperno A, Choi Y, Steinman RM: Recombinant adenovirus is an efficient and non-perturbing genetic vector for human dendritic cells. Eur.J.Immunol. 1999, 29:964–972.

Zitvogel L, Fernandez N, Lozier A, Wolfers J, Regnault A, Raposo G, Amigorena S: Dendritic cells or their exosomes are effective biotherapies of cancer. Eur.J Cancer 1999, 35 Suppl 3:S36-S38.

Zitvogel L, Regnault A, Lozier A, Wolfers J, Flament C, Tenza D, Ricciardi-Castagnoli P, Raposo G, Amigorena S: Eradication of established murine tumors using a novel cell-free vaccine: dendritic cell-derived exosomes. Nat.Med. 1998, 4:594–600.

zur Hausen H: Papillomavirus infections–a major cause of human cancers. Biochim.Biophys.Acta 1996, 1288:F55-F78.

Chapter 9

Dendritic Cell Vaccination and Viral Infection – Animal Models

B. Ludewig

Abstract. Dendritic cells (DCs) play a pivotal role in the initiation and maintenance of immune responses against viruses and other microbial pathogens. Adoptively transferred, in vitro manipulated DCs presenting antigen derived from different viruses have been shown to elicit cytotoxic T cell (CTL) and T helper (Th) cell responses and to induce protective antiviral immunity. Furthermore, DC-based adoptive immunotherapies have the potential to specifically (re)activate antiviral immunity in chronic viral diseases such as HIV or hepatitis virus infections. Cellular dendritic cell vaccines, however, are not suitable for large-scale prophylactic immunization. Strategies for vaccine development should therefore aim at the specific delivery of microbial antigens to DCs in situ. Furthermore, appropriate mobilization and activation of DCs by the vaccine is important for the generation of optimal antimicrobial immune re-

B. Ludewig (✉)
Institute of Experimental Immunology, Department of Pathology,
University of Zürich, Schmelzbergstr. 12, 8091 Zürich, Switzerland

sponses. Here, we discuss recent data on induction of antiviral immunity with various DC-vaccination approaches and outline future directions for the development of specific antigen targeting to DCs.

1 Introduction

Prophylactic vaccines against several viral infections have been developed over the last centuries, leading to the eradication of smallpox and protecting many people from diseases such as measles, rubella, mumps, and polio. However, a number of diseases remain against which vaccines are currently suboptimal or unavailable. Furthermore, there is a growing need to develop therapeutic vaccines that may boost specific immune response to persistent viruses such as HIV or HCV. The critical first step in the development of antiviral vaccines is the identification of the dominant antigens contributing to the different stages of the infection, i.e., initial replication at the site of entry, spread in the host, and establishment of a persistent infection. The methodology for the identification of antigens and the characterization of immunodominant epitopes is well established and has been further advanced by approaches from the fields of proteomics and genomics (Chakravarti et al. 2000). However, the major bottleneck in the development of new and effective vaccines is the delivery of antigens to cellular components of the immune system that initiate protective antiviral immunity. The unmatched capacity of DCs to sample antigen at sites of pathogen entry, transport pathogens and their immunogenic components to secondary lymphoid organs, and initiate activation of T cells makes them the ideal target cell for antimicrobial vaccines.

1.1 DC Functions and Lineages

DCs show a high degree of phenotypical and functional heterogeneity. Different DC lineages have been described that can be found at various anatomical locations and show a distinct set of surface molecules. CD8α-negative, myeloid DCs are found in peripheral nonlymphoid tissues and in secondary lymphoid organs including Langerhans cells of the skin, interstitial DCs of various organs, and marginal zone DCs in the spleen. DCs expressing the CD8α molecule are called lymphoid DCs and are exclusively found in the T cell areas of secondary lymphoid organs and in the thymus (Vremec and Shortman 1997). Whereas the origin, development, and function of myeloid DCs are well documented, both developmental pathways and functions of lymphoid DCs are still a matter of debate.

The key feature in the development of myeloid DCs is the functional switch from the immature to the mature stage. Immature DCs residing in peripheral nonlymphoid organs, and in certain compartments of secondary lymphoid organs, take up antigen via macropinocytosis and endocytosis (Reis et al. 1993; Sallusto et al. 1995). The low expression of costimulatory molecules (Romani et al. 1989; Sallusto and Lanzavecchia 1994) during this stage is accompanied by intracytoplasmic storage of high amounts of MHC class II antigens (Cella et al. 1997; Pierre et al. 1997). DC maturation is characterized by distinct changes in the cytoskeleton, upregulation of costimulatory molecules, display of abundant MHC class II molecules on the cell surface, and secretion of cytokines such as interleukin 12. Furthermore, immature myeloid DCs respond to distinct chemoattractant molecules expressed at sites of inflammation, whereas mature DCs acquire the ability to migrate in response to chemokines expressed in secondary lymphoid organs (Dieu et al. 1998; Forster et al. 1999; Sallusto et al. 1998). Thus myeloid DCs undergo a profound functional switch during their translocation from peripheral nonlymphoid tissues to the T cell areas of secondary lymphoid organs. The special functional properties of myeloid DCs facilitate antigen transport from the periphery to lymphoid organs and, therefore, efficient monitoring of the antigenic environment by the immune system.

It has been suggested that the phenotypical differences in different DC subsets, i.e., lymphoid vs. myeloid, is matched by distinct functional properties and that lymphoid DCs in the periphery induce deletion of potentially self-reactive T cells (de St Groth 1998), e.g., via Fas/FasL-mediated apoptosis during DC-T cell-interactions (Suss and Shortman 1996). However, this hypothesis is challenged by recent reports describing induction of vigorous T cell responses after adoptive transfer of antigen-pulsed lymphoid DCs into naive mice (Maldonado-Lopez et al. 1999; Ruedl and Bachmann 1999). In addition, it may be that CD8α expression by DCs is – like many other surface molecules – developmentally regulated (Merad et al. 2000) and distinguishes the maturation stage rather than a distinct lineage. It appears therefore, that the distinct functions of DCs are closely linked to their developmental stage. For the development of efficient vaccine strategies it is thus important to achieve antigen display on DCs in the optimal maturational stage.

2 Role of DCs in the Initiation and Maintenance of Antiviral Immunity

The excellent capacity of DCs to prime antiviral T cell responses can be readily shown in vitro, i.e., a few DCs can activate large numbers of virus-specific T cells in a mixed lymphocyte culture (Macatonia et al. 1989; Nonacs et al. 1992). The superior stimulatory effect of DCs in vitro can be partly attributed to quantitative aspects of antigen sampling and display: Expression of MHC class I and II products, MHC-peptide complexes, and costimulatory molecules on DCs is 10 – 100 times higher than, for example, on macrophages (Thomas et al. 1993; Inaba et al. 1997). An important prerequisite for the initiation of immune responses in vivo is the translocation of antigens from peripheral sites into secondary lymphoid organs (Zinkernagel et al. 1997). DCs are well equipped to sample antigens from the fluid phase (Sallusto et al. 1995) or via phagocytosis (Reis et al. 1993) and to process these antigens for subsequent presentation in lymphoid tissues. Alternatively, DCs may be infected with viruses in peripheral sites favoring rapid spread and induction of antiviral immune responses in local lymphoid organs.

Susceptibility of DCs to infections with HIV (Knight et al. 1997), influenza virus (Bender et al. 1998), measles virus (Grosjean et al. 1997), and several other viruses (Klagge and Schneider-Schaulies 1999) as been shown. It appears, therefore, that DCs are among the first cells infected after pathogen encounter at peripheral sites. Importantly, virus infection and encounter with other microbial agents (Reis et al. 1999) promote transition of DCs from the immature to the mature stage. Such stimulation can bypass the requirements for $CD4^+$ T cell help during activation of $CD8^+$ T cells (Ridge et al. 1998). Together, early virus infection of DCs triggering activation and migration of DCs to secondary lymphoid organs is most likely a key step in the successful initiation of antiviral T cell responses.

2.1 Induction of Protective Antiviral T Cell Responses by DC Vaccination

Studies in mice facilitate the exact determination of qualitative and quantitative changes in the T cell population during virus infections. The ability of DCs to stimulate primary or secondary antiviral CTL responses in vitro has been shown for herpes simplex virus (Hengel et al. 1987), Sendai virus (Kast et al. 1988), or influenza virus (Nonacs et al. 1992). The high potency of DCs in vivo to induce protective antiviral immunity against a noncytopathic virus has been shown in a study where only 100–1,000 DCs presenting the immunodominant epitope of the lymphocytic choriomeningitis virus (LCMV) were sufficient to elicit a minimal CTL response after direct injection into the spleen (Ludewig et al. 1998). DC-induced CTL responses developed rapidly, i.e., in only 2–4 days and were still detectable after 2 months. DC-immunized mice were protected against acute systemic and peripheral LMCV challenge and cleared LCMV-GP recombinant Vaccinia virus from the ovaries (Ludewig et al. 1998).

Adoptive transfer of DCs pulsed with viral antigens may also protect against cytopathic viruses. Prophylactic vaccination with inactivated equine herpesvirus-1 pulsed DCs protected mice from challenge with lethal doses of the virus (Steinbach et al. 1998). Priming with DCs exogenously loaded with herpesvirus proteins induces rather weak antibody responses but strong T helper (Th) cell response that most likely mediate protection against subsequent lethal challenge via the IFN-γ effector pathway (Schon et al. 2001). Thus, once antigen is displayed appropriately by DC, both protective antiviral Th cells and CTL can be generated in vivo.

It is important to note that DC-induced T cell responses do not confer life-long protection against viral infection. For example, immunological CTL memory against LCMV, induced by DCs, protected mice for prolonged periods of time against a systemic infection that leads to rapid reactivation of T cells in lymphoid tissues and subsequent efficient clearance of the virus (Ludewig et al. 1999). However, protection against peripheral LCMV infection rapidly disappeared after DC priming (Ludewig et al. 1999). These findings support the notion that antiviral memory immune responses against infections of peripheral tissues need reactivation and therefore require reintroduction of antigen into secondary lymphoid organs (Kundig et al. 1996; Lauvau et al. 2001). Interestingly, such restimulation was best achieved when the antigen was delivered by DCs (Ludewig et al. 1999). Thus DCs mediate initiation and maintenance of protective T cell responses against systemic and peripheral virus infections.

2.2 Contribution of DCs to the Induction of Protective Antiviral B Cell Responses

The rapid induction of neutralizing antiviral B cell responses is crucial for protection against infection with cytopathic viruses. The surface structures of cytopathic viruses are often highly organized in a paracrystalline fashion facilitating the induction of T cell-independent IgM responses. Isotype switching to IgG, however, is largely dependent on cognate interaction between antigen-specific B and $CD4^+$ Th cells (Zinkernagel et al. 2001).

Immature DCs residing in the periphery, e.g., Langerhans cells in the epidermis or mucous membranes of the respiratory tract, form a network of sentinel cells. Cutaneously applied antigens are acquired by dendritic cells, transported to regional lymphoid organs for presentation to T cells (Macatonia et al. 1987; Kripke et al. 1990). Similarly, pathogens such as *Leishmania major* or HIV, are specifically routed to lymphoid tissues by DCs (Moll 1993; Spira et al. 1996). Such transport is probably highly efficient as DCs migrate rapidly from the periphery to local lymph nodes, and their turnover may be further enhanced by local inflammatory reactions (DeSmedt et al. 1996). In addition, interferons produced by DCs on viral infection (Ghanekar et al. 1996; Cella et al. 1999) might help to induce resistance of DCs to cytopathic effects of viral infections. Interferon-competent DCs, but not DCs lacking α- and γ-interferon receptors, are able to contain infection with the cytopathic vesicular stomatitis virus so that only a few infectious virus particles are released in local draining lymph nodes (Ludewig et al. 2000b). The rapid transport and release of live virus particles by DCs facilitates induction of efficient B cell responses, characterized by rapid production of IgM, class switch to IgG, and formation of germinal centers. In addition, DC-induced virus-specific Th cell responses were required for the induction of long-lasting B cell memory responses against Th-dependent antigens (Ludewig et al. 2000b). Thus DCs most likely exert two important functions during infections with cytopathic viruses: first, the efficient control of the viral cytopathicity via the interferon system facilitating translocation of the virus to lymphoid organs via DCs and, second, low-level replication of the pathogen in DCs, allowing production of virus particles for direct activation of B cells and offering MHC class II-associated peptides for induction of Th cells.

2.3
Therapeutic Antiviral Vaccination with DCs

Prophylactic immunity against viral and bacterial (Mbow et al. 1997; Su et al. 1998; Demangel et al. 1999) infections can be induced by adoptive transfer of DCs presenting appropriate antigens. Although individualized adoptive DC immunotherapy is feasible and efficient in humans (Dhodapkar et al. 1999), it is clearly not suited for the use in general prophylactic vaccination campaigns. Nevertheless, persistent viral and bacterial infections that are associated with a substantial immunopathology such as HIV, HBV, HCV, and *Mycobacterium tuberculosis* infection, may be potential applications for individualized DC-based treatment strategies. For example, DCs presenting HBV antigen can induce anti-HBV immune responses, even in a situation where HBV-specific T cells are partially tolerant (Shimizu et al. 1998). Moreover, HBV antigen-loaded DCs induced strong primary CTL and Th cell responses in "humanized" Balb/c mice (Bocher et al. 2001). A further important application of DC-based therapeutic antiviral immunization is the protection against or treatment of human papillomavirus (HPV)-associated cervical cancers. Adoptive transfer of DCs pulsed with HPV peptide (Ossevoort et al. 1995) or protein (De Bruijn et al. 1998) induced HPV-specific CTL responses and protected mice from HPV-induced tumors. Furthermore, DCs were found to present HPV-derived peptides after immunization of mice with HPV protein (De Bruijn et al. 1998), indicating that DCs are indeed involved in the activation of anti-HPV T cells in vivo. These and other promising results from animal studies have encouraged clinical studies in humans to treat virus-induced tumors with DC-based immunization strategies (Santin et al. 1999; Adams et al. 2001).

3 Targeting DCs for Antiviral Vaccination

As outlined above, the unique properties of DCs in activating naive T cells in lymphoid tissues are probably of key importance for the initiation of adaptive immune responses against infectious agents. DCs are therefore the crucial target cell population for vaccines. For the discussion of DC-based vaccination strategies one should distinguish between (a) in vitro manipulation of DCs for adoptive immunotherapeutical approaches and (b) targeting of vaccines to DCs in vivo for prophylactic immunization and – for some diseases – immunotherapy of ongoing/chronic disease.

3.1 Manipulating DCs In Vitro

The seminal findings that DCs can be differentiated in vitro from their progenitors found in blood (Caux et al. 1992; Romani et al. 1994) or bone marrow (Inaba et al. 1993a) formed the basis for the development of DC-based vaccination strategies. Because large numbers of highly pure DCs can be obtained with various protocols, specific DC targeting is not needed for antigen delivery. The aim of in vitro manipulation of DCs for immunotherapeutical purposes should be to obtain loading of both MHC class I and II molecules. Antigen display on MHC class II molecules can be achieved by incubation of immature DCs with native or particulate antigen (Inaba et al. 1993b). Although exogenously loaded proteins are optimally processed by immature DCs for the loading of MHC class II molecules, successful exogenous loading of DCs with MHC class I-binding peptides can only be achieved if the peptides bind to the respective MHC class I haplotype. Although narrowing the applicability of MHC class I-binding peptides for adoptive DC-based immunotherapy, this approach has been already used in many clinical studies for treatment of cancer patients. It is important to note that immature DCs, although optimally equipped for antigen processing, are only weakly immunogenic. Therefore, the appropriate maturation status should be induced before adoptive transfer.

New technologies have been developed to further improve in vitro delivery of antigens to DCs. Fusogenic proteins can be used to efficiently deliver antigens to the cytosol of DCs, leading to presentation of exogenous proteins via the MHC class I pathway. Laus and colleagues recently described a strategy facilitating antigen penetration into the cytosol of APC with charge-modified peptides leading to enhanced antigen uptake and improved MHC class I-dependent antigen presentation to CTL (Laus et al. 2000). Likewise, proteins can be delivered to the cytoplasm of DCs by using the antennapedia homeodomain as transfer vehicle (Chikh et al. 2001). Furthermore, proteins and peptides can be delivered to DCs via DNA or RNA transfection, avoiding the need for large-scale production of proteins or peptides (Gilboa et al. 1998; Rea et al. 2001b).

3.2 Targeting DCs In Vivo

Techniques facilitating antigen delivery to DCs in vivo are certainly advantageous for the production of improved prophylactic and therapeutic vaccines. DC targeting can be achieved by using antibodies or ligands binding specifically to DC surface structures. For example, the immunogenicity of antigens coupled to anti-DEC-205 (Hawiger et al. 2001) or anti-CD11c (Wang et al. 2000) is strongly increased. DC-specific vaccines should, in addition, mobilize DCs and induce maturation to the highly immunostimulatory phenotype (Pulendran et al. 2001).

DNA vaccines have been shown the generate both humoral and broad cellular (CTL and Th cells) immune responses. The immunogenicity of DNA vaccines is mediated by bone marrow-derived professional APC, particularly DCs (Condon et al. 1996; Akbari et al. 1999). Importantly, DNA vaccines are able to activate and mobilize DCs via immunostimulatory CpG oligonucleotide sequences that can be found in many plasmid vectors (Sato et al. 1996). DC targeting in vivo can be enhanced by using DC-specific genetic regulatory elements such as the CD11c (Brocker et al. 1997) or the Dectin-2 (Bonkobara et al. 2001) promoter. Delivery of mRNA to the cytoplasm of DCs circumvents problems associated with DNA transfection such as lysosomal degradation of the DNA or insufficient transcription in certain cell types. It has been shown that transfer of mRNA to DCs in vitro is highly efficient (Nair et al. 1998); however, the in vivo application of this potentially powerful immunization method is limited by the very short half-life of RNA in bodily fluids containing RNA-digesting enzymes.

Bacterial and viral vectors facilitate efficient DNA transfection in vivo. For example, oral application of attenuated *Salmonella* leads to transfection of mucosal cells (Levine et al. 1996). Bacterial vectors potentially activate DCs (Reis et al. 1999); however, whether these vectors can be successfully modified to specifically target DCs remains to be established. In contrast, viral vectors can be adapted to target DCs, for example, via specific binding to the CD40 molecule (Tillman et al. 2000) or by selection for high-affinity binding to viral receptors present on DCs (Rea et al. 2001a). In addition, viral vectors may be modified to activate DCs (Tillman et al. 2000) and are therefore well suited for antigen delivery into DCs in vivo.

Viral and bacterial vector systems, however, still bear a risk of potential pathogenicity, particularly in immunodeficient hosts. Nonreplicating virus-like particles (VLPs) are nonenveloped icosahedral particles, composed of viral capsid proteins, that form via self-assembly in the cell nucleus after high-level expression of the respective viral capsid proteins. VLPs resemble capsids from infectious virions morphologically and antigenetically, while lacking encapsidated viral genes. Preclinical animal studies indicate that VLPs can induce both strong humoral and cellular immune responses (Palker et al. 2001). Because DCs can be activated by VLPs (Lenz et al. 2001) and may be specifically modified to enhance DC tropism in vivo, they represent a safe and versatile means for targeting antigens to DCs in vivo.

Other noninfectious delivery systems such as microspheres (Raychaudhuri and Rock 1998) or liposomes can be utilized to deliver peptides, proteins, or nucleic acids to DCs in vivo. In a recent study, the immunogenicity, in vivo distribution, and bioavailability for DCs of a liposomal formulation of antigenic peptides derived from the LCMV glycoprotein was assessed (Ludewig et al. 2000a). Localized in vivo antigen loading of DCs with MHC class I-associated peptides was achieved by intradermal application of the peptide vaccine, leading to induction of protective antiviral CTL responses. The immunogenicity of the vaccine was further enhanced by incorporation of immunostimulatory oligonucleotides mediating activation of DCs (Ludewig et al. 2000a). This and other reports (Nair et al. 1993; Chikh et al. 2001) indicate that liposomal antigen delivery to DCs in vivo may be an easily applicable and, after surface modification to enhance DC tropism, highly efficient approach to vaccinate against viral infections.

4
Concluding Remarks

The primary aim in antimicrobial vaccine development is the generation of a highly immunogenic, inexpensive to produce, and safe off-the-shelf-reagent. Furthermore, therapeutic antiviral and antibacterial vaccines must not interfere with pre-existing pathogen-specific immune responses and should stimulate the host response leading to elimination or – at least – long-term control of the pathogen. The "pole position" of DCs in the initiation of antimicrobial immune responses makes them an important target cell in vaccine development. Several efficient techniques have been established for antigen delivery to DCs in vitro, facilitating individualized adoptive immunotherapy against cancer and chronic infectious diseases. Some of these versatile approaches facilitating specific DC targeting in vivo may lead to the development of widely applicable immunization tools.

Acknowledgements. I thank Betsy Metters for critical reading of the manuscript and Hans Hengartner and Rolf Zinkernagel for discussions and continued support. This work was supported by the Swiss National Science Foundation and the Kanton Zürich.

References

Adams M, Borysiewicz L, Fiander A, Man S, Jasani B, Navabi H, Lipetz C, Evans AS, and Mason M (2001) Clinical studies of human papilloma vaccines in pre-invasive and invasive cancer. Vaccine 19:2549–2556

Akbari O, Panjwani N, Garcia S, Tascon R, Lowrie D, and Stockinger B (1999) DNA vaccination: transfection and activation of dendritic cells as key events for immunity. J.Exp.Med. 189:169–178

Bender A, Albert M, Reddy A, Feldman M, Sauter B, Kaplan G, Hellman W, and Bhardwaj N (1998) The distinctive features of influenza virus infection of dendritic cells. Immunobiology 198:552–567

Bocher WO, Dekel B, Schwerin W, Geissler M, Hoffmann S, Rohwer A, Arditti F, Cooper A, Bernhard H, Berrebi A, Rose-John S, Shaul Y, Galle PR, Lohr HF, and Reisner Y (2001) Induction of strong hepatitis B virus (HBV) specific T helper cell and cytotoxic T lymphocyte responses by therapeutic vaccination in the trimera mouse model of chronic HBV infection. Eur.J.Immunol. 31:2071–2079

Bonkobara M, Zukas PK, Shikano S, Nakamura S, Cruz PDJ, and Ariizumi K (2001) Epidermal langerhans cell-targeted gene expression by a Dectin-2 promoter. J.Immunol. 167:6893–6900

Brocker T, Riedinger M, and Karjalainen K (1997) Targeted expression of major histocompatibility complex (MHC) class II molecules demonstrates that dendritic cells can induce negative but not positive selection of thymocytes in vivo. J.Exp.Med. 185:541–550

Caux C, Dezutter-Dambuyant C, Schmitt D, and Banchereau J (1992) GM-CSF and TNF-alpha cooperate in the generation of dendritic Langerhans cells. Nature 360:258–261

Cella M, Engering A, Pinet V, Pieters J, and Lanzavecchia A (1997) Inflammatory stimuli induce accumulation of MHC class II complexes on dendritic cells. Nature 388:782–787

Cella M, Jarrossay D, Facchetti F, Alebardi O, Nakajima H, Lanzavecchia A, and Colonna M (1999) Plasmacytoid monocytes migrate to inflamed lymph nodes and produce large amounts of type I interferon. Nat.Med. 5:919–923

Chakravarti DN, Fiske MJ, Fletcher LD, and Zagursky RJ (2000) Application of genomics and proteomics for identification of bacterial gene products as potential vaccine candidates. Vaccine 19:601–612

Chikh GG, Kong S, Bally MB, Meunier JC, and Schutze-Redelmeier MP (2001) Efficient delivery of antennapedia homeodomain fused to CTL epitope with liposomes into dendritic cells results in the activation of CD8+ T cells. J.Immunol. 167:6462–6470

Condon C, Watkins SC, Celluzi CM, Thompson K, and Falo LD (1996) DNA-based immunization by in vivo transfection of dendritic cells. Nat.Med. 2:1122–1128

De Bruijn ML, Schuurhuis DH, Vierboom MP, Vermeulen H, de Cock KA, Ooms ME, Ressing ME, Toebes M, Franken KL, Drijfhout JW, Ottenhoff TH, Offringa R, and Melief CJ (1998) Immunization with human papillomavirus type 16 (HPV16) oncoprotein-loaded dendritic cells as well as protein in adjuvant induces MHC class I-restricted protection to HPV16-induced tumor cells. Cancer Res. 58:724–731

de St Groth BF (1998) The evolution of self-tolerance: a new cell arises to meet the challenge of self-reactivity. Immunol.Today 19:448–454

Demangel C, Bean AG, Martin E, Feng CG, Kamath AT, and Britton WJ (1999) Protection against aerosol *Mycobacterium tuberculosis* infection using *Mycobacterium bovis* Bacillus Calmette Guerin-infected dendritic cells. Eur.J.Immunol. 29:1972–1979

DeSmedt T, Pajak B, Muraille E, Lespagnard L, Heinen E, DeBaetselier P, Urbain J, Leo O, and Moser M (1996) Regulation of dendritic cell numbers and maturation by lipopolysaccharide in vivo. J.Exp.Med. 184:1413–1424

Dhodapkar MV, Steinman RM, Sapp M, Desai H, Fossella C, Krasovsky J, Donahoe SM, Dunbar PR, Cerundolo V, Nixon DF, and Bhardwaj N (1999) Rapid generation of broad T-cell immunity in humans after a single injection of mature dendritic cells. J.Clin.Invest. 104:173–180

Dieu MC, Vanbervliet B, Vicari A, Bridon JM, Oldham E, Ait-Yahia S, Briere F, Zlotnik A, Lebecque S, and Caux C (1998) Selective recruitment of immature and mature dendritic cells by distinct chemokines expressed in different anatomic sites. J.Exp.Med. 188:373–386

Forster R, Schubel A, Breitfeld D, Kremmer E, Renner-Muller I, Wolf E, and Lipp M (1999) CCR7 coordinates the primary immune response by establishing functional microenvironments in secondary lymphoid organs. Cell 99:23–33

Ghanekar S, Zheng L, Logar A, Navratil J, Borowski L, Gupta P, and Rinaldo C (1996) Cytokine expression by human peripheral blood dendritic cells stimulated in vitro with HIV-1 and herpes simplex virus. J.Immunol. 157:4028–4036

Gilboa E, Nair SK, and Lyerly HK (1998) Immunotherapy of cancer with dendritic-cell-based vaccines. Cancer Immunol.Immunother. 46:82–87

Grosjean I, Caux C, Bella C, Berger I, Wild F, Banchereau J, and Kaiserlian D (1997) Measles virus infects human dendritic cells and blocks their allostimulatory properties for CD4+ T cells. J.Exp.Med. 186:801–812

Hawiger D, Inaba K, Dorsett Y, Guo M, Mahnke K, Rivera M, Ravetch JV, Steinman RM, and Nussenzweig MC (2001) Dendritic cells induce peripheral T cell unresponsiveness under steady state conditions in vivo. J.Exp.Med. 194:769–779

Hengel H, Lindner M, Wagner H, and Heeg K (1987) Frequency of herpes simplex virus-specific murine cytotoxic T lymphocyte precursors in mitogen- and antigen-driven primary in vitro T cell responses. J.Immunol. 139:4196–4202

Inaba K, Inaba M, Deguchi M, Hagi K, Yasumizu R, Ikehara S, Muramatsu S, and Steinman RM (1993a) Granulocytes, macrophages, and dendritic cells arise from a common major histocompatibility complex class II-negative progenitor in mouse bone marrow. Proc.Natl.Acad.Sci.U.S.A. 90:3038–3042

Inaba K, Inaba M, Naito M, and Steinman RM (1993b) Dendritic cell progenitors phagocytose particulates, including bacillus Calmette-Guerin organisms, and sensitize mice to mycobacterial antigens in vivo. J.Exp.Med. 178:479–488

Inaba K, Pack M, Inaba M, Sakuta H, Isdell F, and Steinman RM (1997) High levels of a major histocompatibility complex II-self peptide complex on dendritic cells from the T cell areas of lymph nodes. J.Exp.Med. 186:665–672

Kast WM, Boog CJ, Roep BO, Voordouw AC, and Melief CJ (1988) Failure or success in the restoration of virus-specific cytotoxic T lymphocyte response defects by dendritic cells. J.Immunol. 140:3186–3193

Klagge IM and Schneider-Schaulies S (1999) Virus interactions with dendritic cells. J.Gen.Virol. 80:823–833

Knight SC, Elsley W, and Wang H (1997) Mechanisms of loss of functional dendritic cells in HIV-1 infection. J.Leukoc.Biol. 62:78–81

Kripke ML, Munn CG, Jeevan A, Tang JM, and Bucana C (1990) Evidence that cutaneous antigen-presenting cells migrate to regional lymph nodes during contact sensitization. J.Immunol. 145:2833–2838

Kundig TM, Bachmann MF, Oehen S, Hoffmann UW, Simard JJL, Kalberer CP, Pircher H, Ohashi PS, Hengartner H, and Zinkernagel RM (1996) On the role of antigen in maintaining cytotoxic T-cell memory. Proc.Natl.Acad.Sci.U.S.A. 93:9716–9723

Laus R, Graddis TJ, Hakim I, and Vidovic D (2000) Enhanced major histocompatibility complex class I-dependent presentation of antigens modified with cationic and fusogenic peptides. Nat.Biotechnol. 18:1269–1272

Lauvau G, Vijh S, Kong P, Horng T, Kerksiek K, Serbina N, Tuma RA, and Pamer EG (2001) Priming of memory but not effector CD8 T cells by a killed bacterial vaccine. Science 294:1735–1739

Lenz P, Day PM, Pang YY, Frye SA, Jensen PN, Lowy DR, and Schiller JT (2001) Papillomavirus-like particles induce acute activation of dendritic cells. J.Immunol. 166:5346–5355

Levine MM, Galen J, Barry E, Noriega F, Chatfield S, Sztein M, Dougan G, and Tacket C (1996) Attenuated *Salmonella* as live oral vaccines against typhoid fever and as live vectors. J.Biotechnol. 44:193–196

Ludewig B, Barchiesi F, Pericin M, Zinkernagel RM, Hengartner H, and Schwendener RA (2000a) In vivo antigen loading and activation of dendritic cells via a liposomal peptide vaccine mediates protective antiviral and anti-tumour immunity. Vaccine 2000 19:23–32

Ludewig B, Ehl S, Karrer U, Odermatt B, Hengartner H, and Zinkernagel RM (1998) Dendritic cells efficiently induce protective antiviral immunity. J.Virol. 72:3812–3818

Ludewig B, Maloy KJ, Lopez-Macias C, Odermatt B, Hengartner H, and Zinkernagel RM (2000b) Induction of optimal anti-viral neutralizing B cell responses by dendritic cells requires transport and release of virus particles in secondary lymphoid organs. Eur.J.Immunol.2000 30:185–196

Ludewig B, Oehen S, Barchiesi F, Schwendener RA, Hengartner H, and Zinkernagel RM (1999) Protective antiviral cytotoxic T cell memory is most efficiently maintained by restimulation via dendritic cells. J.Immunol. 163:1839–1844

Macatonia SE, Knight SC, Edwards AJ, Griffiths S, and Fryer P (1987) Localization of antigen on lymph node dendritic cells after exposure to the contact sensitizer fluorescein isothiocyanate. Functional and morphological studies. J.Exp.Med. 166:1654–1667

Macatonia SE, Taylor PM, Knight SC, and Askonas BA (1989) Primary stimulation by dendritic cells induces antiviral proliferative and cytotoxic T cell responses in vitro. J.Exp.Med. 169:1255–1264

Maldonado-Lopez R, De Smedt T, Pajak B, Heirman C, Thielemans K, Leo O, Urbain J, Maliszewski CR, and Moser M (1999) Role of CD8alpha+ and CD8alpha-dendritic cells in the induction of primary immune responses in vivo. J.Leukoc.Biol. 66:242–246

Mbow ML, Zeidner N, Panella N, Titus RG, Piesman J, and Titus (1997) *Borrelia burgdorferi*-pulsed dendritic cells induce a protective immune response against tick-transmitted spirochetes. Infect.Immun. 65:3386–3390

Merad M, Fong L, Bogenberger J, and Engleman EG (2000) Differentiation of myeloid dendritic cells into CD8alpha-positive dendritic cells in vivo. Blood 96:1865–1872

Moll H (1993) Epidermal Langerhans cells are critical for immunoregulation of cutaneous leishmaniasis. Immunol.Today 14:383–387

Nair S, Babu JS, Dunham RG, Kanda P, Burke RL, and Rouse BT (1993) Induction of primary, antiviral cytotoxic, and proliferative responses with antigens administered via dendritic cells. J.Virol. 67:4062–4069

Nair SK, Boczkowski D, Morse M, Cumming RI, Lyerly HK, and Gilboa E (1998) Induction of primary carcinoembryonic antigen (CEA)-specific cytotoxic T lymphocytes in vitro using human dendritic cells transfected with RNA. Nat.Biotechnol. 16:364–369

Nonacs R, Humborg C, Tam JP, and Steinman RM (1992) Mechanisms of mouse spleen dendritic cell function in the generation of influenza-specific, cytolytic T lymphocytes. J.Exp.Med. 176:519–529

Ossevoort MA, Feltkamp MC, van Veen KJ, Melief CJ, and Kast WM (1995) Dendritic cells as carriers for a cytotoxic T-lymphocyte epitope-based peptide vaccine in

protection against a human papillomavirus type 16-induced tumor. J.Immunother.Emphasis.Tumor Immunol. 18:86–94
Palker TJ, Monteiro JM, Martin MM, Kakareka C, Smith JF, Cook JC, Joyce JG, and Jansen KU (2001) Antibody, cytokine and cytotoxic T lymphocyte responses in chimpanzees immunized with human papillomavirus virus-like particles. Vaccine 19:3733–3743
Pierre P, Turley SJ, Gatti E, Hull M, Meltzer J, Mirza A, Inaba K, Steinman RM, and Mellman I (1997) Developmental regulation of MHC class II transport in mouse dendritic cells [see comments]. Nature 388:787–792
Pulendran B, Banchereau J, Maraskovsky E, and Maliszewski C (2001) Modulating the immune response with dendritic cells and their growth factors. Trends.Immunol. 22:41–47
Raychaudhuri S and Rock KL (1998) Fully mobilizing host defense: building better vaccines. Nat.Biotechnol. 16:1025–1031
Rea D, Havenga MJ, van Den Assem M, Sutmuller RP, Lemckert A, Hoeben RC, Bout A, Melief CJ, and Offringa R (2001a) Highly efficient transduction of human monocyte-derived dendritic cells with subgroup B fiber-modified adenovirus vectors enhances transgene-encoded antigen presentation to cytotoxic T cells. J.Immunol. 166:5236–5244
Rea D, Johnson ME, Havenga MJ, Melief CJ, and Offringa R (2001b) Strategies for improved antigen delivery into dendritic cells. Trends.Mol.Med. 7:91–94
Reis, Sher A, and Kaye P (1999) The role of dendritic cells in the induction and regulation of immunity to microbial infection. Curr.Opin.Immunol. 11:392–399
Reis S-C, Stahl PD, and Austyn JM (1993) Phagocytosis of antigens by Langerhans cells in vitro. J.Exp.Med. 178:509–519
Ridge JP, Di Rosa F, and Matzinger P (1998) A conditioned dendritic cell can be a temporal bridge between a CD4+ T-helper and a T-killer cell. Nature 393:474–478
Romani N, Gruner S, Brang D, Kampgen E, Lenz A, Trockenbacher B, Konwalinka G, Fritsch P, Steinman RM, and Schuler G (1994) Proliferating dendritic cell progenitors in human blood. J.Exp.Med. 180:83–93
Romani N, Lenz A, Glassel H, Stoessel H, Stanzl U, Majdic O, Fritsch P, and Schuler G (1989) Cultured human Langerhans cells resemble lymphoid dendritic cells in phenotype and function. J Invest Dermatol 93:600–609
Ruedl C and Bachmann MF (1999) CTL priming by CD8(+) and CD8(-) dendritic cells in vivo. Eur.J.Immunol. 29:3762–3767
Sallusto F, Cella M, Danieli C, and Lanzavecchia A (1995) Dendritic cells use macropinocytosis and the mannose receptor to concentrate macromolecules in the major histocompatibility complex class II compartment: downregulation by cytokines and bacterial products. J.Exp.Med. 182:389–400
Sallusto F and Lanzavecchia A (1994) Efficient presentation of soluble antigen by cultured human dendritic cells is maintained by granulocyte/macrophage colony-stimulating factor plus interleukin 4 and downregulated by tumor necrosis factor alpha. J.Exp.Med. 179:1109–1118
Sallusto F, Lanzavecchia A, and Mackay CR (1998) Chemokines and chemokine receptors in T-cell priming and Th1/Th2-mediated responses. Immunol.Today 19:568–574
Santin AD, Hermonat PL, Ravaggi A, Chiriva-Internati M, Zhan D, Pecorelli S, Parham GP, and Cannon MJ (1999) Induction of human papillomavirus-specific

CD4(+) and CD8(+) lymphocytes by E7-pulsed autologous dendritic cells in patients with human papillomavirus type 16- and 18-positive cervical cancer. J.Virol. 73:5402–5410

Sato Y, Roman M, Tighe H, Lee D, Corr M, Nguyen MD, Silverman GJ, Lotz M, Carson DA, and Raz E (1996) Immunostimulatory DNA sequences necessary for effective intradermal gene immunization. Science 273:352–354

Schon E, Harandi AM, Nordstrom I, Holmgren J, and Eriksson K (2001) Dendritic cell vaccination protects mice against lethality caused by genital herpes simplex virus type 2 infection. J.Reprod.Immunol. 50:87–104

Shimizu Y, Guidotti LG, Fowler P, and Chisari FV (1998) Dendritic cell immunization breaks cytotoxic T lymphocyte tolerance in hepatitis B virus transgenic mice. J.Immunol. 161:4520–4529

Spira AI, Marx PA, Patterson BK, Mahoney J, Koup RA, Wolinsky SM, and Ho DD (1996) Cellular targets of infection and route of viral dissemination after an intravaginal inoculation of simian immunodeficiency virus into rhesus macaques. J.Exp.Med. 183:215–225

Steinbach F, Borchers K, Ricciardi-Castagnoli P, Ludwig H, Stingl G, and Elbe-Burger A (1998) Dendritic cells presenting equine herpesvirus-1 antigens induce protective anti-viral immunity. J.Gen.Virol. 79:3005–3014

Su H, Messer R, Whitmire W, Fischer E, Portis JC, and Caldwell HD (1998) Vaccination against chlamydial genital tract infection after immunization with dendritic cells pulsed ex vivo with nonviable *Chlamydiae*. J.Exp.Med. 188:809–818

Suss G and Shortman K (1996) A subclass of dendritic cells kills CD4 T cells via Fas Fas-ligand-induced apoptosis. J.Exp.Med. 183:1789–1796

Thomas R, Davis LS, and Lipsky PE (1993) Comparative accessory cell function of human peripheral blood dendritic cells and monocytes. J.Immunol. 151:6840–6852

Tillman BW, Hayes TL, DeGruijl TD, Douglas JT, and Curiel DT (2000) Adenoviral vectors targeted to CD40 enhance the efficacy of dendritic cell-based vaccination against human papillomavirus 16-induced tumor cells in a murine model. Cancer Res. 60:5456–5463

Vremec D and Shortman K (1997) Dendritic cell subtypes in mouse lymphoid organs: cross-correlation of surface markers, changes with incubation, and differences among thymus, spleen, and lymph nodes. J.Immunol. 159:565–573

Wang H, Griffiths MN, Burton DR, and Ghazal P (2000) Rapid antibody responses by low-dose, single-step, dendritic cell-targeted immunization. Proc.Natl.Acad. Sci.U.S.A. 97:847–852

Zinkernagel RM, Ehl S, Aichele P, Oehen S, Kundig T, and Hengartner H (1997) Antigen localisation regulates immune responses in a dose- and time-dependent fashion: a geographical view of immune reactivity. Immunol.Rev. 156:199–209:199–209

Zinkernagel RM, Lamarre A, Ciurea A, Hunziker L, Ochsenbein AF, McCoy KD, Fehr T, Bachmann MF, Kalinke U, and Hengartner H (2001) Neutralizing antiviral antibody responses. Adv.Immunol. 79:1–53.:1–53

Chapter 10

Interplay Between Human Papillomaviruses and Dendritic Cells

R. Offringa, A. de Jong, R. E. M. Toes, S. H. van der Burg, C. J. M. Melief

Abstract. The design of the human papillomavirus (HPV) infection cycle is tightly fitted to the differentiation program of its natural host, the keratinocyte. This has important consequences for the role of antigen-presenting cells in the priming of antiviral immunity. The confinement of HPV infection to epithelia puts the epithelial dendritic cell, the Langerhans cell (LC), in charge of the induction of T cell-dependent immunity. Because HPV-infected keratinocytes cannot reach the regional lymphoid organs, and HPV-infection of LCs does not result in viral gene expression, priming of antiviral T cells exclusively depends on cross-presentation of viral antigens by the LC. Sensitization of the immune system in the regional lymphoid organs elicits systemic anti-HPV immunity as well as intraepithelial immune surveillance by memory-type intraepi-

R. Offringa (✉), A. de Jong, R.E.M. Toes, S.H. van der Burg, C.J.M. Melief
Tumor Immunology Group,
Department of Immunohematology and Blood Transfusion,
Leiden University Medical Center, Albinusdreef 2, 2333 ZA Leiden, The Netherlands

thelial T cells and locally produced antibodies. The high rate of spontaneous rejections of high-risk HPV-infections and HPV-positive premalignant lesions indicates that in general the LC-driven antigen presentation machinery is capable of raising an effective immune defense against HPV. Epidemiological studies also reveal that a decrease in the vigilance of the immune system is readily exploited by HPV to escape immune destruction, resulting in persistent infections and development of HPV-positive cancers. In view of the inherent antigenicity of HPV, immune intervention strategies constitute a promising approach for both the prevention and the therapeutic treatment of HPV-induced diseases. Importantly, the mechanisms that govern the induction and effector phases of the intraepithelial immune surveillance against HPV must be taken into account when designing such strategies.

1 Introduction

Human papillomaviruses (HPVs) have an important causative role in a variety of epithelial disorders in the skin, respiratory tract, and anogenital mucosa. Although the majority of HPV types induce relatively harmless skin warts, other types are associated with epithelial tumors. Of the HPV types capable of inducing neoplasia, the so-called low-risk types are primarily found in anogenital warts and respiratory papillomas (e.g., HPV 6 and -11), whereas the high-risk types are found in non-melanoma skin cancers, head and neck cancers, and anogenital cancers (e.g., HPV 16 and -18). In particular, the association of high-risk HPV types with cervical cancer, the second most common cancer in women worldwide, has attracted the attention of tumor biologists and immunologists (Stoler 2000; zur Hausen 2000). Research on HPV therefore focuses on two major questions: (1) By which mechanism do HPVs cause cancers? and (2) How can virus-positive tumors arise in the face of the immune defense?

Crucial information with respect to these questions lies in the sophisticated design of the HPV infection cycle, which is closely tuned to the physiology of its epithelial target, the keratinocyte. Successful initiation of the HPV infection cycle requires the virus to reach the basal layers of keratinocytes, implicating that epithelial damage will facilitate HPV infection by allowing the virus to breach this multilayered physical barrier. Most cervical neoplastic lesions originate from the squamocolumnar junction. Because of its architecture, this connection between the columnar epithelium of the endocervix and the squamous epithelium of the ectocervix is particularly vulnerable to HPV infection (Fig. 1). Initiation of keratinocyte differentiation in the immediate suprabasal zone unleashes the full program of HPV gene expression, which starts with expression of the immediate-early gene products E6 and E7. The major function of these proteins is to prepare the keratinocyte for viral DNA replication by disconnecting the DNA synthesis machinery of this cell from cell cycle control. This process involves inactivation of key regulatory switches in cell cycle regulation and apoptosis, such as the p105Rb (E7) and p53 (E6) tumor suppressor proteins. Once the cellular DNA synthesis machinery is made available, other HPV early proteins, in particular E1 and E2, will ensure that this machinery is geared toward the replication of HPV DNA instead of cellular DNA. At least initially, productive HPV infection does not induce cell division, nor does it grossly disturb the process of keratinocyte differentiation that shapes stratified epithelia. In fact, the program of viral gene expression is tightly linked to the differentiation stage of the keratinocyte and develops in a step-by-step fashion while the keratinocyte moves up in the epithelium. During the final stage of the infection cycle, viral gene expression is dominated by synthesis of the structural components of the viral particle (L1 and L2). HPV infection does not induce cell destruction, but the new viral particles are simply released through wear and tear of the top layer of the epithelium (Fig. 1). It is clear that the minimal disturbance of the epithelial architecture and function during HPV infection increases the chance of this virus to evade immediate detection and elimination by the host immune defense.

HPV infection interrupts the checkpoints of cell cycle and DNA synthesis, and, therefore, infected cells are predisposed to accumulation of DNA damage and to increased rates of cellular proliferation. Aberrancies in the infection process can result in abortion of the productive infection cycle and in sustained expression of the immediate-early genes. For HPV, like other DNA tumor viruses, a hallmark of this event is the stable integration of viral DNA sequences into the cellular genome. Analysis of integrated viral DNA in HPV-positive cancers has shown that integration associated with transformation generally results in disruption of the E2 gene, while leaving E6 and E7 intact. The absence of functional E2 gene product, which otherwise keeps E6 and E7 gene expression in check, results in sustained and elevated levels of these oncoproteins. Consequent-

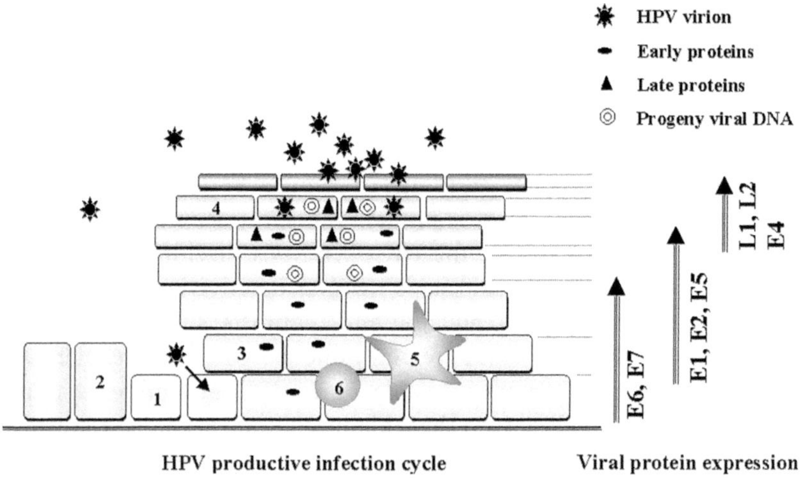

Fig. 1. Productive HPV infection at the cervical squamocolumnar junction. The "reserve, basal, or stem" cells (*1*) at the squamocolumnar junction are capable of differentiating along neuroendocrine, glandular (*2*), or squamous (*3*) lines and are responsible for maintenance of both the endo- (*2*) and exocervical (*3*) epithelia. It is conceivable that the architecture of this junction allows the virus to reach these basal cells more easily than those in other parts of the cervical epithelium. Furthermore, features associated with the relatively undifferentiated state of these cells, such as expression of virus receptor molecules and sensitivity to estrogen hormones, appear to contribute to their permissiveness to HPV infection (Elson et al. 2000; Payne et al. 2001). Commitment of infected stem cells to the squamous lineage allows full activation of the HPV gene expression program, which unfolds in a stepwise fashion in parallel to the keratinocyte differentiation path. Viral expression in the suprabasal layers is restricted to the early genes (*3*), whereas late gene expression dominates in the upper layers of keratinocytes (*4*). Newly produced virus is released from the top layers of the epithelium as part of the natural desquamification process. The Langerhans cell (*LC*) (*5*) is the key APC involved in induction of HPV-specific immunity. It can take up viral antigens derived from HPV-infected keratinocytes or from free viral particles. Innate immune effector cells such as the TCRγ/δ intraepithelial lymphocytes (*IEL*) (*6*) can increase the availability of viral antigens by killing HPV-infected keratinocytes. Furthermore, they can emit, together with the infected keratinocytes, inflammatory signals that will stimulate LC migration and maturation. This will enable the LCs to prime TCRα/β T cells in the regional lymphoid organs, resulting in the establishment of T-cell dependent cellular and humoral antiviral immunity

ly, cell cycle progression driven by the immediate-early genes will be geared toward replication of the cellular DNA and cell transformation. With respect to HPV infection of cervical epithelium, prolonged and elevated expression of the E6 and E7 oncoproteins sets the stage for a multistep transformation process starting with cervical intraepithelial neoplasia (CIN) and potentially evolving to cervical cancer. Because functional integration of viral DNA into the cellular genome is a rare event, viral persistence is critically required for this to happen.

2
Immune Surveillance Against HPV

Epidemiological studies have shown that the likelihood of persistent HPV infections and HPV-positive epithelial lesions is primarily determined by the frequency of viral encounters (Ho et al. 1998; Koutsky 1997) and by the time window that the immune defense allows for infectious virus to persist in the host. The common clearance of HPV infections and frequent spontaneous regression of low-grade CIN lesions indicate that in most healthy persons the immune system succeeds in eliminating the virus before malignant disease can develop (Evander et al. 1995; Koutsky 1997). However, in immunosuppressed subjects, such as transplant recipients and human immunodeficiency virus (HIV)-positive patients, the prevalence of HPV-positive anogenital and skin cancers is greatly increased, demonstrating that failure of immune surveillance can indeed result in HPV persistence and HPV-induced disease (Harwood et al. 2000; Palefsky 2000; Sasadeusz et al. 2001; Serraino et al. 1999). This latter point is further illustrated by the fact that downregulation of MHC class I and overexpression of antiapoptotic proteins that block the cytolytic impact of effector T cells have been frequently found in cervical carcinomas (Garrido et al. 1997; Medema et al. 2001). Direct evidence for immune surveillance against HPV is provided by the detection in HPV-infected patients of serum immunoglobulin (primarily IgG and IgA type), T helper (Th), and cytotoxic T lymphocyte (CTL) responses against various HPV early and late gene products. (reviewed in Konya and Dillner 2001; Man and Fiander 2001; Stern et al. 2000; Tjiong et al. 2001). Notably, in several cases these responses were found associated with progressive tumor development, implying that the immune reactivity observed was incapable of eliminating the HPV-positive lesions and, therefore, had come too late. In fact, in contrast to HPV infections and low-grade CIN lesions, more progressed cervical lesions are rarely cleared in the absence of surgical intervention, suggesting that the immune defense needs to kick in during the premalignant phase of HPV-infection to be effective against this virus.

The striking prevalence of HPV infection in the human population together with the frequent clearance of HPV infections imply that successful clearance of this virus, like that of many other viruses, must install virus-specific immunological memory. In accordance with this notion, healthy individuals frequently show systemic and/or local IgG and IgA-type antibody responses against HPV capsid proteins. Similarly, antibody memory has been observed against the HPV E2 antigen (Konya and Dillner 2001). Because antibody isotype switching largely depends on antigen-specific $CD4^+$ Th cell responses, these observations point toward the presence of underlying HPV-specific Th immunity. Indeed, we have recently found memory-type Th responses, so far against both HPV16 E2 and E6, in a large fraction of healthy subjects (de Jong et al. 2002; our additional unpublished observations). Further charting of HPV-specific T cell responses in healthy subjects versus patients with HPV-positive lesions will be required to determine key parameters that are decisive in the success or failure of HPV-specific immune surveillance.

3
APC Function During HPV Infection

At present, a major debate with respect to the priming of tumor-specific T cell immunity focuses on the question of whether this primarily involves direct priming by tumor cells that travel to the regional lymph nodes or cross-priming of tumor-derived antigens by professional antigen-presenting cells (APCs) (Melief et al. 2000; Ochsenbein et al. 2001; Pardoll 2001). The relative contribution of both processes is conceivably influenced by characteristics of the cell type and the tissue from which the tumor originates. In the case of HPV infection and HPV-positive premalignant lesions, it is highly unlikely that HPV-positive keratinocytes will be able to traverse the basement membrane and travel to the regional lymph nodes for exposure of their antigenic load to naive T cells. Consequently, induction of T cell immunity against HPV infections is exclusively dependent on professional APCs, in particular dendritic cells (DCs), which are specialized in transferring peripheral antigenic signals to the lymphoid organs. Because HPV infection is confined to the epidermis, the DC subset primarily responsible for this process is the epidermal DC or Langerhans cell (LC) (Nestle and Burg 1999). Importantly, the HPV infection cycle is specifically adapted to the keratinocyte differentiation program, rendering it extremely unlikely that HPV infection of LCs will result in the expression of HPV antigens by the LCs themselves. Together, the characteristics of the HPV infection process dictate that priming of T cell immunity against HPV must completely rely on the cross-presentation of HPV antigensby LCs. This implies that the failure or success of the HPV-specific immune defense will critically depend on the outcome of the interplay between HPV and LCs.

During subclinical manifestation of HPV infection no premature death of HPV-infected keratinocytes takes place and the architecture of the epithelium is only minimally disturbed. Furthermore, the availability of viral particles for uptake by DCs is limited because the virus is released at the top of the stratified epithelium. Consequently, the supply of antigen and of inflammatory signals required for DC activation and migration is very restricted, making it conceivable that the immune system can, at least for some time, remain ignorant of the virus. If the immune system has not yet encountered the virus, initial detection of HPV can, however, be enhanced through subtle aberrancies in the behavior of HPV-infected keratinocytes. Secretion of proinflammatory cytokines, including type I interferons (IFN), interleukin (IL)-1, transforming growth factor α (TGF-α), and macrophage inflammatory protein 3α (MIP-3α/CCL20), by keratinocytes has been reported under certain conditions (Dieu-Nosjean et al. 2000; Kondo 1999). This could enhance the recruitment and activation of LCs as well as the activation of innate immune effector cells, such as the TCR γ/δ^+ subset of intraepithelial lymphocytes (IELs) (Havran et al. 1991; Hayday et al. 2001; Figs. 1 and 2). Furthermore, interference of the E7 protein with class I MHC-restricted antigen presentation may make the cells more "visible" to natural killer cells (Evans et al. 2001; Vambutas et al. 2001). Finally, binding and uptake of HPV viruslike particles was recently shown to induce DC activation (Rudolf et al. 2001), suggesting that the virus itself may constitute a direct inflammatory signal for LCs.

Once immunological memory has been installed through previous encounter(s) with the infecting HPV type, or possibly other HPV types with similar antigenic determinants, local immune surveillance will be heightened through the presence of HPV-specific antibodies (Konya and

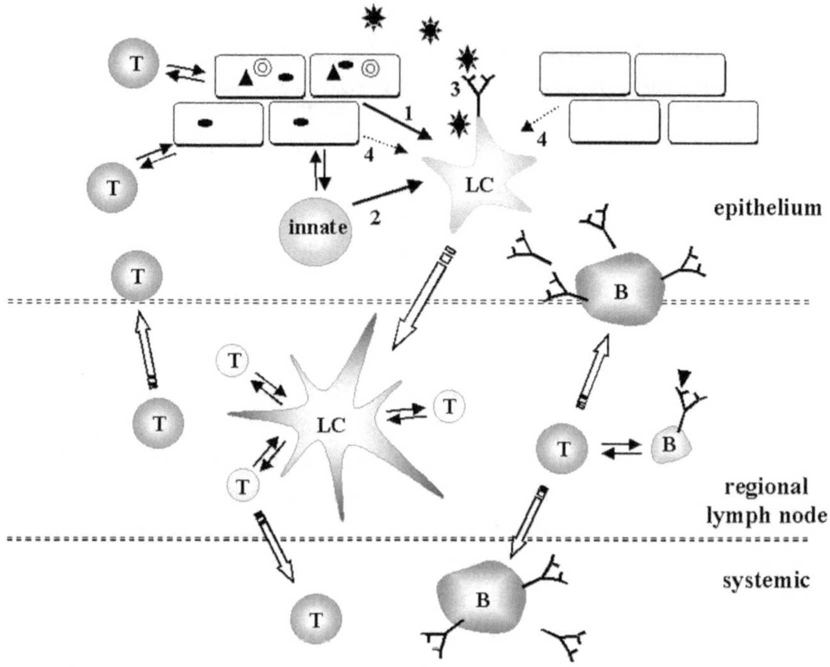

Fig. 2. Priming of the adaptive immune response during HPV infection. The Langerhans cell (*LC*) can acquire viral antigen through uptake of viral particles or (debris of) HPV-infected keratinocytes. Proinflammatory signals that can drive LC maturation are emitted by infected keratinocytes (*1*) and innate immune cells (*2*). Direct and antibody-dependent binding of virus particles to LC (*3*) also constitutes a maturation signal. Importantly, efficient LC migration as well as full maturation of LCs to immunostimulatory antigen-presenting cells requires these proinflammatory signals to dominate over the homeostatic signals (*4*) derived from (noninfected) keratinocytes. On arrival of fully matured LCs in the regional lymphoid organs, antigen-specific T and B cells are primed. This results in homing of activated TCRα/β $CD4^+$ and $CD8^+$ T cells as well as neutralizing antibody-producing B cells to the epithelia. After the infection has been cleared, local HPV-specific T and B cell memory will persist. Priming of the HPV-specific immune response also results in systemic cellular and humoral immunity

Dillner 2001) as well as by a second type of IEL, skin-homing effector-type memory TCRα/β^+ T cells (Masopust et al. 2001). The efficacy of anti-HPV antibodies is largely restricted to the L1 and L2 capsid antigens, involving neutralization of virus particles and uptake by DC. Elimination of virus-infected cells will require HPV-specific T cells. Although

T cells directed against late HPV antigens can definitely play a role in this process, these are only expressed at a time when the viral infection cycle is almost complete and the differentiating keratinocyte has moved up in the epithelium. Early intervention with the infection process and virus production is therefore likely to benefit from T cell immunity directed against (immediate) early antigens. From this perspective, the intraepithelial memory T cells are expected to play a major role in the efficient suppression of HPV infection. Keratinocytes keep these memory T cells in their vicinity through the secretion of the chemokine CCL27, which selectively attracts skin-homing memory T cells. Exposure of keratinocytes to proinflammatory stimuli increases their CCL27 secretion and can thereby result in the enhanced recruitment of skin-homing T cells to the epidermis (Morales et al. 1999; Reiss et al. 2001). However, presentation of cognate antigen by epithelial cells does not necessarily suffice for functional activity of memory T cells or even effector T cells. Experiments in transgenic mice that express ovalbumin (OVA) as an intestinal epithelium-specific antigen showed that adoptively transferred OVA-specific CTL initially expanded in the mucosal lymphoid tissues and homed to the OVA-expressing epithelia. Despite this clear evidence of CTL activation, no signs of autoimmune tissue destruction were observed. Autoimmunity was observed when CTL transfer was accompanied by a systemic challenge with OVA antigen, indicating that inflammatory mediators could prevent the induction of functional tolerance. Nevertheless, in transgenic mice with low epithelial OVA expression levels autoimmune disease subsided rapidly. Interestingly, the OVA-expressing epithelia of these mice displayed a persisting population of memory-type OVA-specific IELs that was functionally anergic, in that these CTL could not be induced to produce the inflammatory cytokines IFN-γ and TNF-α (Vezys et al. 2000). The notion that the interplay between epithelial cells and the immune system can be highly restrictive for CTL activity is supported by experiments with transgenic mice expressing the HPV16 E7 antigen under control of a keratinocyte-specific promoter. E7-transgenic skin grafted onto nontransgenic littermates was not rejected, even if the mice received an E7-specific immunization that resulted in effective E7-specific immunity capable of rejecting E7-expressing transplantable tumors (Frazer et al. 1999, 2001). This showed that even in recently grafted skin sections, which exhibit a certain level of inflammatory activity because of the grafting process, (1) antigen presentation was insufficient for mobilizing E7-specific immunity and (2) preestablished E7-specific immunity was incapable of attacking E7-positive keratinocytes. Thus, the epithelial site of E7 antigen expression, as well as the nature of the keratinocyte and its context, set strict requirements for the immune system at both the induction and the effector level.

It should be noted that the relative contributions of the different arms of the immune response are likely to be different between the nonmucosal skin epithelium and the mucosal epithelia of the respiratory and genital tracts. This is illustrated by the fact that the repertoires of IELs show striking differences between mucosal and nonmucosal epithelia, and even between different mucosal epithelia (Hayday et al. 2001). Furthermore, it is conceivable that neutralizing antiviral antibodies play a more prominent role in the defense against mucosal HPV infections than against HPV infections in the skin.

4 APC Function in HPV-Associated Cancers

The pivotal role of LCs in immune surveillance against HPV implies that immune evasion by HPV is likely to involve interference with LC function. The ability of the virus to persist for at least some time before elimination, as well as the lag phase between infection and the appearance of detectable anti-HPV immunity (Konya and Dillner 2001), suggests that the virus is indeed capable of doing so. The manner in which this interference takes place is most likely related to the physiology of the infected keratinocyte. Although this cell has been reported to secrete proinflammatory cytokines under certain conditions, it also produces immunoregulatory factors such as transforming growth factor β (TGF-β) and IL-10. In fact, both of these latter factors play an essential role in the differentiation of LCs as well as in their retention in the epidermis (Jakob et al. 2001). Furthermore, these factors inhibit the induction of LC maturation by proinflammatory factors. It is evident that LC-mediated activation of T cells in the regional lymph nodes requires the proinflammatory signals to dominate over the homeostasis of LCs as controlled by TGF-β and IL-10 and that lack of sufficient proinflammatory signals will result in immunological ignorance or tolerance. The experiments in the aforementioned transgenic mouse models argue that the lack of effective immunity in this setting is not merely a result of immunological ignorance caused by a lack of antigen presentation, but instead involves tolerization of antigen-specific T cells. In vivo experiments in murine models as well as in vitro experiments with human DCs have indeed shown that under certain conditions DCs can induce antigen-specific tolerance instead of immune activation (Hackstein et al. 2001; Heath and Carbone 2001; Jonuleit et al. 2001; Lanzavecchia and Sallusto 2001; Roncarolo et al. 2001). An essential aspect of this mechanism for tolerance induction is the presentation of antigen by DCs that, because of a lack of inflammatory signals, have not properly matured. As a consequence, the antigen is presented to T cells, but not in an appropriate costimulatory context. In the E7-transgenic skin graft model one could indeed envision that the grafting procedure would result in a weak inflammatory signal combined with limited death of E7-expressing keratinocytes and, thereby, in cross-presentation of E7 antigen by improperly matured DCs. The situation during natural HPV infection in humans is probably much more favorable toward the induction of effective immunity, rather than tolerance, as indicated by the fact that the vast majority of HPV infections are eventually eliminated. However, it is clear that a period of decreased vigilance of the immune system can result in minor disturbance of the balance between LC homeostasis and LC activation and thereby allow the virus a sufficient time window for establishment of a status quo in which HPV and/or HPV-transformed cells are difficult to eliminate.

The capacity of keratinocytes to secrete immunomodulatory cytokines is very pronounced in tumors. In particular, cervical carcinoma cells were found to secrete IL-10 (de Gruijl et al. 1999; Giannini et al. 1998; Mota et al. 1999) vascular endothelium growth factor (VEGF) (Dobbs et al. 1997; Fujimoto et al. 1999; Le Buanec et al. 1999), and PGE_2 (Sales et al. 2001). In addition, considerable levels of TGF-β (Comerci et al. 1996; Mota et al. 1999; Santin et al. 1997; Tervahauta et al. 1994) and IL-6 (Pao et al. 1995; Takano et al. 1996; Tartour et al. 1994) were detected in these tumors, although it should be noted that current data suggest that these cytokines were produced by hematopoietic cells in the surrounding stroma rather than by the tumor cells themselves. These cytokines have been found to downmodulate the APC function of human monocytes (Bogdan et al. 1991; Bonham et al. 1996; de Waal Malefyt et al. 1991; Fiorentino et al. 1991; Gabrilovich et al. 1998; Menetrier-Caux et al. 1998; Ohm et al. 1999; Peguet-Navarro et al. 1994; Takeuchi et al. 1998) or to even convert these into suppressor cells (Alleva et al. 1994). Furthermore, such cytokines were shown to inhibit differentiation of human monocytes into immature DCs (Allavena et al. 1998; Buelens et al.

1997). The inhibitory effect of tumor-produced IL-10 on APC recruitment to the tumor site was also observed in a murine tumor model (Qin et al. 1997), and IL-10 was shown to inhibit the APC function of LCs (Beissert et al. 1995; Enk et al. 1993). Furthermore, activation of DCs or LCs in the presence of IL-10, TGF-β, VEGF, or PGE_2 was shown to result in APCs with decreased costimulatory capacity, lacking IL-12 secretion, which are poor stimulators of Th1/CTL immunity and which can even induce T cell tolerance (Buelens et al. 1997; De Smedt et al. 1997; Enk et al. 1997; Gabrilovich et al. 1996; Ishida et al. 1998; Kalinski et al. 1998; Steinbrink et al. 1997, 1999). In several cases such APCs were shown rather to induce Th2-type immunity (D'Orazio and Niederkorn 1998; King et al. 1998; Liu et al. 1998), which is relatively ineffective against solid tumors and which seems to prevail in patients with progressed cancers such as cervical cancer (al-Saleh et al. 1998; Clerici et al. 1998). Domination of Th2-type immunity is thought to induce a paracrine loop between Th2 cells and APCs that will further incapacitate the APC system with respect to the induction of Th1/CTL-type immunity (Ria et al. 1998).

A considerable number of reports have indicated that the failure of the immune system to clear progressed HPV-positive cervical lesions is indeed the result of insufficient and/or improper APC activity. In particular, analysis of LCs in normal and neoplastic cervical tissue has revealed that the number of LCs in high-grade CIN lesions and cervical tumors is significantly decreased (al-Saleh et al. 1998; Hughes et al. 1988; McArdle and Muller 1986; Mota et al. 1998; Poppe et al. 1996; Puts et al. 1986; Tay et al. 1987; Viac et al. 1990). A recent study has shown that this may be related to the secretion of IL-10 by the neoplastic cells (Mota et al. 1999). Alternatively, the reduced numbers of LC in neoplastic epithelia could be related to decreased production by transformed keratinocytes of chemotactic factors essential for recruitment of LC precursors to the epithelium. In this respect, it would be highly worthwhile to investigate the levels of MIP-3α/CCL20 in normal versus HPV-infected/transformed epithelia (Dieu-Nosjean et al. 2000; Tohyama et al. 2001). In conclusion, the (so far merely descriptive) observations concerning APC function in HPV-positive human tumors, as well as the experiments in transgenic mouse models, argue that failure of APC function during HPV infection indeed plays a dominant role in the occasional escape of HPV from the immune system and that this escape can lead to a situation of sustained failure of the antiviral immune response.

5
Vaccination Strategies Against HPV

The force that must be applied for the induction of effective HPV-specific immunity, in particular Th1/CTL-type responses capable of eradicating HPV-positive cells, will largely depend on the immunological status of the infected person. Prophylactic vaccination of healthy subjects does not need to overcome pre-existing unfavorable immune conditions, such as HPV-specific tolerance or the more general suppression of the immune system frequently found associated with progressed cancers. It should "merely" aim at enhancing the vigilance of the immune defense against HPV. Although at first sight this seems rather straightforward, several requirements must be met. Because it is unclear when a vaccinated subject will encounter HPV, and repeated periodic vaccination is not feasible from financial and logistic perspectives, prophylactic vaccination must install long-term HPV-specific memory. Several of the costimulatory signals are known to be required for the induction of lasting T cell immunity (Marrack et al. 2000). Because only fully activated DCs can provide these signals, management of DC function is an essential feature of such immune intervention schemes. A key parameter is the efficient codelivery of both antigenic and proinflammatory signals to the DCs, as this will ensure that the epitopes of interest are only presented on properly activated DCs. Our views concerning this issue have recently been described elsewhere (Melief et al. 2000; Offringa et al. 2000; Rea et al. 2001).

The confinement of HPV infection to epithelia implies that at least part of the immune memory must be installed at the sites where it can immediately encounter the virus. With respect to prophylactic vaccination against HPV-induced cancers in mucosal epithelia, this embodies the induction of memory-type TCRα/β IELs and mucosal B cells. The currently prevailing view on mucosal immunity is that homing of antigen-experienced B and T lymphocytes to mucosal tissues involves conditioning of these lymphocytes by mucosa-associated lymphoid tissue (MALT). Through this conditioning, the lymphocytes acquire a pattern of surface molecules, including the expression of certain integrins and chemokine receptors, which assist them to adhere to – and traverse – the endothelium at mucosal sites. Notably, the mucosal immune system is regionalized. Lymphocytes primed in the gut-associated lymphoid tissue (GALT) preferentially home to the intestinal mucosa, whereas lymphocytes primed in the airways or urogenital tract preferentially home back to these latter sites (Brandtzaeg et al. 1999). Likewise, there are marked differences in the requirements for lymphocyte homing to the skin versus mucosal epithelia. Although the molecular nature of the signals that install this homing behavior is not yet known, it is likely that the LC plays a key role in this programming (Campbell and Butcher 2002). In view of this, the delivery of a vaccine dose at a mucosal site in the airways or urogenital tract should be considered when designing vaccination schemes against mucosal HPV infection. Furthermore, one should be aware that this approach may not optimally protect against HPV infection in the skin.

Several laboratories are developing mucosal anti-HPV vaccines, primarily based on viruslike particles (VLPs) formed by the in vitro assembly of recombinant HPV capsid proteins, and have shown that intranasal delivery of such vaccines can elicit both mucosal and systemic B and T cell immunity in mice (Gerber et al. 2001; Revaz et al. 2001). The antigenicity of VLP-based vaccines is largely restricted to the L1 and L2 proteins, because chimeric VLPs can only accommodate small fragments of HPV early proteins (Schiller and Lowy 2001). Recombinant adenoviral vectors, which can be made to encode any desired HPV antigen and for which the airways constitute the natural target tissue for infection, constitute a promising alternative as a mucosal delivery tool. Administration of vaccines through the airways must, however, be performed in a very careful manner. On one hand, strong inflammatory responses in the airways can be dangerous for the subject, thereby putting constraints on vaccine dose and formulation. On the other hand, potent adjuvants are usually required for the induction of mucosal responses and to prevent the induction of tolerance instead (Czerkinsky et al. 1999). In this respect, an interesting concept for vaccination against HPV can be derived from studies concerning vaccination against HIV, which is also transmitted through the mucosal route. In a study in macaques, the animals first received several parenteral doses of SIV-specific vaccine, after which they were boosted with a mucosal dose of this vaccine (Marx et al. 1993). The concept of this vaccination strategy is compelling, in that systemic priming can be performed to ensure sufficient virus-specific immune potential, after which a single mucosal boost can be used to focus the antiviral immune defense toward the sites of viral entry. For high-risk HPV16, several clinical-grade vaccines are available in which formulation and/or vector type are clearly distinct. This allows the application of a second concept, heterologous prime-boost protocols. These are superior to homologous prime-boost protocols because they focus immune reactivity toward the HPV-specific component rather than the vector components (van der Burg et al. 2001 and references therein). We propose that a combination of both prime-boost concepts is likely to result in a highly powerful vaccination protocol.

6
Management of DC Function During Therapeutic Immune Intervention

Immune intervention in therapeutic settings primarily applies to clinical settings in which patients exhibit HPV-positive lesions that are unlikely to regress spontaneously. The aim of therapeutic vaccination is to mobilize an effector-type of immune response that will immediately launch an attack against the HPV-positive lesion. If this effector phase develops in a successful fashion, tumor-specific immune memory will most likely ensue. Notably, the presence of the lesion is evidence of the failure of the spontaneous immune response. It is highly unlikely that in the case of an antigenic virus-positive tumor such failure is a result of ignorance. More likely, HPV-specific priming of the immune system has taken place, as witnessed by the frequent detection of HPV-specific immunity in patients with HPV-positive cervical lesions (see above), but this response is unable to deal with the tumor. If this is primarily related to the magnitude of this response, antigen-specific immune intervention should be performed along guidelines similar to those described for prophylactic vaccination. However, when failure of the immune response is related to improper polarization of the HPV-specific immune response (Clerici et al. 1998), or even the induction of HPV-specific tolerance (Frazer et al. 1999, 2001), the immune system will need to be redirected rather than merely boosted. In such settings, mobilization of potent APC function requires repeated delivery of strong Th1-polarizing adjuvants. We have recently found that mice bearing established HPV16-positive tumors were capable of rejecting these tumors if repeatedly immunized with HPV16E7 peptide vaccine in combination with a strong Th1-stimulating CpG adjuvant but not when immunized with this formulation only once or when repeatedly immunized with this peptide in combination with incomplete Freund's adjuvant (Zwaveling et al. 2002). In severe cases, breaching of immunosuppressive mechanisms may even require a "hard reset" of the T cell immune system, for instance, by eliminating all CD25-expressing lymphocytes, a population that at least in the mouse includes the highly potent CD4$^+$ T-regulatory subset (Sakaguchi et al. 2001; Shevach et al. 2001).

Although such profound perturbations of the immune defense are likely to be found in patients with progressed, metastatic disease, it is conceivable that in patients with premalignant lesions, or in those who may exhibit minimal residual disease after resection of the primary tumor, the immune system has not been incapacitated to such an extent. Especially in the case of antigenic virus-positive tumors, one could envision that presentation of tumor antigens to the T cells in the regional lymph nodes does take place, but that insufficient inflammatory signals result in antigenic and costimulatory signals that are below the thres hold for effective mobilization of antitumor T cell immunity. In such a setting, provision of a potent (pro-)inflammatory signal without codelivery of an antigen-specific vaccine could be sufficient to trigger this immune response. In a murine tumor model featuring human adenovirus type 5 (Ad5)-transformed cells, we have recently demonstrated that such therapeutic approaches may be feasible. Mice carrying pre-established subcutaneous Ad5 tumors were capable of rejecting these tumors if treated by a systemic or peritumoral injection of an agonistic antibody against CD40 (van Mierlo et al. 2002). We previously demonstrated that application of this antibody results in full activation of DC in vitro as well as in vivo (Diehl et al. 1999; Schuurhuis et al. 2000). Comparison of treated and nontreated mice showed that (1) both exhibited presentation of the Ad5 tumor antigen in their tumor-draining lymph nodes and (2) both showed a population of tumor-specific CD8$^+$ T cells that stained with an MHC/tetramer specific for this tumor antigen. Importantly, the numbers of tumor-specific CD8$^+$ T cells were greatly increased in the anti-CD40-treated mice. Furthermore, these mice showed antitumor CD8$^+$ T cells in all lymphoid compartments, whereas the control group only contained such T cells in the tumor-draining lymph node. The existence of systemic effector-type antitumor immunity was supported by the fact that mice bearing Ad5 positive tumors in both flanks were capable of rejecting both tumors on peritumoral injection of anti-CD40 antibody in only one of the tumors. Together, these data show that spontaneous presentation of antigen derived from virus-positive tumors may very well result in the induction of a low level of tumor-specific T lymphocytes but a strong inflammatory signal such as the DC-activating anti-CD40 antibody is required to transform this local "tumor-observing" T cell response into a systemic effector response.

From a more general perspective, these findings show that both the recruitment and the activation state of peritumoral DC are decisive factors in the failure or success of the immune defense against virus-positive tumors. Especially in the case of highly antigenic, poorly inflammatory HPV-positive premalignant lesions, delivery of a molecularly defined (pro-)inflammatory signal is an essential aspect of therapeutic interventions. A possible additional advantage of such local treatment would be that the T cell responses will be induced through the correct afferent lymphoid organs and will thereby acquire the potential for homing to the relevant epithelial sites (see above). In fact, in the aforementioned experiments with nontransgenic mice that were grafted with E7-transgenic skin, a local antigenic signal in combination with a strong inflammatory signal consisting of systemic administration of *Listeria monocytogenes*-induced E7-specific T cell immunity capable of rejecting recently transplanted grafts (Frazer et al. 2001).

7
Perspectives for Preclinical Research and Development of Clinical Immune Intervention Schemes

At present, a wide variety of clinically applicable vaccine formulations against HPV and molecularly defined adjuvants are available. Furthermore, our insights on the physiology of the epithelial LC and on the requirements for induction of effective epithelial antiviral immunity are increasing rapidly. It is now imperative to integrate the tools and knowledge into a unifying concept for vaccination against HPV. Unfortunately, the testing of such vaccination schemes in a physiological preclinical model for HPV infection is precluded by the fact that HPV cannot be propagated and/or isolated in large quantities. However, even if purified HPV were available, it is questionable whether the design of the HPV infection cycle, which is so delicately tuned to the human keratinocyte differentiation program, would allow productive infection, and therefore meaningful experimentation, in animal experiments. Cottontail rabbit papillomavirus does offer such an animal model (Breitburd et al. 1997), but in contrast to the mouse immune system that of rabbits has not been mapped at the molecular level. As a consequence, the extensive repertoire of sophisticated tools required for incisive experimentation is lacking. Although mouse models for HPV-induced carcinogenesis can only represent an approximation of HPV-induced carcinogenesis in humans, essential improvements can be made to the currently available experimental systems. Most work has so far been performed with preclinical models in which tumor growth is initiated through introduction into mice of in vitro expanded HPV-positive tumor cells. In contrast to naturally arising HPV-positive lesions, these tumors do not arise stealthily within the context of skin or mucosal epithelium, but rather from transplanted cells that immediately form a foreign body outside the context of a natural somatic tissue. This difference will certainly and profoundly affect the APC population involved, the type of immunity induced, as well as the quantitative and qualitative requirements of the effector response required for elimination of the HPV-positive tumor cells. As an alternative, several groups have generated transgenic mice that constitutively express HPV oncoproteins under control of keratinocyte-specific promoters. A major drawback of this approach is that the HPV proteins are expressed very early in life by epithelia throughout the body and therefore automatically regarded by the immune system as epithelial autoantigens. Furthermore, oncogene-induced neoplastic cell growth is spread over many epithelia throughout the body, making effective therapeutic intervention hard to achieve. The best solution to these problems so far, as reported by Frazer and coworkers, was the grafting of HPVE7-transgenic skin onto nontransgenic mice (Frazer et al. 2001). A more sophisticated approach involves the generation of mouse models for controlled sporadic virus-induced cancer, in which the transgenic mice harbor conditionally active oncogenes, the expression of which can be unleashed in a spatiotemporally controlled fashion through Cre/lox-mediated DNA recombination (Meuwissen et al. 2001a,b). Cre/lox-controlled mice that express the conditionally active HPV16 E6 and E7 oncogenes can be exploited to study the consequences of local HPV oncogene expression both on the immune response and on tumorigenesis. Furthermore, the effects of immune suppression versus HPVE6/E7-specific vaccination strategies (see above) can be evaluated. With respect to the latter, it will be of particular interest to compare the effects of vaccine delivery at mucosal or skin epithelia or at parenteral sites.

Notwithstanding the fact that such animal models will soon be available, it is important to acknowledge that several different clinical HPV-specific vaccine formulations have already been tested for toxicity in safety trials. This should allow the in depth analysis and comparison of (heterologous) prime-boost protocols and vaccine delivery strategies in humans.

Note added in proof. An HPV16 VLP vaccine was recently shown to have prophylactic efficacy in young women (Koutsky et al. 2002).

Acknowledgements. Our work as discussed in this manuscript was supported through grants provided by the Dutch Cancer Research fund, in particular grants 99–2024 and 99–2025 provided to R.O, R.E.M.T., and C.J.M.M.

References

Allavena P, Piemonti L, Longoni D, Bernasconi S, Stoppacciaro A, Ruco L. and Mantovani A, IL-10 prevents the differentiation of monocytes to dendritic cells but promotes their maturation to macrophages. Eur J Immunol 28 359–69. (1998)

Alleva D.G, Burger C.J. and Elgert K.D, Tumor-induced regulation of suppressor macrophage nitric oxide and TNF-alpha production. Role of tumor-derived IL-10 TGF-beta and prostaglandin E2. J Immunol 153 1674–86. (1994)

al-Saleh W, Giannini S.L, Jacobs N, Moutschen M, Doyen J, Boniver J. and Delvenne P, Correlation of T-helper secretory differentiation and types of antigen-presenting cells in squamous intraepithelial lesions of the uterine cervix. J Pathol 184 283–90. (1998)

Beissert S, Hosoi J, Grabbe S, Asahina A. and Granstein R.D, IL-10 inhibits tumor antigen presentation by epidermal antigen-presenting cells. J Immunol 154 1280–6. (1995)

Bogdan C, Vodovotz Y. and Nathan C, Macrophage deactivation by interleukin 10. J Exp Med 174 1549–55. (1991)

Bonham C.A, Lu L, Banas R.A, Fontes P, Rao A.S, Starzl T.E, Zeevi A. and Thomson A.W, TGF-beta 1 pretreatment impairs the allostimulatory function of human bone marrow-derived antigen-presenting cells for both naive and primed T cells. Transpl Immunol 4 186–91. (1996)

Brandtzaeg P, Farstad I.N. and Haraldsen G, Regional specialization in the mucosal immune system: primed cells do not always home along the same track. Immunol Today 20 267–77. (1999)

Breitburd F, Salmon J. and Orth G, The rabbit viral skin papillomas and carcinomas: a model for the immunogenetics of HPV-associated carcinogenesis. Clin Dermatol 15 237–47. (1997)

Buelens C, Verhasselt V, De Groote D, Thielemans K, Goldman M. and Willems F, Interleukin-10 prevents the generation of dendritic cells from human peripheral blood mononuclear cells cultured with interleukin-4 and granulocyte/macrophage-colony-stimulating factor. Eur J Immunol 27 756–62. (1997)

Campbell D.J. and Butcher E.C, Rapid acquisition of tissue-specific homing phenotypes by CD4(+) T cells activated in cutaneous or mucosal lymphoid tissues. J Exp Med 195 135–41. (2002)

Clerici M, Shearer G.M. and Clerici E, Cytokine dysregulation in invasive cervical carcinoma and other human neoplasias: time to consider the TH1/TH2 paradigm. J Natl Cancer Inst 90 261–3. (1998)

Comerci J.T, Jr, Runowicz C.D, Flanders K.C, De Victoria C, Fields A.L, Kadish A.S. and Goldberg G.L, Altered expression of transforming growth factor-beta 1 in cervical neoplasia as an early biomarker in carcinogenesis of the uterine cervix. Cancer 77 1107–14. (1996)

Czerkinsky C, Anjuere F, McGhee J.R, George-Chandy A, Holmgren J, Kieny M.P, Fujiyashi K, Mestecky J.F, Pierrefite-Carle V, Rask C. and Sun J.B, Mucosal immunity and tolerance: relevance to vaccine development. Immunol Rev 170 197–222. (1999)

de Gruijl T.D, Bontkes H.J, van den Muysenberg A.J, van Oostveen J.W, Stukart M.J, Verheijen R.H, van der Vange N, Snijders P.J, Meijer C.J, Walboomers J.M. and Scheper R.J, Differences in cytokine mRNA profiles between premalignant and malignant lesions of the uterine cervix. Eur J Cancer 35 490–7. (1999)

de Jong A, van der Burg S.H, Kwappenberg K.M, van der Hulst J.M, Franken K.L, Geluk A, van Meijgaarden K.E, Drijfhout J.W, Kenter G, Vermeij P, Melief C.J. and Offringa R, Frequent detection of human papillomavirus 16 E2-specific T-helper immunity in healthy subjects. Cancer Res 62 472–9. (2002)

De Smedt T, Van Mechelen M, De Becker G, Urbain J, Leo O. and Moser M, Effect of interleukin-10 on dendritic cell maturation and function. Eur J Immunol 27 1229–35. (1997)

de Waal Malefyt R, Abrams J, Bennett B, Figdor C.G. and de Vries J.E, Interleukin 10(IL-10) inhibits cytokine synthesis by human monocytes: an autoregulatory role of IL-10 produced by monocytes. J Exp Med 174 1209–20. (1991)

Diehl L, den Boer A.T, Schoenberger S.P, van der Voort E.I, Schumacher T.N, Melief C.J, Offringa R. and Toes R.E, CD40 activation in vivo overcomes peptide-induced peripheral cytotoxic T-lymphocyte tolerance and augments anti-tumor vaccine efficacy. Nat Med 5 774–9. (1999)

Dieu-Nosjean M.C, Massacrier C, Homey B, Vanbervliet B, Pin J.J, Vicari A, Lebecque S, Dezutter-Dambuyant C, Schmitt D, Zlotnik A. and Caux C, Macrophage inflammatory protein 3alpha is expressed at inflamed epithelial surfaces and is the most potent chemokine known in attracting Langerhans cell precursors. J Exp Med 192 705–18. (2000)

Dobbs S.P, Hewett P.W, Johnson I.R, Carmichael J. and Murray J.C, Angiogenesis is associated with vascular endothelial growth factor expression in cervical intraepithelial neoplasia. Br J Cancer 76 1410–5 (1997)

D'Orazio T.J. and Niederkorn J.Y, A novel role for TGF-beta and IL-10 in the induction of immune privilege. J Immunol 160 2089–98. (1998)

Elson D.A, Riley R.R, Lacey A, Thordarson G, Talamantes F.J. and Arbeit J.M, Sensitivity of the cervical transformation zone to estrogen-induced squamous carcinogenesis. Cancer Res 60 1267–75. (2000)

Enk A.H, Angeloni V.L, Udey M.C. and Katz S.I, Inhibition of Langerhans cell antigen-presenting function by IL-10. A role for IL-10 in induction of tolerance. J Immunol 151 2390–8. (1993)

Enk A.H, Jonuleit H, Saloga J. and Knop J, Dendritic cells as mediators of tumor-induced tolerance in metastatic melanoma. Int J Cancer 73 309–16. (1997)

Evander M, Edlund K, Gustafsson A, Jonsson M, Karlsson R, Rylander E. and Wadell G, Human papillomavirus infection is transient in young women: a population-based cohort study. J Infect Dis 171 1026–30. (1995)

Evans M, Borysiewicz L.K, Evans A.S, Rowe M, Jones M, Gileadi U, Cerundolo V. and Man S, Antigen processing defects in cervical carcinomas limit the presentation of a CTL epitope from human papillomavirus 16 E6. J Immunol 167 5420–8. (2001)

Fiorentino D.F, Zlotnik A, Mosmann T.R, Howard M. and O'Garra A, IL-10 inhibits cytokine production by activated macrophages. J Immunol 147 3815–22. (1991)

Frazer I.H, De Kluyver R, Leggatt G.R, Guo H.Y, Dunn L, White O, Harris C, Liem A. and Lambert P, Tolerance or immunity to a tumor antigen expressed in somatic cells can be determined by systemic proinflammatory signals at the time of first antigen exposure. J Immunol 167 6180–7. (2001)

Frazer I.H, Thomas R, Zhou J, Leggatt G.R, Dunn L, McMillan N, Tindle R.W, Filgueira L, Manders P, Barnard P. and Sharkey M, Potential strategies utilised by papillomavirus to evade host immunity. Immunol Rev 168 131–42. (1999)

Fujimoto J, Sakaguchi H, Hirose R, Ichigo S. and Tamaya T, Expression of vascular endothelial growth factor (VEGF) and its mRNA in uterine cervical cancers. Br J Cancer 80 827–33. (1999)

Gabrilovich D, Ishida T, Oyama T, Ran S, Kravtsov V, Nadaf S. and Carbone D.P, Vascular endothelial growth factor inhibits the development of dendritic cells and dramatically affects the differentiation of multiple hematopoietic lineages in vivo. Blood 92 4150–66. (1998)

Gabrilovich D.I, Chen H.L, Girgis K.R, Cunningham H.T, Meny G.M, Nadaf S, Kavanaugh D. and Carbone D.P, Production of vascular endothelial growth factor by human tumors inhibits the functional maturation of dendritic cells. Nat Med 2 1096–103. (1996)

Garrido F, Ruiz-Cabello F, Cabrera T, Perez-Villar J.J, Lopez-Botet M, Duggan-Keen M. and Stern P.L, Implications for immunosurveillance of altered HLA class I phenotypes in human tumours. Immunol Today 18 89–95. (1997)

Gerber S, Lane C, Brown D.M, Lord E, DiLorenzo M, Clements J.D, Rybicki E, Williamson A.L. and Rose R.C, Human papillomavirus virus-like particles are efficient oral immunogens when coadministered with *Escherichia coli* heat-labile enterotoxin mutant R192G or CpG DNA. J Virol 75 4752–60. (2001)

Giannini S.L, Al-Saleh W, Piron H, Jacobs N, Doyen J, Boniver J. and Delvenne P, Cytokine expression in squamous intraepithelial lesions of the uterine cervix: implications for the generation of local immunosuppression. Clin Exp Immunol 113 183–9. (1998)

Hackstein H, Morelli A.E. and Thomson A.W, Designer dendritic cells for tolerance induction: guided not misguided missiles. Trends Immunol 22 437–42. (2001)

Harwood C.A, Surentheran T, McGregor J.M, Spink P.J, Leigh I.M, Breuer J. and Proby C.M, Human papillomavirus infection and non-melanoma skin cancer in immunosuppressed and immunocompetent individuals. J Med Virol 61 289–97. (2000)

Havran W.L, Chien Y.H. and Allison J.P, Recognition of self antigens by skin-derived T cells with invariant gamma delta antigen receptors. Science 252 1430–2. (1991)

Hayday A, Theodoridis E, Ramsburg E. and Shires J, Intraepithelial lymphocytes: exploring the Third Way in immunology. Nat Immunol 2 997–1003. (2001)

Heath W.R. and Carbone F.R, Cross-presentation dendritic cells tolerance and immunity. Annu Rev Immunol 19 47–64 (2001)

Ho G.Y, Bierman R, Beardsley L, Chang C.J. and Burk R.D, Natural history of cervicovaginal papillomavirus infection in young women. N Engl J Med 338 423–8. (1998)

Hughes R.G, Norval M. and Howie S.E, Expression of major histocompatibility class II antigens by Langerhans' cells in cervical intraepithelial neoplasia. J Clin Pathol 41 253–9. (1988)

Ishida T, Oyama T, Carbone D.P. and Gabrilovich D.I, Defective function of Langerhans cells in tumor-bearing animals is the result of defective maturation from hemopoietic progenitors. J Immunol 161 4842–51. (1998)

Jakob T, Ring J. and Udey M.C, Multistep navigation of Langerhans/dendritic cells in and out of the skin. J Allergy Clin Immunol 108 688–96. (2001)

Jonuleit H, Schmitt E, Steinbrink K. and Enk A.H, Dendritic cells as a tool to induce anergic and regulatory T cells. Trends Immunol 22 394–400. (2001)

Kalinski P, Schuitemaker J.H, Hilkens C.M. and Kapsenberg M.L, Prostaglandin E2 induces the final maturation of IL-12-deficient CD1a+CD83+ dendritic cells: the levels of IL-12 are determined during the final dendritic cell maturation and are resistant to further modulation. J Immunol 161 2804–9. (1998)

King C, Davies J, Mueller R, Lee M.S, Krahl T, Yeung B, O'Connor E. and Sarvetnick N, TGF-beta1 alters APC preference polarizing islet antigen responses toward a Th2 phenotype. Immunity 8 601–13. (1998)

Kondo S, The roles of keratinocyte-derived cytokines in the epidermis and their possible responses to UVA-irradiation. J Investig Dermatol Symp Proc 4 177–83. (1999)

Konya J. and Dillner J, Immunity to oncogenic human papillomaviruses. Adv Cancer Res 82 205–38. (2001)

Koutsky L, Epidemiology of genital human papillomavirus infection. Am J Med 102 3–8. (1997)

Koutsky L.A, Ault K.A, Wheeler C.M, Brown D.R, Barr E, Alvarez F.B, Chiacchierini L.M, Jansen K.U, A controlled trial of a human papillomavirus type 16 vaccine. N Engl J Med 347 1645-51. (2002)

Lanzavecchia A. and Sallusto F, The instructive role of dendritic cells on T cell responses: lineages plasticity and kinetics. Curr Opin Immunol 13 291–8. (2001)

Le Buanec H, D'Anna R, Lachgar A, Zagury J.F, Bernard J, Ittele D, d'Alessio P, Hallez S, Giannouli C, Burny A, Bizzini B, Gallo R.C. and Zagury D, HPV-16 E7 but not E6 oncogenic protein triggers both cellular immunosuppression and angiogenic processes. Biomed Pharmacother 53 424–31. (1999)

Liu L, Rich B.E, Inobe J, Chen W. and Weiner H.L, Induction of Th2 cell differentiation in the primary immune response: dendritic cells isolated from adherent cell culture treated with IL-10 prime naive CD4+ T cells to secrete IL-4. Int Immunol 10 1017–26. (1998)

Man S. and Fiander A, Immunology of human papillomavirus infection in lower genital tract neoplasia. Best Pract Res Clin Obstet Gynaecol 15 701–14. (2001)

Marrack P, Bender J, Hildeman D, Jordan M, Mitchell T, Murakami M, Sakamoto A, Schaefer B.C, Swanson B. and Kappler J, Homeostasis of alpha beta TCR+ T cells. Nat Immunol 1 107–11. (2000)

Marx P.A, Compans R.W, Gettie A, Staas J.K, Gilley R.M, Mulligan M.J, Yamshchikov G.V, Chen D. and Eldridge J.H, Protection against vaginal SIV transmission with microencapsulated vaccine. Science 260 1323–7. (1993)

Masopust D, Vezys V, Marzo A.L. and Lefrancois L, Preferential localization of effector memory cells in nonlymphoid tissue. Science 291 2413–7. (2001)

McArdle J.P. and Muller H.K, Quantitative assessment of Langerhans' cells in human cervical intraepithelial neoplasia and wart virus infection. Am J Obstet Gynecol 154 509–15. (1986)

Medema J.P, de Jong J, Peltenburg L.T, Verdegaal E.M, Gorter A, Bres S.A, Franken K.L, Hahne M, Albar J.P, Melief C.J. and Offringa R, Blockade of the granzyme B/perforin pathway through overexpression of the serine protease inhibitor PI-9/SPI-6 constitutes a mechanism for immune escape by tumors. Proc Natl Acad Sci U S A 98 11515–20. (2001)

Melief C.J, Toes R.E, Medema J.P, van der Burg S.H, Ossendorp F. and Offringa R, Strategies for immunotherapy of cancer. Adv Immunol 75 235–82 (2000)

Menetrier-Caux C, Montmain G, Dieu M.C, Bain C, Favrot M.C, Caux C. and Blay J.Y, Inhibition of the differentiation of dendritic cells from CD34(+) progenitors by tumor cells: role of interleukin-6 and macrophage colony-stimulating factor. Blood 92 4778–91. (1998)

Meuwissen R, Jonkers J. and Berns A, Mouse models for sporadic cancer. Exp Cell Res 264 100–10. (2001a)

Meuwissen R, Linn S.C, van der Valk M, Mooi W.J. and Berns A, Mouse model for lung tumorigenesis through Cre/lox controlled sporadic activation of the K-Ras oncogene. Oncogene 20 6551–8. (2001b)

van Mierlo G.J, den Boer A.T, Medema J.P, van Der Voort E.I, Fransen M.F, Offringa R, Melief C.J, and Toes R.E. CD40 stimulation leads to effective therapy of CD40- tumors through induction of strong systemic cytotoxic T lymphocyte immunity Proc Natl Acad Sci USA 99 5561-6. (2002)

Morales J, Homey B, Vicari A.P, Hudak S, Oldham E, Hedrick J, Orozco R, Copeland N.G, Jenkins N.A, McEvoy L.M. and Zlotnik A, CTACK a skin-associated chemokine that preferentially attracts skin-homing memory T cells. Proc Natl Acad Sci U S A 96 14470–5. (1999)

Mota F, Rayment N, Chong S, Singer A. and Chain B, The antigen-presenting environment in normal and human papillomavirus (HPV)-related premalignant cervical epithelium. Clin Exp Immunol 116 33–40. (1999)

Mota F.F, Rayment N.B, Kanan J.H, Singer A. and Chain B.M, Differential regulation of HLA-DQ expression by keratinocytes and Langerhans cells in normal and premalignant cervical epithelium. Tissue Antigens 52 286–93. (1998)

Nestle F.O. and Burg G, Dendritic cells: role in skin diseases and therapeutic applications. Clin Exp Dermatol 24 204–7. (1999)

Ochsenbein A.F, Sierro S, Odermatt B, Pericin M, Karrer U, Hermans J, Hemmi S, Hengartner H. and Zinkernagel R.M, Roles of tumour localization second signals and cross priming in cytotoxic T-cell induction. Nature 411 1058–64. (2001)

Offringa R, van der Burg S.H, Ossendorp F, Toes R.E. and Melief C.J, Design and evaluation of antigen-specific vaccination strategies against cancer. Curr Opin Immunol 12 576–82. (2000)

Ohm J.E, Shurin M.R, Esche C, Lotze M.T, Carbone D.P. and Gabrilovich D.I, Effect of vascular endothelial growth factor and FLT3 ligand on dendritic cell generation in vivo. J Immunol 163 3260–8. (1999)

Palefsky J.M, Human papillomavirus-related tumors. AIDS 14 S189–95. (2000)

Pao C.C, Lin C.Y, Yao D.S. and Tseng C.J, Differential expression of cytokine genes in cervical cancer tissues. Biochem Biophys Res Commun 214 1146–51. (1995)

Pardoll D, T cells and tumours. Nature 411 1010–2. (2001)

Payne E, Bowles M.R, Don A, Hancock J.F. and McMillan N.A, Human papillomavirus type 6b virus-like particles are able to activate the Ras-MAP kinase pathway and induce cell proliferation. J Virol 75 4150–7. (2001)

Peguet-Navarro J, Moulon C, Caux C, Dalbiez-Gauthier C, Banchereau J. and Schmitt D, Interleukin-10 inhibits the primary allogeneic T cell response to human epidermal Langerhans cells. Eur J Immunol 24 884–91. (1994)

Poppe W.A, Drijkoningen M, Ide P.S, Lauweryns J.M. and Van Assche F.A, Langerhans' cells and L1 antigen expression in normal and abnormal squamous epithelium of the cervical transformation zone. Gynecol Obstet Invest 41 207–13 (1996)

Puts J.J, Moesker O, de Waal R.M, Kenemans P, Vooijs G.P. and Ramaekers F.C, Immunohistochemical identification of Langerhans cells in normal epithelium and in epithelial lesions of the uterine cervix. Int J Gynecol Pathol 5 151–62 (1986)

Qin Z, Noffz G, Mohaupt M. and Blankenstein T, Interleukin-10 prevents dendritic cell accumulation and vaccination with granulocyte-macrophage colony-stimulating factor gene-modified tumor cells. J Immunol 159 770–6. (1997)

Rea D, Johnson M.E, Havenga M.J, Melief C.J. and Offringa R, Strategies for improved antigen delivery into dendritic cells. Trends Mol Med 7 91–4. (2001)

Reiss Y, Proudfoot A.E, Power C.A, Campbell J.J. and Butcher E.C, CC chemokine receptor (CCR)4 and the CCR10 ligand cutaneous T cell-attracting chemokine (CTACK) in lymphocyte trafficking to inflamed skin. J Exp Med 194 1541–7. (2001)

Revaz V, Benyacoub J, Kast W.M, Schiller J.T, De Grandi P. and Nardelli-Haefliger D, Mucosal vaccination with a recombinant *Salmonella typhimurium* expressing human papillomavirus type 16 (HPV16) L1 virus-like particles (VLPs) or HPV16 VLPs purified from insect cells inhibits the growth of HPV16-expressing tumor cells in mice. Virology 279 354–60. (2001)

Ria F, Penna G. and Adorini L, Th1 cells induce and Th2 inhibit antigen-dependent IL-12 secretion by dendritic cells. Eur J Immunol 28 2003–16. (1998)

Roncarolo M.G, Levings M.K. and Traversari C, Differentiation of T regulatory cells by immature dendritic cells. J Exp Med 193 F5–9. (2001)

Rudolf M.P, Fausch S.C, Da Silva D.M. and Kast W.M, Human dendritic cells are activated by chimeric human papillomavirus type-16 virus-like particles and induce epitope-specific human T cell responses in vitro. J Immunol 166 5917–24. (2001)

Sakaguchi S, Sakaguchi N, Shimizu J, Yamazaki S, Sakihama T, Itoh M, Kuniyasu Y, Nomura T, Toda M. and Takahashi T, Immunologic tolerance maintained by CD25+ CD4+ regulatory T cells: their common role in controlling autoimmunity tumor immunity and transplantation tolerance. Immunol Rev 182 18–32. (2001)

Sales K.J, Katz A.A, Davis M, Hinz S, Soeters R.P, Hofmeyr M.D, Millar R.P. and Jabbour H.N, Cyclooxygenase-2 expression and prostaglandin E(2) synthesis are up-regulated in carcinomas of the cervix: a possible autocrine/paracrine regulation of neoplastic cell function via EP2/EP4 receptors. J Clin Endocrinol Metab 86 2243–9. (2001)

Santin A.D, Hermonat P.L, Hiserodt J.C, Fruehauf J, Schranz V, Barclay D, Pecorelli S. and Parham G.P, Differential transforming growth factor-beta secretion in adenocarcinoma and squamous cell carcinoma of the uterine cervix. Gynecol Oncol 64 477–80. (1997)

Sasadeusz J, Kelly H, Szer J, Schwarer A.P, Mitchell H. and Grigg A, Abnormal cervical cytology in bone marrow transplant recipients. Bone Marrow Transplant 28 393–7. (2001)

Schiller J. and Lowy D, Papillomavirus-like particle vaccines. J Natl Cancer Inst Monogr 28 50–4 (2001)

Schuurhuis D.H, Laban S, Toes R.E, Ricciardi-Castagnoli P, Kleijmeer M.J, van der Voort E.I, Rea D, Offringa R, Geuze H.J, Melief C.J. and Ossendorp F, Immature dendritic cells acquire CD8(+) cytotoxic T lymphocyte priming capacity upon activation by T helper cell-independent or -dependent stimuli. J Exp Med 192 145–50. (2000)

Serraino D, Carrieri P, Pradier C, Bidoli E, Dorrucci M, Ghetti E, Schiesari A, Zucconi R, Pezzotti P, Dellamonica P, Franceschi S. and Rezza G, Risk of invasive cervical cancer among women with or at risk for HIV infection. Int J Cancer 82 334–7. (1999)

Shevach E.M, McIlugh R.S, Piccirillo C.A. and Thornton A.M, Control of T-cell activation by CD4+ CD25+ suppressor T cells. Immunol Rev 182 58–67. (2001)

Steinbrink K, Jonuleit H, Muller G, Schuler G, Knop J. and Enk A.H, Interleukin-10-treated human dendritic cells induce a melanoma-antigen-specific anergy in CD8(+) T cells resulting in a failure to lyse tumor cells. Blood 93 1634–42. (1999)

Steinbrink K, Wolfl M, Jonuleit H, Knop J. and Enk A.H, Induction of tolerance by IL-10-treated dendritic cells. J Immunol 159 4772–80. (1997)

Stern P.L, Brown M, Stacey S.N, Kitchener H.C, Hampson I, Abdel-Hady E.S. and Moore J.V, Natural HPV immunity and vaccination strategies. J Clin Virol 19 57–66. (2000)

Stoler M.H, Human papillomaviruses and cervical neoplasia: a model for carcinogenesis. Int J Gynecol Pathol 19 16–28. (2000)

Takano H, Harigaya K, Ishii G, Sugaya Y, Soeta S, Nunoyama T, Shirasawa H, Shimizu K, Tokita H, Simizu B, Mikata A. and Sekiya S, Interleukin-6 (IL-6) production in carcinoma of the cervix. Arch Gynecol Obstet 258 25–33 (1996)

Takeuchi M, Alard P. and Streilein J.W, TGF-beta promotes immune deviation by altering accessory signals of antigen-presenting cells. J Immunol 160 1589–97. (1998)

Tartour E, Gey A, Sastre-Garau X, Pannetier C, Mosseri V, Kourilsky P. and Fridman W.H, Analysis of interleukin 6 gene expression in cervical neoplasia using a quantitative polymerase chain reaction assay: evidence for enhanced interleukin 6 gene expression in invasive carcinoma. Cancer Res 54 6243–8. (1994)

Tay S.K, Jenkins D, Maddox P, Campion M. and Singer A, Subpopulations of Langerhans' cells in cervical neoplasia. Br J Obstet Gynaecol 94 10–5. (1987)

Tervahauta A, Syrjanen S, Yliskoski M, Gold L.I. and Syrjanen K, Expression of transforming growth factor-beta 1 and -beta 2 in human papillomavirus (HPV)-associated lesions of the uterine cervix. Gynecol Oncol 54 349–56. (1994)

Tjiong M.Y, Out T.A, Ter Schegget J, Burger M.P. and Van Der Vange N, Epidemiologic and mucosal immunologic aspects of HPV infection and HPV-related cervical neoplasia in the lower female genital tract: a review. Int J Gynecol Cancer 11 9–17. (2001)

Tohyama M, Shirakara Y, Yamasaki K, Sayama K. and Hashimoto K, Differentiated keratinocytes are responsible for TNF-alpha regulated production of macrophage inflammatory protein 3alpha/CCL20 a potent chemokine for Langerhans cells. J Dermatol Sci 27 130–9. (2001)

Vambutas A, DeVoti J, Pinn W, Steinberg B.M. and Bonagura V.R, Interaction of human papillomavirus type 11 E7 protein with TAP-1 results in the reduction of ATP-dependent peptide transport. Clin Immunol 101 94–9. (2001)

van der Burg S.H, Kwappenberg K.M, O'Neill T, Brandt R.M, Melief C.J, Hickling J.K. and Offringa R, Pre-clinical safety and efficacy of TA-CIN a recombinant HPV16 L2E6E7 fusion protein vaccine in homologous and heterologous prime-boost regimens. Vaccine 19 3652–60. (2001)

Vezys V, Olson S. and Lefrancois L, Expression of intestine-specific antigen reveals novel pathways of CD8 T cell tolerance induction. Immunity 12 505–14. (2000)

Viac J, Guerin-Reverchon I, Chardonnet Y. and Bremond A, Langerhans cells and epithelial cell modifications in cervical intraepithelial neoplasia: correlation with human papillomavirus infection. Immunobiology 180 328–38. (1990)

zur Hausen H, Papillomaviruses causing cancer: evasion from host-cell control in early events in carcinogenesis. J Natl Cancer Inst 92 690–8. (2000)

Zwaveling S, Ferreira-Mota S.C, Nouta J, Johnson M, Lipford G.B, Offringa R, Van der Burg S.H, Melief C.J.M, Established human papillomavirus type 16-expressing tumors are effectively eradicated following vaccination with long peptides. J Immunol 169 350-8. (2002)

Chapter 11

Viral Vectors for Dendritic Cell-Based Immunotherapy

J. Humrich, L. Jenne

Abstract. Dendritic cells (DCs) constitute a specialised system of antigen-presenting cells with a high capacity to induce and to modulate the immune response against microbial, tumour and self-antigens. New techniques to generate large amounts of DCs together with the molecular identification of human tumour-associated antigens (TAA) have opened new ways for antigen-specific cancer immunotherapies. DCs loaded either with TAA-derived MHC class I-specific synthetic peptides or with whole tumour cell preparations have been used in numerous clinical trials evaluating the efficacy of DCs in patients with cancer. However, the disadvantages of DCs pulsed with synthetic peptides from TAA include the uncertainty regarding the longevity of antigen presentation, the re-

J. Humrich, L. Jenne (✉)
Department of Dermatology, University of Erlangen,
Hartmannstrasse 14, 91052 Erlangen, Germany

striction by the patient's haplotype and the relatively low number of known MHC class I and in particular of MHC class II helper cell-related epitopes. Whole tumour cell preparations are difficult to standardise, and they depend on the availability of tumour cells. Thus the utilisation of viral vectors genetically modified to express TAA for the ex vivo transduction of DCs is an attractive alternative to achieve a MHC I- and MHC II-restricted presentation of tumoural antigens. To induce protective anti-tumoural immune response an increasing number of modified viral vectors have been used to transduce DCs. Although high transduction efficacies were reported for several viruses, analysis of the interaction of viral vectors with DCs has revealed several viral mechanisms that interfere with main functions of DCs, dampening somewhat the initial optimism in the field of DC transduction. However, promising results with different vectors have been achieved. In this review we summarise available data and discuss advantages and drawbacks of currently available vectors.

1
Introduction

Dendritic cells (DCs) constitute a specialised system of antigen-presenting cells (APCs) that are initiators and modulators of the immune response against microbial, tumour and self-antigens (Banchereau et al. 2001). Their unique capacity to induce or boost immunity makes them an attractive tool for immunotherapy, in particular for the induction of anti-tumoural immunity. The development of techniques to generate clinical-grade DCs together with molecular identification of human cancer antigens in the last decade has ushered in a new era of antigen-specific cancer immunotherapy specifically targeting these antigens. Numerous phase I and II clinical trials evaluating the efficacy of DCs in patients with cancer have been initiated, mostly in the post-adjuvant setting (for review see Jenne and Bhardwaj 2001). In these trials, DCs were loaded either with MHC class I-specific synthetic peptides or with whole tumour cell preparations. The disadvantages of using DCs pulsed with synthetic peptides from TAA include the uncertainty regarding the longevity of antigen presentation, the restriction by the patient's haplotype and the relatively low number of known MHC class I and in particular of MHC class II helper cell-related epitopes. Whole tumour cell preparations such as tumour lysates, apoptotic tumour cells or DC tumour cell fusions are difficult to standardise, and they depend on the availability of tumour cells. Thus the utilisation of viral vectors genetically modified to express TAA for the ex vivo transduction of DCs seems to be an attractive alternative to achieve a MHC I- and MHC II-restricted presentation of tumoural antigens in order to induce protective anti-tumoural immune response. An increasing number of modified viral vectors initially designed for gene therapy purposes have been used to transduce DCs. Most of the vectors have been adopted from the field of gene therapy. However, the requirements of vectors for ex vivo transduction of DCs differ considerably from gene therapy vector types, which have been designed so as to achieve the long-term expression of modified genes in selected tissues or organs. The aim of DC-based immunotherapy is not the long-term expression of genes but rather the induction of strong cellular and humoral immunity against an inserted TAA, whereas the immunological response against a given vector should be almost absent in gene therapy vectors. Analysis of the interaction of viral vectors with DCs has revealed several new viral mechanisms that perturb the main function of DCs (for review see Jenne et al. 2001), dampening somewhat the initial optimism in the field of DC transduction. However, promising results with different vectors have been achieved.

2
Dendritic Cells in Immunotherapy

The application of DCs ex vivo-transduced with a virus coding for one or multiple TAA might have several potential advantages over the technically easier direct vaccination. First, endogenous antigen synthesis within DCs ensures direct and simultaneous access to the MHC I and MHC II antigen processing pathway, the latter thought to be essential for the generation of cellular anti-tumoural immunity (Toes et al. 1999). Second, ex vivo transduction allows for the control of DC quality (i.e. maturation status) and expression level of recombinant TAA. Third, vaccination with DCs transduced ex vivo by recombinant vectors might circumvent the problem of anti-viral antibodies known to limit the effect of viruses used for vaccination purposes (Brossart et al. 1997). Fourth, ex vivo-generated and transduced DCs can be injected at anatomical sites of interest (i.e. lymph nodes or tumours). Nevertheless, to the best of our knowledge, no scientific work in humans has focused on the direct comparison of immune responses induced by conventional vaccination with those generated with ex vivo virus-transduced DCs, and thus the advantages/disadvantages of each approach remain to be proven.

3 Viral Vectors Used to Transduce DCs

3.1 Poxviridae

Three members of the Poxviridae family have been developed as vector systems for gene therapy purposes. One is a member of the Orthopoxvirus family, Vaccinia virus (VV), which has been used as poxvirus vaccination with remarkable efficacy and safety in the poxvirus eradication campaign, and two are avipoxviruses (usually bird restricted), the canarypox virus (ALVAC) and the fowlpoxvirus (FWPV).

It has been demonstrated that FWPVs are able to infect CD83-positive DCs at a MOI of 10 (Brown et al. 1999) and that a FWPV coding for recombinant CD80, CD54 and LFA-3 can be used to transduce DCs, thereby increasing their capacity to activate T cells (Zhu et al. 2001). Nevertheless, no further investigations regarding the phenotypical and functional properties of these infected DCs have been performed so far.

ALVAC has been used in phase I clinical vaccination trials as a carrier for carcinoembryonic antigen without clinical side effects, suggesting a favourable safety profile (Horig et al. 2000). However, ALVAC seems to be a less than perfect vector for the transduction of DCs, because it induces apoptosis in immature and mature DCs (Ignatius et al. 2000).

Here we focus on VV as a representative viral vector of the poxviridae. The VV system has been developed into a versatile vector system for gene therapy purposes, able to carry at least 25 kb of heterologous DNA (Carroll and Moss 1997), as it is technically relatively easy to produce stable recombinant VV by homologous recombination (Carroll and Moss 1997; Mackett and Smith 1986). Although the mechanism of entry of VV into DCs is not completely understood, a receptor-mediated process seems likely as MOIs of 2.5–5 are sufficient for the transduction of both immature and mature DCs (Jenne et al. 2000; Drillien et al. 2000). After transduction, only genes regulated by early promoters, although to a far lesser degree, and early-intermediate genes are expressed in monocyte-derived DCs (Jenne et al. 2000). Because of the lack of late gene expression, no virus assembly and consequently no cell lysis can be observed in infected DCs. Nevertheless, the viability of infected DCs is reduced because of the induction of apoptosis (Jenne et al. 2000), presumably representing yet another mechanism of viral immune evasion. Interestingly, further maturation of immature, VV-infected DCs cannot be induced by either inflammatory maturation cytokines or other maturation stimuli such as LPS or Poly-I:C (Jenne et al., unpublished observation). Thus other known VV escape mechanisms, such as the syntheses of viral proteins that bind TNF or IL-1β (Alcami and Smith 1992), do not account for this maturation blockade.

Mature DCs can be infected with an equal or slightly better efficacy than immature DCs. However, mature DCs infected with the western reserve (WR) strain are unable to stimulate the proliferation of allogeneic T cells (Jenne et al. 2000), one of the key features of DCs. Thus, apart from numerous immune evasion strategies aimed at the inhibition of anti-viral immunity (recently reviewed by Alcami and Koszinowski 2000), VV has the capacity to specifically target DC function. In search of mechanisms accounting for this inhibition it was demonstrated that, comparable to HSV, VV induces the loss of CD83 cell surface expression on mature DCs. This effect is initially observed only on infected DC, but after more than 48 h partially also on non-infected bystander cells (Jenne et al. 2000). Even more striking is the VV-induced effect regarding CD80, a costimulatory molecule of vital importance for the induction of potent immune responses (Harris and Ronchese 1999). CD80 is down-regulated on both infected DCs and uninfected bystander DCs (Jenne et al. 2000; Drillien et al. 2000), while other costimulatory molecules such as CD86 and CD40 remain unaltered. Strategies to avoid the immunosuppression of VV include the inactivation of the viral particles by UV light. This allowed for the generation of an anti-EBV immunity (Subklewe et al. 1999) and the utilisation of attenuated virus strains such as the modified virus Ankara (MVA) (Meyer et al. 1991; Sutter and Moss 1992) and strains deficient for late gene expression (Holzer et al. 1999). In vitro, the allostimulatory capacity of MVA-infected DCs is only partially inhibited, and the generation of cellular immunity with MVA-infected mature DCs in vitro has been demonstrated against MUC1 (Trevor et al. 2001) and tyrosinase (Drexler et al. 1999).

For future development of VV-derived vectors it appears important to identify and characterise the viral gene products which interfere with DC functions by down-regulating relevant surface molecules such as CD83 and CD80 and to generate vectors deleted of such genes.

4
Herpesviridae

The family of Herpesviridae consists of eight human herpesviruses: herpes simplex viruses (HSV-1 and HSV-2), cytomegalovirus (CMV), Epstein-Barr virus (EBV), varicella zoster virus (VZV), and human herpesviruses 6/7/8. These viruses are, as is the case for VV, double-stranded DNA viruses. So far only HSVs have been developed as a versatile vector for gene therapy, in particular for targeting the CNS (Fink and Glorioso 1997). The HSV genome has a molecular weight of approximately 150 kb, encodes for about 70 viral proteins and is surrounded by an icosahedral nucleocapsid and a lipid layer envelope. HSV has the ability to package large genetic insertions, thus enabling the simultaneous expression of multiple TAA transcripts even in combination with immunomodulators such as IL-2, IL-12 or CD80 and CD86.

Both immature and mature DCs can be infected with HSV as DCs express two of the necessary entry receptors, Prr-1, an Ig superfamily member, and herpesvirus entry mediator (Hve-A), a member of the TNF/nerve growth factor (NGF) family, to a sufficient degree (Salio et al. 1999). As a consequence of the receptor-mediated entry a relatively low MOI of 1–5 is sufficient to transduce up to 90% of cultured DCs (Salio et al. 1999; Kruse et al. 2000b). Viral replication has only been demonstrated in immature DCs (Mikloska et al. 2001), whereas mature DCs do not support the generation of infectious viral particles (Kruse et al. 2000b). The lytic properties of replicating HSV have a severe implication for the viability of the infected DCs. Although immature DCs are prone to cell death (Mikloska et al. 2001), mature DCs are stable for a relatively long time period (Kruse et al. 2000b).

Generally, HSV has the capacity to hamper host cell protein synthesis by disruption of polysomes and degradation of mRNA (Sydiskis and Roizman 1966). This function is exhibited by the viral protein *vhs*, a product of the UL41 gene (Kwong and Frenkel 1989). The up-regulation of several DC surface molecules such as MHC II, CD86 and CD1a (Salio et al. 1999) after HSV infection argues against a complete host protein shut-off in DCs. Similarly, the inhibition of monocytic cell function by HSV-1 infection seems to be independent of *vhs* expression (Hayward et al. 1993). Thus the inability to induce maturation in HSV-infected immature DCs (Salio et al. 1999) may be induced by a different, as yet unidentified, viral gene product. In addition, apoptosis could also be partially responsible for this maturation block.

Furthermore, VP16 mediates the activation of JNK/SAPK and p38 in Vero cells (Zachos et al. 1999), indicating active interference with signal transduction pathways. However, it remains to be determined as to whether or not these events contribute to the maturation blockade. With regard to the cytokine synthesis, it has been shown that HSV-1-infected immature DCs do not produce any cytokine with or without stimulation (Salio et al. 1999). Hayward et al. reported the reduced expression of costimulatory and adhesion molecules on the surface of infected monocytes (Hayward et al. 1993), whereas Mikloska et al. reported the down-regulation of CD1a, CD40, CD54, CD80 and CD86 on HSV-1-infected immature DCs (Mikloska et al. 2001). Nevertheless, because the HSV-1 infection induces a high degree of cytopathicity in immature DCs (20%–45% of DCs die within the first 24–48 h) the observed down-regulation could be due to cell death rather than a consequence of a specific viral mechanism. Together, the work on the interaction of immature DCs with HSV reveals an immuno-evasive strategy of the virus because only mature DCs can induce an anti-viral immunity.

Because only fully mature DCs induce potent immune responses (Jonuleit et al. 2001), it seems crucial to transduce fully mature DCs with genetically modified HSV vectors. Mature DCs are characterised by the expression of a specific array of surface molecules. These include costimulatory molecules such as CD40, CD80 and CD86, the high expression of MHC I and MHC II molecules and CD83. Unfortunately, wild-type HSV-1, used in investigations to date, severely impairs the allostimulatory capacity of mature DCs (Kruse et al. 2000b). It is noted that costimulatory molecules such as CD80 and CD86 as well as MHC I and MHC II molecules remain unaltered and are highly expressed on HSV-transduced DCs.

On the other hand, CD83, the best-known marker for mature DC, is completely degraded 10 h after infection. Although the precise biological function of CD83 has not been defined as yet, it was demonstrated that the inhibition of the nuclear export of CD83 mRNA from the nucleus into the cytoplasm leads to a reduction of the CD83 cell surface expression on DCs, which is paralleled by a reduced allostimulatory capacity of these DCs (Kruse et al. 2000a). The CD83 degradation in HSV-1-infected DCs is probably due to a virus-induced protein degradation process

(Kruse et al. 2000a). Interestingly, CD83-degradation occurs only in infected DCs; uninfected bystander DCs are not affected.

Together these data suggest that wild-type HSV-1 vectors are not suitable for DC transduction. Furthermore, there are major obstacles to overcome, which include safety problems, long-term gene expression, cytotoxicity, immunological competence of the host and in vivo control of the vector (Fink and Glorioso 1997). For the future development of HSV vectors for ex vivo transduction of DCs, it is crucial to develop viral vectors with a better safety profile than the wild-type viruses and, in addition, to generate mutants from which the gene responsible for the CD83 degradation is deleted.

Several second-generation vectors have already been developed but have not yet been used for DC transduction. These include HSV deletion mutants, displaying reduced neurovirulence (Chou et al. 1990) and increased safety (Mineta et al. 1995), which have been designed and used for phase I clinical trials to treat malignant glioma (Markert et al. 2000). Furthermore, two vector systems with reduced lytic activity have also been developed: first, strains with deletion of the IE gene VP16, necessary for the initiation of the cascade of lytic gene expression (Ace et al. 1989), and second, a vector system based on the cotransfection of a plasmid containing an HSV origin of replication and packaging sequence with cosmids containing HSV genome but with a defective packaging sequence, termed HSV-1 amplicons (Geller 1997; Sena-Esteves 2000). The latter system is currently under investigation in our laboratory.

5
Adenovirus

Adenoviruses (AVs) are naked DNA viruses consisting of 51 different serotypes divided into 6 subgroups designated A to F (Horwitz 1996). Serotypes 2 and 5, to which most adults have been exposed, have been developed as vector systems, initially as a vehicle to transfer genes into the respiratory epithelium of patients with cystic fibrosis but later as a means to transfer genes into a variety of tissues (Kovesdi et al. 1997) with a total capacity of 28–32 kb. The viral dsDNA genome itself consists of approximately 36 kb encoding for over 50 polypeptides. AVs can be either replication competent or replication deficient, depending on the location into which foreign DNA is inserted in the genome (Imler 1995). First-generation AV vectors have deleted early region 1 (E1) genes, which serve together with E2 and E4 genes as transcriptional regulators for viral replication. Because even in the absence of E1 gene products low-level transcription of remaining genes led to the generation of AV immune responses, reducing the lifespan of the transduced cells (Yang and Wilson 1995), second- and third-generation AVs with deleted E1 and E2 and/or E1 and E4 were constructed (Lusky et al. 1998). Genomic deletions in the E3 and E4 region of the virus lead to an increase in the otherwise limited cloning capacity without affecting viral viability (Imler 1995). Infusion of an E1/E4-deleted AV into the hepatic artery to treat partial OTC deficiency unfortunately caused the death of one patient, and consequently further trials with this virus were suspended (Ready 2001). The further deletion of AV genes led to the generation of high-capacity or gutless AV vectors (Kochanek et al. 2001); however, to the best of our knowledge no report of DCs infected with such AV is published.

The different requirements of vectors suitable for gene therapy and DC transduction are highlighted by the fact that in gene therapy approaches, the insertion of genes from the E3 cassette into other viral vectors has been used to reduce anti-vector immunity (Ilan et al. 1997) as the E3 cassette contains genes that regulate the virus-cell interaction and in particular the host immunosurveillance (Wold and Gooding 1991). Deletion of this cassette increases the capacity to insert foreign DNA into the viral genome by 8 kb and appears valuable for DC transduction to avoid undesired viral immunosuppressions.

The entry of AV into cells and thus the cellular tropism is mediated by a two-step process consisting of the virus attachment to the cellular membrane via Ad fibre knobs (Stevenson et al. 1995) followed by internalisation upon binding of the penton base RGD motifs to cellular integrins (Wickham et al. 1993). The coxsackie-adenovirus receptor (CAR) is used by most subgroups (A, C, D and F) as a high-affinity cellular ligand, whereas subgroup B fibres do not bind to CAR (Roelvink et al. 1998). As DCs do not express CAR, high viral titres are required to achieve significant gene transfer (Zhong et al. 1999; Linette et al. 2000; Rea et al. 1999). Three approaches have been applied to improve the DC tropism of AV. First, fibre knobs of serotype B were used to replace Ad5 fibres, which resulted in the reduction of AV titres needed to transduce DCs by an unidentified mechanism (Rea et al. 2001). However, MOIs of 10,000 were still needed to infect immature DCs and a MOI of 1,000 to infect mature DCs, a surprising reversion of the usually better infection of mature DCs. Second, a bi-specific antibody directed against CD40 allowed for both the targeting of mature DCs and the simultaneous further activation of the DCs (Tillman et al. 1999). Third, immature DCs were pre-treated with a CAR-expressing AV, leading to the increased transduction efficacy of the following AV (Stockwin et al. 2002).

Immature DCs preferentially express the $\alpha v\beta^5$ integrin responsible for the final uptake of the AV. Thus immature DCs were generally infected with higher efficacy by AV (Zhong et al. 1999). Differences in the AV strain or virus production procedure might account for discrepancies in the analysis as to whether or not immature DCs are activated by AV. Whereas some groups found no activation (Zhong et al. 1999; Dietz et al. 1998; Diao et al. 1999), others reported a change of DC features towards a more mature profile (Rea et al. 1999; Rouard et al. 2000; Morelli et al. 2000) or even the induction of $CD83^+$ cells from $CD14^+$ monocytes by adding a replication deficient AV at an MOI of 200 (Lyakh et al. 2002). Immature AV-infected DCs could be matured by standard procedures without affecting the transgene expression (Dietz et al. 2001).

The efficient transgene expression of DCs infected with AV and the consecutive induction of transgene-specific T cells has been demonstrated for MUC1 (Pecher et al. 2001; Maruyama et al. 2001), MAGE-3 (Schultz et al. 2002) and MART-1 (Philip et al. 2000). In the latter study, comparable anti-tumoural T-cell effects were achieved with AV-infected DCs and MART-1 peptide-pulsed DC, validating the AV vectors for future immunotherapy studies. On the basis of these promising data, AV-based DC therapy is expected to play a role in future clinical studies.

6
Adeno-Associated Virus

Adeno-associated viruses (AAVs) are non-pathogenic parvoviruses that usually require a helper virus, such as AV, to mediate a productive infection (Muzyczka 1992). Because the AAV vector genome lacks viral coding sequences, the vector has not been associated with toxicity and many different organs can be targeted via in vivo administration (Monahan and Samulski 2000). The major limitation of the AAV vector is the limited cloning capacity, which is only approximately 5 kb.

After initial reports on the infection of DCs with an AAV vector (Ponnazhagan et al. 2001), which demonstrated unaltered DC properties and sound transgene expression, the efficacy of AAV-infected DCs to generate an anti-papillomavirus immunity was compared with that of other loading methods (Chiriva-Internati et al. 2002) and showed some superiority. These data raise hope for AAV vectors as efficient ex vivo DC vector system.

7 Influenza Virus

Negative-stranded RNA viruses comprise several important human pathogens such as measles, mumps, rabies, respiratory syncytial, Ebola, hantavirus and influenza. New techniques to manipulate the genome of such viruses have led to research testing these viruses for potential applications (Palese et al. 1996). From this group of viruses, apart from a rabies virus deleted of its own single glycoprotein and expressing a chimeric human immunodeficiency virus type 1 envelope protein (Foley et al. 2002)], only the capacity of influenza viruses (IVs) to serve as vector for DC transduction has been tested extensively. Unlike DNA viruses, which integrate into the host genome or have a DNA replication intermediate and thus pose the risk of adverse mutations, IV RNA does not integrate into the host genome. Furthermore, IV has a good safety profile, as IV strains are sensitive to anti-viral pharmaceuticals and fail to establish persistent infections. The efficient generation of a plasmodium T cell immunity by vaccinating mice with a recombinant IV containing an epitope of the circumsporozoite protein (Li et al. 1993) was the first experiment demonstrating the efficacy of such viruses. Here the surface glycoprotein haemagglutinin was genetically modified. Bender et al. subsequently analysed the interaction of IV with DCs (Bender et al. 1995). They demonstrated that, in contrast to monocytes, DCs infected with (live or inactivated) IV do not undergo apoptosis and can generate a strong anti-influenza $CD8^{+}$ T cell response even with inactivated virus particles at extremely low levels of infection (Bender et al. 1998). Furthermore, the efficient MHC I presentation (cross-presentation) of viral antigen by DCs that had phagocytosed IV-infected, and thus apoptotic, macrophages was demonstrated (Albert et al. 1998).

With regard to the interaction of IV with DCs, Cella et al. reported that viral infection with the IV strain PR-8 resulted in the maturation of immature DCs, increased antigen presentation and T cell stimulatory capacity (Cella et al. 1999). This was not observed when an avian, so far non-human, pathogenic influenza virus vector was used (Strobel et al. 2000). The latter strain was designed to include an additional gene segment for displaying a recombinant protein. It was shown to be a very efficient vector for delivery of foreign genes into mature human monocyte-derived DCs with a transduction efficacy of over 90% at a MOI of only 1 (Strobel et al. 2000). The transduced DCs retained their characteristic phenotype and were potent immune stimulators, indicating the versatility of the influenza virus vector in the development of genetically modified DCs for immunotherapy.

8
Lentivirus

Lentiviruses (LVs) are a genus in the Retroviridae family, which comprises HIV-1/2, the simian immunodeficiency virus (SIV) and a few non-primate viruses. In addition to the *gag*, *pol* and *env* proteins of simple retroviruses, the LV genome contains accessory regulatory proteins (Frankel and Young 1998). In contrast to retroviruses, LVs are capable of integrating their provirus in certain non-dividing cells (Lewis and Emerman 1994), a feature that makes this virus attractive for both gene therapy studies and ex vivo transduction of DCs derived from non-proliferating $CD14^+$ progenitors. As the genetic information required to package a functional LV core is only a fraction of the genome, the deletion of genes critical for viral pathogenesis was feasible and led to the generation of vectors with high bio-safety (Dull et al. 1998). Progress in the field of vector development has significantly increased the feasibility of using LVs in clinical trials (Trono 2000; Kafri 2001).

DCs represent a natural target for the LVs HIV and SIV. Thus,an attenuated/modified variant might constitute a suitable vector to genetically modify DCs. Several groups have reported the successful transduction of both immature and mature DC. By using a HIV-1ΔEN vector at a MOI of 5, 26% of immature DCs and 5% of mature DCs were infected (Gruber et al. 2000). Our own group found a transduction efficacy of around 50% on using a LV vector to infect mature DCs (Jenne, Piguet, unpublished data) Furthermore, a HIV-1 derived LV, with deletions of all structural and accessory genes, infected 40% of the mature DCs at a MOI of 20 and was able to induce an anti-influenza immunity in the absence of $CD4^+$ helper cells (Dyall et al. 2001). The insertion of the sequence coding for the central triple-stranded DNA structure, usually created during HIV-1 reverse transcription, increased the transduction efficacies of the LV vector (Firat et al. 2002), which infected 95% of immature human and mouse DCs. Importantly, in all these reports, immature DCs infected with LV were still able to mature. Furthermore, mature LV-infected DCs stimulated allogeneic T cells as well as CTL clones. DCs infected in vivo with a LV encoding for the murine tyrosinase-related protein 2 were efficient as therapeutic vaccination in treating a pre-established melanoma in B16 mice (Metharom et al. 2001).

Also, SIVmac251-derived LV vectors have been reported to transduce DCs (Negre et al. 2000; Mangeot et al. 2002). At an MOI of 10, 10% of immature and 32% of mature DCs were reported GFP positive. As for HIV-derived vectors, SIV-infected DCs were comparable to uninfected DCs.

These data indicate that LV-based vector systems represent useful tools to genetically modify DCs. The door to trials with LV-infected DCs in tumour therapy will be opened once first trials in the field of gene therapy prove the safety of LV vectors.

9
Conclusions and Future Outlook for Viral Vectors

To date, viral vectors utilised for DC transduction have been adopted from gene therapy strategies somewhat ignoring the specific effects of viral vectors on DCs and their antigen presentation capacity. It should be a prime goal of future research in the field of DC transduction to further characterise viral immune evasion mechanisms to design viral vectors, which allow for non-perturbating DC transduction.

Regarding the problems associated with vector-specific immunity, although possibly less important than in gene therapy, prime-boost strategies will have to be tested. To minimise anti-vector responses and to booster therapeutic anti-tumour reactions, it might be advantageous to combine different vector systems expressing the same TAA in sequential immunisation protocols or to boost with DCs loaded with peptide or whole tumour cell preparations for antigen presentation.

Acknowledgements. The authors' work was supported by the Marohn Fond. We thank Armin Bender and Vincent Piguet for valuable scientific discussions.

References

Ace CI, McKee TA, Ryan JM, Cameron JM, Preston CM: Construction and characterization of a herpes simplex virus type 1 mutant unable to transinduce immediate-early gene expression. J.Virol. 1989, 63:2260–2269.

Albert ML, Sauter B, Bhardwaj N: Dendritic cells acquire antigen from apoptotic cells and induce class I-restricted CTLs. Nature 1998, 392:86–89.

Alcami A, Koszinowski UH: Viral mechanisms of immune evasion. Immunol.Today 2000, 21:447–455.

Alcami A, Smith GL: A soluble receptor for interleukin-1 beta encoded by vaccinia virus: A novel mechanism of virus modulation of the host response to infection. Cell 1992, 71:153–167.

Banchereau J, Schuler-Thurner B, Palucka AK, Schuler G: Dendritic cells as vectors for therapy. Cell 2001, 106:271–274.

Bender A, Albert M, Reddy A, Feldman M, Sauter B, Kaplan G, Hellman W, Bhardwaj N: The distinctive features of influenza virus infection of dendritic cells. Immunobiology 1998, 198:552–567.

Bender A, Bui LK, Feldman MA, Larsson M, Bhardwaj N: Inactivated influenza virus, when presented on dendritic cells, elicits human CD8+ cytolytic T cell responses. J.Exp.Med. 1995, 182:1663–1671.

Brossart P, Goldrath AW, Butz EA, Martin S, Bevan MJ: Virus-mediated delivery of antigenic epitopes into dendritic cells as a means to induce CTL. J Immunol 1997, 158:3270–3276.

Brown M, Davies DH, Skinner MA, Bowen G, Hollingsworth SJ, Mufti GJ, Arrand JR, Stacey SN: Antigen gene transfer to cultured human dendritic cells using recombinant avipoxvirus vectors. Cancer Gene Ther. 1999, 6:238–245.

Carroll MW, Moss B: Poxviruses as expression vectors. Curr.Opin.Biotechnol. 1997, 8:573–577.

Cella M, Salio M, Sakakibara Y, Langen H, Julkunen I, Lanzavecchia A: Maturation, activation, and protection of dendritic cells induced by double-stranded RNA. J.Exp.Med. 1999, 189:821–829.

Chiriva-Internati M, Liu Y, Salati E, Zhou W, Wang Z, Grizzi F, Roman JJ, Lim SH, Hermonat PL: Efficient generation of cytotoxic T lymphocytes against cervical cancer cells by adeno-associated virus/human papillomavirus type 16 E7 antigen gene transduction into dendritic cells. Eur.J.Immunol. 2002, 32:30–38.

Chou J, Kern ER, Whitley RJ, Roizman B: Mapping of herpes simplex virus-1 neurovirulence to gamma 134.5, a gene nonessential for growth in culture. Science 1990, 250:1262–1266.

Diao J, Smythe JA, Smyth C, Rowe PB, Alexander IE: Human PBMC-derived dendritic cells transduced with an adenovirus vector induce cytotoxic T-lymphocyte responses against a vector-encoded antigen in vitro. Gene Ther. 1999, 6:845–853.

Dietz AB, Bulur PA, Brown CA, Pankratz VS, Vuk-Pavlovic S: Maturation of dendritic cells infected by recombinant adenovirus can be delayed without impact on transgene expression. Gene Ther. 2001, 8:419–423.

Dietz AB, Vuk-Pavlovic S: High efficiency adenovirus-mediated gene transfer to human dendritic cells. Blood 1998, 91:392–398.

Drexler I, Antunes E, Schmitz M, Wolfel T, Huber C, Erfle V, Rieber P, Theobald M, Sutter G: Modified vaccinia virus Ankara for delivery of human tyrosinase as melanoma-associated antigen: induction of tyrosinase- and melanoma-specific human leukocyte antigen A*0201-restricted cytotoxic T cells in vitro and in vivo. Cancer Res. 1999, 59:4955–4963.

Drillien R, Spehner D, Bohbot A, Hanau D: Vaccinia virus-related events and phenotypic changes after infection of dendritic cells derived from human monocytes. Virology 2000, 268:471–81.

Dull T, Zufferey R, Kelly M, Mandel RJ, Nguyen M, Trono D, Naldini L: A third-generation lentivirus vector with a conditional packaging system. J.Virol. 1998, 72:8463–8471.

Dyall J, Latouche JB, Schnell S, Sadelain M: Lentivirus-transduced human monocyte-derived dendritic cells efficiently stimulate antigen-specific cytotoxic T lymphocytes. Blood 2001, 97:114–121.

Fink DJ, Glorioso JC: Herpes simplex virus-based vectors: problems and some solutions. Adv.Neurol. 1997, 72:149–156.

Firat H, Zennou V, Garcia-Pons F, Ginhoux F, Cochet M, Danos O, Lemonnier FA, Langlade-Demoyen P, Charneau P: Use of a lentiviral flap vector for induction of CTL immunity against melanoma. Perspectives for immunotherapy. J.Gene Med. 2002, 4:38–45.

Foley HD, Otero M, Orenstein JM, Pomerantz RJ, Schnell MJ: Rhabdovirus-based vectors with human immunodeficiency virus type 1 (HIV-1) envelopes display HIV-1-like tropism and target human dendritic cells. J.Virol. 2002, 76:19–31.

Frankel AD, Young JA: HIV-1: fifteen proteins and an RNA. Annu.Rev.Biochem. 1998, 67:1–25.

Geller AI: Herpes simplex virus-1 plasmid vectors for gene transfer into neurons. Adv.Neurol. 1997, 72:143–148.

Gruber A, Kan-Mitchell J, Kuhen KL, Mukai T, Wong-Staal F: Dendritic cells transduced by multiply deleted HIV-1 vectors exhibit normal phenotypes and functions and elicit an HIV-specific cytotoxic T-lymphocyte response in vitro. Blood 2000, 96:1327–1333.

Harris NL, Ronchese F: The role of B7 costimulation in T-cell immunity. Immunol.Cell Biol. 1999, 77:304–311.

Hayward AR, Read GS, Cosyns M: Herpes simplex virus interferes with monocyte accessory cell function. J.Immunol. 1993, 150:190–196.

Holzer GW, Remp G, Antoine G, Pfleiderer M, Enzersberger OM, Emsenhuber W, Hammerle T, Gruber F, Urban C, Falkner FG, Dorner F: Highly efficient induction of protective immunity by a vaccinia virus vector defective in late gene expression. J.Virol. 1999, 73:4536–4542.

Horig H, Lee DS, Conkright W, Divito J, Hasson H, LaMare M, Rivera A, Park D, Tine J, Guito K, Tsang KW, Schlom J, Kaufman HL: Phase I clinical trial of a recombinant canarypoxvirus (ALVAC) vaccine expressing human carcinoembryonic antigen and the B7.1 co-stimulatory molecule. Cancer Immunol.Immunother. 2000, 49:504–514.

Horwitz MS: Adenoviruses. In *Fields Virology*. Edited by Fields BN, Knipe DM, Howley PM, Chancrock RM. Philadelphia: 1996:2149–2171.

Ignatius R, Marovich M, Mehlhop E, Villamide L, Mahnke K, Cox WI, Isdell F, Frankel S, Mascola JR, Steinman RM, Pope M: Canarypox-induced maturation of dendritic cells is mediated by apoptotic cell death and tumor necrosis factor-a secretion. J.Virol. 2000, 74:11329–11338.

Ilan Y, Droguett G, Chowdhury NR, Li Y, Sengupta K, Thummala NR, Davidson A, Chowdhury JR, Horwitz MS: Insertion of the adenoviral E3 region into a recombinant viral vector prevents antiviral humoral and cellular immune responses and permits long-term gene expression. Proc.Natl.Acad.Sci.U.S.A 1997, 94:2587–2592.

Imler JL: Adenovirus vectors as recombinant viral vaccines. Vaccine 1995, 13:1143–1151.

Jenne L, Bhardwaj N: Perspectives of DC based immunotherapies. In *Principles and Practice of Oncology*, 6th Edition, Edited by DeVita VT, Hellman S, Rosenberg SA. Lippincott Williams & Wilkins; 2001:1–15.

Jenne L, Hauser C, Arrighi JF, Saurat JH, Hugin AW: Poxvirus as a vector to transduce human dendritic cells for immunotherapy: abortive infection but reduced APC function. Gene Ther 2000, 7:1575–83.

Jenne L, Schuler G, Steinkasserer A: Viral vectors for dendritic cell-based immunotherapy. Trends Immunol 2001, 22:102–107.

Jonuleit H, Giesecke-Tuettenberg A, Tuting T, Thurner-Schuler B, Stuge TB, Paragnik L, Kandemir A, Lee PP, Schuler G, Knop J, Enk AH: A comparison of two types of dendritic cell as adjuvants for the induction of melanoma-specific T-cell responses in humans following intranodal injection. Int.J.Cancer 2001, 93:243–251.

Kafri T: Lentivirus vectors: difficulties and hopes before clinical trials. Curr.Opin.-Mol.Ther. 2001, 3:316–326.

Kochanek S, Schiedner G, Volpers C: High-capacity 'gutless' adenoviral vectors. Curr.Opin.Mol.Ther. 2001, 3:454–463.

Kovesdi I, Brough DE, Bruder JT, Wickham TJ: Adenoviral vectors for gene transfer. Curr.Opin.Biotechnol. 1997, 8:583–589.

Kruse M, Rosorius O, Kratzer F, Bevec D, Kuhnt C, Steinkasserer A, Schuler G, Hauber J: Inhibition of CD83 cell surface expression during dendritic cell maturation by interference with nuclear export of CD83 mRNA. J.Exp.Med. 2000a, 191:1581–1590.

Kruse M, Rosorius O, Kratzer F, Stelz G, Kuhnt C, Schuler G, Hauber J, Steinkasserer A: Mature dendritic cells infected with herpes simplex virus type I exhibit inhibited T cell stimulatory capacity. J.Virol. 2000b, 74:7127–7136.

Kwong AD, Frenkel N: The herpes simplex virus virion host shutoff function. J.Virol. 1989, 63:4834–4839.

Lewis PF, Emerman M: Passage through mitosis is required for oncoretroviruses but not for the human immunodeficiency virus. J.Virol. 1994, 68:510–516.

Li S, Rodrigues M, Rodriguez D, Rodriguez JR, Esteban M, Palese P, Nussenzweig RS, Zavala F: Priming with recombinant influenza virus followed by administration of recombinant vaccinia virus induces CD8+ T-cell-mediated protective immunity against malaria. Proc.Natl.Acad.Sci.U.S.A 1993, 90:5214–5218.

Linette GP, Shankara S, Longerich S, Yang S, Doll R, Nicolette C, Preffer FI, Roberts BL, Haluska FG: In vitro priming with adenovirus/gp100 antigen-transduced dendritic cells reveals the epitope specificity of HLA-A*0201-restricted CD8+ T cells in patients with melanoma. J.Immunol. 2000, 164: 3402–3412.

Lusky M, Christ M, Rittner K, Dieterle A, Dreyer D, Mourot B, Schultz H, Stoeckel F, Pavirani A, Mehtali M: In vitro and in vivo biology of recombinant adenovirus vectors with E1, E1/E2A, or E1/E4 deleted. J.Virol. 1998, 72:2022–2032.

Lyakh LA, Koski GK, Young HA, Spence SE, Cohen PA, Rice NR: Adenovirus type 5 vectors induce dendritic cell differentiation in human CD14(+) monocytes cultured under serum-free conditions. Blood 2002, 99:600–608.

Mackett M, Smith GL: Vaccinia virus expression vectors. J.Gen.Virol. 1986, 67:2067–2082.

Mangeot PE, Duperrier K, Negre D, Boson B, Rigal D, Cosset FL, Darlix JL: High levels of transduction of human dendritic cells with optimized SIV vectors. Mol.Ther. 2002, 5:283–290.

Markert JM, Medlock MD, Rabkin SD, Gillespie GY, Todo T, Hunter WD, Palmer CA, Feigenbaum F, Tornatore C, Tufaro F, Martuza RL: Conditionally replicating herpes simplex virus mutant, G207 for the treatment of malignant glioma: results of a phase I trial. Gene Ther. 2000, 7:867–874.

Maruyama K, Akiyama Y, Nara-Ashizawa N, Hojo T, Cheng JY, Mizuguchi H, Hayakawa T, Yamaguchi K: Adenovirus-mediated MUC1 gene transduction into human blood-derived dendritic cells. J.Immunother. 2001, 24:345–353.

Metharom P, Ellem KA, Schmidt C, Wei MQ: Lentiviral vector-mediated tyrosinase-related protein 2 gene transfer to dendritic cells for the therapy of melanoma. Hum.Gene Ther. 2001, 12:2203–2213.

Meyer H, Sutter G, Mayr A: Mapping of deletions in the genome of the highly attenuated vaccinia virus MVA and their influence on virulence. J.Gen.Virol. 1991, 72 (Pt 5):1031–1038.

Mikloska Z, Bosnjak L, Cunningham AL: Immature monocyte-derived dendritic cells are productively infected with herpes simplex virus type 1. J.Virol. 2001, 75:5958–5964.

Mineta T, Rabkin SD, Yazaki T, Hunter WD, Martuza RL: Attenuated multi-mutated herpes simplex virus-1 for the treatment of malignant gliomas. Nat.Med. 1995, 1:938–943.

Monahan PE, Samulski RJ: AAV vectors: is clinical success on the horizon? Gene Ther. 2000, 7:24–30.

Morelli AE, Larregina AT, Ganster RW, Zahorchak AF, Plowey JM, Takayama T, Logar AJ, Robbins PD, Falo LD, Thomson AW: Recombinant adenovirus induces maturation of dendritic cells via an NF-kappaB-dependent pathway. J.Virol. 2000, 74:9617–9628.

Muzyczka N: Use of adeno-associated virus as a general transduction vector for mammalian cells. Curr.Top.Microbiol.Immunol. 1992, 158:97–129.

Negre D, Mangeot PE, Duisit G, Blanchard S, Vidalain PO, Leissner P, Winter AJ, Rabourdin-Combe C, Mehtali M, Moullier P, Darlix JL, Cosset FL: Characterization of novel safe lentiviral vectors derived from simian immunodeficiency virus (SIVmac251) that efficiently transduce mature human dendritic cells. Gene Ther. 2000, 7:1613–1623.

Palese P, Zheng H, Engelhardt OG, Pleschka S, Garcia-Sastre A: Negative-strand RNA viruses: genetic engineering and applications. Proc.Natl.Acad.Sci.U.S.A 1996, 93:11354–11358.
Pecher G, Spahn G, Schirrmann T, Kulbe H, Ziegner M, Schenk JA, Sandig V: Mucin gene (MUC1) transfer into human dendritic cells by cationic liposomes and recombinant adenovirus. Anticancer Res. 2001, 21:2591–2596.
Philip R, Alters SE, Brunette E, Ashton J, Gadea J, Yau J, Lebkowski J, Philip M: Dendritic cells loaded with MART-1 peptide or infected with adenoviral construct are functionally equivalent in the induction of tumor-specific cytotoxic T lymphocyte responses in patients with melanoma. J.Immunother. 2000, 23:168–176.
Ponnazhagan S, Mahendra G, Curiel DT, Shaw DR: Adeno-associated virus type 2-mediated transduction of human monocyte-derived dendritic cells: implications for ex vivo immunotherapy. J.Virol. 2001, 75:9493–9501.
Rea D, Havenga MJ, van Den AM, Sutmuller RP, Lemckert A, Hoeben RC, Bout A, Melief CJ, Offringa R: Highly efficient transduction of human monocyte-derived dendritic cells with subgroup B fiber-modified adenovirus vectors enhances transgene-encoded antigen presentation to cytotoxic T cells. J.Immunol. 2001, 166:5236–5244.
Rea D, Schagen FH, Hoeben RC, Mehtali M, Havenga MJ, Toes RE, Melief CJ, Offringa R: Adenoviruses activate human dendritic cells without polarization toward a T-helper type 1-inducing subset. J Virol. 1999, 73:10245–10253.
Ready T: Trials suspended due to death at Hopkins. Nat.Med. 2001, 7:877.
Roelvink PW, Lizonova A, Lee JG, Li Y, Bergelson JM, Finberg RW, Brough DE, Kovesdi I, Wickham TJ: The coxsackievirus-adenovirus receptor protein can function as a cellular attachment protein for adenovirus serotypes from subgroups A, C, D, E, and F. J.Virol. 1998, 72:7909–7915.
Rouard H, Leon A, Klonjkowski B, Marquet J, Tenneze L, Plonquet A, Agrawal SG, Abastado JP, Eloit M, Farcet JP, Delfau-Larue MH: Adenoviral transduction of human 'clinical grade' immature dendritic cells enhances costimulatory molecule expression and T-cell stimulatory capacity. J Immunol Methods 2000, 241:69–81.
Salio M, Cella M, Suter M, Lanzavecchia A: Inhibition of dendritic cell maturation by herpes simplex virus. Eur.J.Immunol. 1999, 29:3245–53.
Schultz ES, Chapiro J, Lurquin C, Claverol S, Burlet-Schiltz O, Warnier G, Russo V, Morel S, Levy F, Boon T, Van den Eynde BJ, van der BP: The production of a new MAGE-3 peptide presented to cytolytic T lymphocytes by HLA-B40 requires the immunoproteasome. J.Exp.Med. 2002, 195:391–399.
Sena-Esteves M, Saeki Y, Fraefel C, Breakefield XO: HSV-1 amplicon vectors – simplicity and versatility. Mol.Ther. 2000, 2:9–15.
Stevenson SC, Rollence M, White B, Weaver L, McClelland A: Human adenovirus serotypes 3 and 5 bind to two different cellular receptors via the fiber head domain. J.Virol. 1995, 69:2850–2857.
Stockwin LH, Matzow T, Georgopoulos NT, Stanbridge LJ, Jones SV, Martin IG, Blair-Zajdel ME, Blair GE: Engineered expression of the Coxsackie B and adenovirus receptor (CAR) in human dendritic cells enhances recombinant adenovirus-mediated gene transfer. J.Immunol.Methods 2002, 259:205–215.
Strobel I, Krumbholz M, Menke A, Hoffmann E, Dunbar PR, Bender A, Hobom G, Steinkasserer A, Schuler G, Grassmann R: Efficient expression of the tumor-asso-

ciated antigen MAGE-3 in human dendritic cells, using an avian influenza virus vector. Hum.Gene Ther. 2000, 11:2207–2218.

Subklewe M, Chahroudi A, Schmaljohn A, Kurilla MG, Bhardwaj N, Steinman RM: Induction of Epstein-Barr virus-specific cytotoxic T-lymphocyte responses using dendritic cells pulsed with EBNA-3A peptides or UV- inactivated, recombinant EBNA-3A vaccinia virus. Blood 1999, 94:1372–1381.

Sutter G, Moss B: Nonreplicating vaccinia vector efficiently expresses recombinant genes. Proc.Natl.Acad.Sci.U.S.A 1992, 89:10847–10851.

Sydiskis RJ, Roizman B: Polysomes and protein synthesis in cells infected with a DNA virus. Science 1966, 153:76–78.

Tillman BW, de Gruijl TD, Luykx-de Bakker SA, Scheper RJ, Pinedo HM, Curiel TJ, Gerritsen WR, Curiel DT: Maturation of dendritic cells accompanies high-efficiency gene transfer by a CD40-targeted adenoviral vector. J.Immunol. 1999, 162:6378–6383.

Toes RE, Ossendorp F, Offringa R, Melief CJ: CD4 T cells and their role in antitumor immune responses. J.Exp.Med. 1999, 189:753–756.

Trevor KT, Hersh EM, Brailey J, Balloul JM, Acres B: Transduction of human dendritic cells with a recombinant modified vaccinia Ankara virus encoding MUC1 and IL-2. Cancer Immunol.Immunother. 2001, 50:397–407.

Trono D: Lentiviral vectors: turning a deadly foe into a therapeutic agent. Gene Ther. 2000, 7:20–23.

Wickham TJ, Mathias P, Cheresh DA, Nemerow GR: Integrins alpha v beta 3 and alpha v beta 5 promote adenovirus internalization but not virus attachment. Cell 1993, 73:309–319.

Wold WS, Gooding LR: Region E3 of adenovirus: a cassette of genes involved in host immunosurveillance and virus-cell interactions. Virology 1991, 184:1–8.

Yang Y, Wilson JM: Clearance of adenovirus-infected hepatocytes by MHC class I-restricted CD4+ CTLs in vivo. J.Immunol. 1995, 155:2564–2570.

Zachos G, Clements B, Conner J: Herpes simplex virus type 1 infection stimulates p38/c-Jun N-terminal mitogen-activated protein kinase pathways and activates transcription factor AP-1. J.Biol.Chem. 1999, 274:5097–5103.

Zhong L, Granelli-Piperno A, Choi Y, Steinman RM: Recombinant adenovirus is an efficient and non-perturbing genetic vector for human dendritic cells. Eur.J.Immunol. 1999, 29:964–972.

Zhu M, Terasawa H, Gulley J, Panicali D, Arlen P, Schlom J, Tsang KY: Enhanced activation of human T cells via avipox vector-mediated hyperexpression of a triad of costimulatory molecules in human dendritic cells. Cancer Res. 2001, 61:3725–3734.

Chapter 12

Cross-Presentation of Cell-Associated Antigens by Dendritic Cells

M. Larsson, J. F. Fonteneau, N. Bhardwaj

Abstract. There is a strict requirement for professional antigen-presenting cells (APCs) in the generation of immunity toward most viruses. Exogenous pathways of MHC class I-restricted antigen presentation play an important role in the generation of antiviral immunity, particularly in the immune surveillance of virus-infected tissues of nonhematopoietic origin, and to bypass the detrimental effects of direct virus infection on professional APCs. The mechanisms underlying generation of antiviral immunity under these circumstances are discussed.

1 Introduction

$CD8^+$ T cells develop into cytotoxic effector cells after recognition of peptides on MHC class I molecules on the surface of professional antigen-presenting cells (pAPCs). Proteins synthesized endogenously within the cytoplasm of pAPCs are generally the source of MHC class I-restrict-

M. Larsson, J.F. Fonteneau, N. Bhardwaj (✉)
The Laboratory of Molecular Neuro-Oncology, The Rockefeller University, 1230 York Avenue, New York, NY 10021, USA

ed peptides. Most peptides arise through proteosomal processing and transported into the ER via transporter associated with antigen presentation (TAP) molecules for delivery onto newly synthesized MHC class I molecules. In contrast, antigens presented to CD4$^+$ T cells are acquired exogenously via endocytosis, macropinocytosis, or phagocytosis and processed into peptides within late endosomal compartments where they access nascent MHC class II molecules. After acquisition of peptides, both MHC class I and II molecules are exocytosed to the cell surface. (Yewdell et al. 1999a; Villadangos and Ploegh 2000). Because CD8 immunity is efficiently generated to viruses that do not infect pAPCs or that undermine pAPC function and viability, alternative MHC class I-restricted antigen processing pathways exist.

Michael Bevan first proposed the concept that antigens could be acquired from exogenous cell-associated sources by pAPC to be presented on MHC class I molecules (Bevan 1976). This phenomenon, referred to as "cross-priming," was subsequently supported by several findings (reviewed in Heath and Carbone 2001). Many antigens were shown to be "cross-presented" by APCs, e.g., viral, bacterial, tumor, and self-antigens, foreign histocompatibility antigens, and H-Y antigens, and in various forms, e.g., live and dead cells, antibody-opsonized cells, immune complexes, exosomes, heat shock proteins, DNA, and soluble, non-cell-associated proteins (reviewed in Larsson et al. 2001a). In this review we limit our discussion to the cross-presentation of cell-associated antigens by dendritic cells (DCs), with particular attention to viral antigens.

2
Identity of the APC Mediating Cross-Presentation

Bone marrow (BM)-derived APCs are essential for priming cytotoxic T cell responses to intracellular pathogens such as *Listeria*, vaccinia, and polio viruses (Lenz et al. 2000; Sigal and Rock 2000; Sigal et al. 1999). In vivo studies utilizing BM chimeras demonstrated that cross-presentation of cell-associated, viral antigens was a property of BM-derived APCs (Sigal and Rock 2000; Heath et al. 1998; Huang et al. 2000; Sotomayor et al. 2001; Bennett et al. 1998). Sigal et al. showed that the exogenous MHC class I pathway is the obligatory mechanism for the initiation of CTL responses to viruses that infect only nonhematopoietic cells. They created a series of BM chimeras (B6 to cPVR mice, the latter a transgenic strain expressing the polio virus receptor) that permitted the infection of only non-BM-derived cells. Strong anti-polio virus CTLs were generated in a TAP-dependent manner as no responses were seen when TAP -/- mice were used as donors of the BM (Heath et al. 1998).

In vitro studies suggested that DCs, macrophages, B cells, and even liver sinusoidal epithelial cells had the capacity to cross-present antigens from exogenous sources to $CD8^+$ T cells (reviewed in Heath and Carbone 2001 and Larsson et al. 2001a). den Haan et al., in an elegant set of experiments, unequivocally showed that DCs but not macrophages cross-primed CTLs in vivo. Mice primed i.v. with irradiated OVA-loaded splenocytes developed OVA-specific, MHC class I-restricted responses. The cells responsible for the in vivo cross-presentation were DCs expressing CD11c, CD8α, DEC-205, and low levels of CD11b (formerly referred to as the lymphoid DCs) (den Haan et al. 2000). In contrast, CD8α-, $CD11b^+$ DCs and macrophages that also internalized OVA-loaded cells failed to mediate cross-presentation. Similar results were reported by Pooley et al. Mice were injected i.v. with FITC-conjugated OVA (Pooley et al. 2001). All DC subsets from the spleen ($CD4^+/CD8^+$; $CD4^+/CD8^-$; $CD4^-/CD8^-$) showed uptake of OVA within 1 h after injection of antigen (Pooley et al. 2001). However, effective stimulation of class I MHC-restricted $CD8^+$ T cells was obtained only with the $CD8^+$ DCs, whereas class II MHC-restricted stimulation was most potent with the $CD4^-/CD8^-$ subset. With the concomitant administration of LPS to the mice, the various DC subsets became more efficient, although the relative preference by $CD8^+$ and $CD8^-$ DCs for MHC class I- and II-restricted antigens, respectively, remained. This observation that DC subsets are not equal in their cross-priming capacity is strengthened by studies performed in a rat model (Huang et al. 2000). Only the $CD4^-/OX40^-$ DCs in mesenteric afferent lymphatic vessels contain apoptotic gut epithelial cells. They exhibit poor T cell stimulation capacity compared with the other subset found in the afferent lymph ($CD4^+/OX41^+$) (Huang et al. 2000). A primary role of DCs in cross-presentation of exogenous self antigens (cross-tolerance) was also reported by Kurts et al. (Kurts et al. 2001). These data suggest that different DC lineages have specialized functions. Whether this is related to differences in antigen processing and/or molecular signaling is currently not known. It remains to be seen whether cross-priming and cross-tolerance are mediated by the same DC subsets and/or by different DC subsets.

3
Viral Antigens Undergoing Cross-Presentation

Because a shared outcome of infection by many viruses is apoptotic cell death (and possibly secondary necrosis), "resurrection" of antigens via phagocytosis of infected dead cells or cell debris has been proposed as one mechanism by which DCs cross-prime T cells (Albert et al. 1998b). Substantial evidence exists that cross-presentation of cell-associated antigen and particulate antigens is more efficient than presentation of soluble antigens, up to 10^5-fold in some cases (Rodriguez et al. 1999; Li et al. 2001; Machy et al. 2000).

Cross-presentation of exogenous antigens derived from infected cells has been confirmed in vitro for a variety of viruses, influenza, HIV-1, vaccinia, EBV, and CMV, which leads to $CD8^+$ T cell, and in some cases to $CD4^+$ T cell responses (Albert et al. 1998b; Larsson et al. 2001c; Tabi et al. 2001; Arrode et al. 2000; Herr et al. 2000; Subklewe et al. 2001; Ferlazzo et al. 2000) and are reviewed below.

The first studies were done with influenza virus that established itself as a model virus to define basic elements of cross-presentation in vitro. Albert et al. first demonstrated that human DCs rather than monocytes were the pAPCs responsible for cross-presentation (Albert et al. 1998b). Virus-infected syngeneic, allogeneic, and xenogeneic cells of myeloid, lymphoid, thymic, epithelial, or fibroblast origin were then shown to be effective sources of antigens for DCs after apoptotic or necrotic cell death (Albert et al. 1998b; Larsson et al. 2001c; Sauter et al. 2000). Cross-presentation was shown to require two steps: phagocytosis by immature DCs of apoptotic cells utilizing receptors such as CD36 and $\alpha v\beta 5$ and probably others, followed by delivery of a maturation signal to DCs to optimize both antigen processing and presentation (Larsson et al. 2001c; Sauter et al. 2000; Albert et al. 1998a). This scenario leads to the activation of both $CD4^+$ and $CD8^+$ effector T cells. Interestingly, the efficiency of presentation was extraordinary in that just one influenza-infected apoptotic cell per 100 DCs was sufficient to activate memory $CD8^+$ T cells from primed donors (Albert et al. 1998b).

As with influenza virus, CTL immunity to vaccinia virus in vivo is TAP dependent and requires BM-derived APCs (Sigal and Rock 2000). In vitro, vaccinia virus readily infects immature human DCs, but the infection is abortive, induces apoptosis, inhibits maturation, and impairs the immunostimulatory capacity of DCs (Engelmayer et al. 1999; Jenne et al. 2000). This raises the issue of how one generates such effective antivaccinia immunity after vaccination (Demkowitz et al. 1996). Larsson et al. demonstrated that human DCs rapidly elicited CD8 effector cells from blood memory cells after the phagocytosis of either apoptotic or necrotic vaccinia-infected cells (Larsson et al. 2001c). Cross-presentation was mediated most efficiently by CD40-L matured DCs but not by monocytes, even if the latter were activated with CD40-L. Traditionally, vaccinia is delivered for vaccination by scarification, and local epidermal DCs, i.e., Langerhans cells in the skin may provide a vehicle to traffic vaccinia to draining lymph nodes. In vivo, one mechanism for engaging and activating T cells may be cross-presentation by bystander DCs that have ingested vaccinia-infected dying cells and subsequently underwent maturation as a result of the inflammation associated with infection and/or CD40L-CD40 interactions with intranodal $CD4^+$ T cells (Larsson et al. 2001c; Engelmayer et al. 1999).

EBV, a gamma herpesvirus, is largely restricted to cells of the B lymphoid lineage in which it establishes a lytic or latent infection. $CD8^+$ T cell responses have been described to antigens expressed in both stages. Cross-presentation has been proposed as one mechanism by which EBV responses are generated. Acquisition of apoptotic or necrotic BLCLs by DCs activated $CD4^+$ and/or $CD8^+$ T cells toward EBNA-1, EBNA-3A, and latent membrane protein 2 (Herr et al. 2000; Subklewe et al. 2001). Interestingly, $CD8^+$ responses have been identified in some donors at relative high frequencies toward EBNA-1, which is a surprising observation because this antigen in its endogenously expressed form is not directly presented by infected B cells to $CD8^+$ T cells because of an internal glycine-alanine repeat that inhibits proteosomal processing (reviewed in Rickinson 1996). Cross-presentation by DCs is probably the mechanism allowing such a $CD8^+$ response, as Blake et al. have shown that full-length EBNA-1 can be processed by DCs, (Blake et al. 2000). In this case, one would presume that clinically effective immunity would be mediated by $CD4^+$ rather than $CD8^+$ T cells, because infected B cells would not express relevant epitopes for the latter.

Human cytomegalovirus (HCMV) latently infects $CD34^+$ hematopoietic precursors and DCs. Reactivation can occur as macrophage/DC precursors differentiate (Hahn et al. 1998). In vitro, human monocyte-derived immature DCs are susceptible to productive CMV infection by strains propagated in endothelial cells, producing syncytia that ultimately die (Riegler et al. 2000). These findings, as with vaccinia, raise the issue of how immunity to HCMV is generated, because DC function is compromised. HCMV-infected fibroblast lines can be cross-presented by human DCs to $CD8^+$ T cells, while simultaneously providing a maturation signal to the DCs (Tabi et al. 2001; Arrode et al. 2000). Exogenous CMV lysates and pp65 protein are also cross-presented to $CD8^+$ T cells (Maecker et al. 2001). Cross-presentation may enable DCs to bypass immune evasion tactics used by HCMV to avoid T cell recognition.

HSV-1, a member of the alpha-herpesvirus family, normally infects cells in mucosal surfaces or squamous epithelia and enters a latent period in neurons. $CD8^+$ T cell immunity is necessary to clear HSV-1 infection; however, the virus persists throughout the lifetime of the infected host through evolution of multiple strategies of immune evasion, including inhibition of antigen processing (Ploegh 1998). HSV-1 destroys cells it infects and induces apoptosis of PBMCs, $CD4^+$ T cells, and hepatocytes (Daheshia et al. 1998). Infection of immature DCs inhibits their maturation and adversely affects their T cell stimulatory capacity (Salio et al. 1999: Kruse et al. 2000). However, dying DCs can induce bystander maturation of uninfected DCs. Therefore, phagocytosis of virus-infected cells or DCs by immature DCs that simultaneously undergo bystander maturation may be a significant mechanism by which antiviral immunity to HSV-1 is generated.

Life-long immunity develops after measles virus (MV) infection, but acute infection can be accompanied by transient immunosuppression. A productive infection of DCs and possibly T cells with ensuing cell death is speculated to be the cause. In vitro, direct infection of DCs with MV induces syncytia formation and cell death (Fugier-Vivier et al. 1997; Servet-Delprat et al. 2000b). Bystander maturation of uninfected DCs takes place after contact and/or engulfment of MV-infected apoptotic DCs. The addition of T cells to the cultures augments virus replication and results in apoptosis of both infected and uninfected T cells (Servet-Delprat et al. 2000a). These observations provide an explanation for the immunosuppression seen during MV infection while indicating cross-presentation as a mechanism by which immunity is generated (Servet-Delprat et al. 2000b).

Extensive viral replication, syncytia formation, and cell death follow the interaction between $CD4^+$ T cells and DCs previously exposed to HIV-1 virus. Furthermore, DC blood levels decline after acute infection. These effects, should they occur in vivo, would be likely to compromise potent stimulation of antiviral immune responses against HIV-1. DCs from HIV$^+$ donors can present nonreplicating, chemically inactivated HIV-1 to autologous T cells and induce full activation. Furthermore, DCs acquire apoptotic CD4$^+$ cells (infected with infectious or chemically inactivated HIV-1) and activate viral antigen-specific CD4$^+$ and CD8$^+$ T cells. Defective virus or infected dead cells, rather than live virus, may be critical sources of antigen for the generation and maintenance of anti-HIV-1 immunity by DCs (Larsson et al., submitted).

These studies confirm that human DCs are the primary APCs mediating cross-presentation, at least in vitro, and suggest an important role for cross-priming in vivo for the initiation of antiviral immune responses. Given that many viruses infect DCs, often with detrimental consequences, cross-priming allows induction of CTLs even when APC function is compromised.

4 Role of CD4$^+$ T Cells and DC Maturation in Cross-Presentation

Cross-priming of CD8$^+$ T cells to foreign antigens requires helper CD4$^+$ T cells, which recognize cognate antigen on the same APC (Bennett et al. 1997). Help is mediated primarily via engagement of CD40 on APCs, as its ligation completely restores CD8$^+$ CTL activities in CD4$^+$ T cell-depleted mice (Bennett et al. 1998). Similarly, cross-tolerance to self- or tumor antigens is converted to priming after exogenous delivery of antigen-specific CD4$^+$ T cells (Kurts et al. 1997) or triggering of CD40 with antibodies (Sotomayor et al. 1999). CD40-independent components for cross-priming of CD8$^+$ CTLs toward recombinant influenza hemagglutinin (HA) or lysate of a tumor line expressing HA have been identified (Lu et al. 2000). A model has been proposed to integrate all these findings, in which primary CD4 help is provided to DCs via CD40 but can be replaced by CD4$^+$ T cell-derived cytokines (Lu et al. 2000). The precise nature of this CD40-independent help, which occurs after DC sensitization, remains to be determined, but other TNF family members, costimulatory molecules, and cytokines will need to be evaluated. It is interesting to speculate whether the loss of CD4 help in diseases such as HIV infection leads to the decline of cross-presentation and generation of CTL effectors.

The differentiation or activation state of the APC mediating cross-presentation appears to be a central determinant of T cell priming versus tolerance. In the absence of activating signals, immature DCs constitutively capturing tissue antigens would induce tolerogenic $CD8^+$ response, whereas simultaneous exposure to factors that mature DCs would induce stimulatory $CD8^+$ responses. Such factors would include inflammatory cytokines that can be produced by virus- or bacteria-infected cells undergoing caspase-dependent apoptosis (Restifo 2000). In addition, viruses or bacteria releasing dsRNA, LPS or CpG-expressing DNA constructs and infected dead and dying cells releasing these products and/or heat shock proteins would promote DC maturation (reviewed in Larsson et al. 2001b). These factors can bind to pattern recognition molecules, e.g., Toll-like receptors and the conserved molecule CD91. The concomitant induction of CD4 helper cell responses within the draining lymph nodes would further enhance cross-presentation to microbial antigens.

In vitro and in vivo studies support the concept that DC maturation controls elements of the immune response. For example, in healthy volunteers, injection of influenza matrix protein peptide (MP)-pulsed immature DCs induces immunoregulatory $CD8^+$ T cells, which inhibit the function of antigen-specific cytolytic effectors (Dhodapkar et al. 2001). In contrast, MCM-matured DCs induce cytolytic effectors (Dhodapkar and Bhardwaj 2000). In vitro, immature DCs have been shown to induce immunoregulatory $CD4^+$ T cells whereas mature DCs induce conventional $CD4^+$ effectors (Jonuleit et al. 2000). This model has recently been challenged. Albert et al. found that mature DCs are required for *both* cross-tolerance and cross-priming (Albert et al. 2001). In an in vitro cross-presentation system using apoptotic influenza-infected EL4 cells, they showed that monocyte-conditioned medium or TNF-α plus PGE_2-matured human DCs caused the proliferation and eventual deletion of influenza-specific $CD8^+$ T cells. The generation of IFN-γ-secreting effectors was only achieved when CD40L, an agonistic anti-CD40 antibody, $CD4^+$ T cells, or a $CD4^+$ T cell derived supernatant was added to the "mature" DCs. The authors interpreted this result to mean that "mature" DCs cross-presenting antigens require additional signals to induce the full activation of $CD8^+$ T cells into effectors. However, other studies have shown that mature DCs can cross-present viral antigens (vaccinia, influenza, or HIV-1) to purified $CD8^+$ T cells and induce IFN-γ production and proliferation (Larsson et al. 2001c). More recently, Menges et al. showed that TNF-α-matured DCs pulsed with an autoantigenic peptide prevented the induction of EAE in mice through the induction of IL-10-producing immunoregulatory $CD4^+$ T cells. In contrast, LPS- and anti-

CD40-matured DCs induced Th1 responses and had no effect on disease progression (Menges et al. 2002). Both sets of DCs were "mature" by phenotypic standards but differed in that the LPS/CD40-matured DCs produced IL-12 and IL-10. This study is a good example of the need to establish a consensus on what maturation means, as results on T cell activation may differ depending on what maturation stimuli are used.

5 Intracellular Pathways for Cross-Presentation

Kovacsovics-Bankowski and Rock (Kovacsovics-Bankowski and Rock 1995) suggested the existence of a common final pathway for class I presentation of exogenous and endogenous antigens. Sensitivity to proteosomal inhibitors, mutations in TAP-1 and TAP-2 and to Brefeldin A, indicated that soluble or particulate antigens from exogenous sources accessed the cytoplasm for further processing and exited the Golgi for recognition by T cells (Yewdell et al. 1999b). In vivo murine models confirmed that TAP proteins are essential for cross-priming MHC class I-restricted responses to protein, tumor, or viral antigens (Sigal and Rock 2000; Huang et al. 1996).

Rodriguez et al. (Rodriguez et al. 1999) used immune-complexed antigens and a mouse dendritic cell line (D1) to demonstrate that antigens could access the cytosol from endosomes. This phenomenon was specific to DCs but not evident within macrophages. They showed that acidification of endosomes, functional proton-pump ATPases, functional proteosome, and TAP were needed for cytosolic transport and efficient processing and presentation of antigens. Furthermore, molecules of 3–20 kDa but no larger could be localized within the cytosol. The authors speculated that preliminary processing occurs in the phagosome and active transporters or "leaky membranes" permit the transfer of antigens into the cytosol (Rodriguez et al. 1999).

Because some antiviral CTL responses are dependent on BM-derived APCs but independent of TAP function alternative pathways must exist by which these CTLs are generated (Lenz et al. 2000; Sigal and Rock 2000). Certain exogenous antigens are internalized and loaded on MHC class I molecules independent of functional proteasome and TAP and are resistant to Brefeldin A treatment (Harding and Song 1994; Bachmann et al. 1996; Singh-Jasuja et al. 2000; Norbury et al. 2001). For instance, Hsp-70-mediated presentation of OVA epitopes can be TAP independent, depending on the orientation of the peptide in the context of Hsps (Castellino et al. 2000). Processing within endosomal/phagosomal vesicles and access to recycling MHC class I molecules have been proposed to account for these findings (Ramachandra et al. 1999). Investigators believe that endosomal processing would be a less dominant pathway given that it is less efficient than the TAP-dependent pathways (Lenz et al. 2000; Sigal and Rock 2000).

Regurgitation of peptides as a result of endosomal processing from phagosomes into extracellular fluids where peptides bind surface MHC class I molecules has been proposed as a third pathway for the cross-presentation of exogenous antigens (Harding and Song 1994; Pfeifer et al. 1993). It is unlikely that this pathway plays a dominant role, if any, in cross-presentation.

6
Summary

The field of cross-presentation by DCs is relatively young but growing. The next decade should shed light on remaining unanswered questions such as the identities of DC subsets that mediate cross-presentation for tolerance versus priming and cellular and molecular signals and routes that define the cross-presentation pathway in these APCs.

Note: Due to publication delays the authors appologize for not being able to include recent publications relevant to this topic.

References

Albert ML, Jegathesan M, Darnell RB: Dendritic cell maturation is required for the cross-tolerization of CD8+ T cells. Nat Immunol 2001, 2:1010–1017.

Albert ML, Pearce SFA, Francisco LM, Sauter B, Roy P, Silverstein RL, Bhardwaj N: Immature dendritic cells phagocytose apoptotic cells via $\alpha v\beta 5$ and CD36, and cross-present antigens to cytotoxic T lymphocytes. J. Exp. Med. 1998a, 188:1359–1368.

Albert ML, Sauter B, Bhardwaj N: Dendritic cells acquire antigen from apoptotic cells and induce class I-restricted CTLs. Nature 1998b, 392:86–89.

Arrode G, Boccaccio C, Lule J, Allart S, Moinard N, Abastado JP, Alam A, Davrinche C: Incoming human cytomegalovirus pp65 (UL83) contained in apoptotic infected fibroblasts is cross-presented to CD8(+) T cells by dendritic cells. J Virol 2000, 74:10018–10024.

Bachmann MF, Lutz MB, Layton GT, Harris SJ, Fehr T, Rescigno M, Ricciardi-Castagnoli P: Dendritic cells process exogenous viral proteins and virus-like particles for class I presentation to CD8+ cytotoxic T lymphocytes. Eur.J. Immunol. 1996, 26:1–7.

Bennett SRM, Carbone FR, Karamalis F, Flavell RA, Miller JFAP, Heath WR: Help for cytotoxic-T-cell responses is mediated by CD40 signalling. Nature 1998, 393:478–480.

Bennett SRM, Carbone FR, Karamalis F, Miller JJAP, Heath WR: Induction of a CD8+ cytotoxic T lymphocyte response by cross-priming requires cognate CD4+ T cell help. J. Exp. Med. 1997, 186:65–70.

Bevan MJ: Cross-priming for a secondary cytotoxic response to minor H antigens with H-2 congenic cells which do not cross-react in the cytotoxic assay. J. Exp. Med. 1976, 143:1283–1288.

Blake N, Haigh T, Shaka'a G, Croom-Carter D, Rickinson A: The importance of exogenous antigen in priming the human CD8(+) T cell response: lessons from the EBV nuclear antigen EBNA1. J Immunol 2000, 165:7078–7087. High frequency of CTLs against EBNA1 was detected in the blood of EBV-seropositive donors. The authors conclude that CTL responses toward EBNA1 are initiated by an exogenous HLA class I presentation pathway, as epitopes from EBNA1 are poorly generated by the classic endogenous pathway of HLA class I presentation. Furthermore, the observation that DCs cocultured with soluble EBNA1 are able to activate EBNA1-specific CTL supports their conclusion.

Castellino F, Boucher PE, Eichelberg K, Mayhew M, Rothman JE, Houghton AN, Germain RN: Receptor-mediated uptake of antigen/heat shock protein complexes results in Major Histocompatibility Complex class I antigen presentation via two distinct processing pathways. J. Exp. Med. 2000, 191:1957–1964.

Daheshia M, Feldman L, Rouse B: Heres simplex virus latency and immune response. Current Opinion in Microbiology 1998, 1:430–435.

Demkowitz WE, Jr. Littaua RA, Wang J, Ennis FA: Human cytotoxic T-cell memory: long-lived responses to vaccinia virus. J. Virol. 1996, 70:2627–2631.

den Haan J, Lehar S, Bevan M: CD8+ but not CD8- dendritic cells cross-prime cytotoxic T cells in vivo. J. Exp. Med. 2000, 192:1685–1696. The authors used OVA-loaded splenocytes incapable of direct MHC class I antigen presentation to identify the APC responsible for cross-presenting antigens to CTL. Whereas the $CD8^+$ and $CD8^-$ DC subsets are equally able to acquire cellular associated components from the injected splenocytes, only the $CD8^+$/$CD11c^+$/$DEC205^+$/$CD4^-$ DCs are able to cross-present OVA epitopes.

Dhodapkar MV, Bhardwaj N: Active immunization of humans with dendritic cells. J. Clin. Immunol. 2000, 20:167–174.

Dhodapkar MV, Steinman RM, Krasovsky J, Munz C, Bhardwaj N: Antigen specific inhibition of effector T cell function in humans after injection of immature dendritic cells. J. Exp. Med. 2001, 193:233–238.

Engelmayer J, Larsson M, Subklewe M, Chahroudi A, Schmaljohn A, William C, Steinman RM, Bhardwaj N: Vaccinia virus inhibits the maturation of human dendritic cells: a novel mechanism of immune evasion. J. Immunol. 1999, 163:6762–6768.

Ferlazzo G, Semino C, Spaggiari GM, Meta M, Mingari MC, Melioli G: Dendritic cells efficiently cross-prime HLA class I-restricted cytolytic T lymphocytes when pulsed with both apoptotic and necrotic cells but not with soluble cell-derived lysates. Int Immunol 2000, 12:1741–1747.

Fugier-Vivier I, Servet-Delprat C, Rivailler P, Rissoan M-C, Liu Y-J, Rabourdin-Combe C: Measles virus suppresses cell-mediated immunity by interfering with the survival and functions of DC and T cells. Submitted 1997.

Hahn G, Jores R, Mocarski E: Cytomegalovirus remains latent in a common precursor of dendritic and myeloid cells. Proc Natl Acad Sci USA 1998, 95:3937–3942.

Harding CV, Song R: Phagocytic processing of exogenous particulate antigens by macrophages for presentation by class I MHC molcules. J. Immunol. 1994, 153:4925–4933.

Heath WR, Carbone FR: Cross-presentation, dendritic cells, tolerance and immunity. Annu. Rev. Immunol. 2001, 19:47–64.

Heath WR, Kurts C, Miller JFAP, Carbone F: Cross-tolerance: a pathway for inducing tolerance to peripheral tissue antigens. J. Exp. Med. 1998, 187:1549–1553.

Herr W, Ranieri E, Olson W, Zarour H, Gesualdo L, Storkus WJ: Mature dendritic cells pulsed with freeze-thaw cell lysates define an effective in vitro vaccine designed to elicit EBV-specific CD4(+) and CD8(+) T lymphocyte responses. Blood 2000, 96:1857–1864.

Huang AYC, Bruce AT, Pardoll DM, Levitsky HI: In vivo cross-priming of MHC class I-restricted antigens requires the TAP transporter. Immunity 1996, 4:349–355.

Huang F-P, Platt N, Wykes M, Major JR, Powell TJ, Jenkins CD, MacPherson GG: A discrete subpopulation of dendritic cells transports apoptotic intestinal epithelial cells to T cell areas of mesenteric lymph nodes. J. Exp. Med. 2000, 191:435–444. In a rat model, a subset of DCs migrating to secondary lymph node from the gut were shown to contain apoptotic gut epithelial cells. These DCs are $CD4^{-}OX40^{-}$ and exhibit in vitro low capacity to enhance T cell proliferation compared with the other subset found in the afferent lymph ($CD4^{+}OX41^{+}$). This subset of rat DCs may represent the BM-derived APCs responsible for peripheral tolerance.

Jenne L, Hauser C, Arrighi JF, Saurat JH, Hugin AW: Poxvirus as a vector to transduce human dendritic cells for immunotherapy: abortive infection but reduced APC function. Gene Ther 2000, 7:1575–1583.

Jonuleit H, Schmitt E, Schuler G, Knop J, Enk AH: Induction of interleukin 10-producing, nonproliferating CD4(+) T cells with regulatory properties by repetitive stimulation with allogeneic immature human dendritic cells. J Exp Med 2000, 192:1213–1222.

Kovacsovics-Bankowski M, Rock KL: A phagosome-to-cytosol pathway for exogenous antigens presented on MHC class I molecules. Science 1995, 267:243–246.

Kruse M, Rosorius O, Kratzer F, Stelz G, Kuhnt C, Schuler G, Hauber J, Steinkasserer A: Mature dendritic cells infected with herpes simplex virus type I exhibit inhibited T cell stimulatory capacity. J. Virol. 2000, 74:7127–7136.

Kurts C, Cannarile M, Klebba I, Brocker T: Cutting Edge: Dendritic cells are sufficient to cross-present self-antigens to CD8 T cells in vivo. J Immunol 2001, 166:1439–1442. BM from transgenic mice with MHC class I expression driven selectively in DCs under the CD11c promoter was used to reconstitute irradiated transgenic mice that express membrane-bound OVA in the pancreas. An antigen-specific $CD8^+$ T cell division, restricted only by the donor HLA allele, was observed, demonstrating that DCs were sufficient to cross-present self-antigens.

Kurts C, Carbone FR, Barnden M, Blanas E, Allison J, Heath WR, Miller JF: CD4+ T cell help impairs CD8+ T cell deletion induced by cross-presentation of self-antigens and favors autoimmunity. J. Exp. Med. 1997, 186:2057–2062.

Larsson M, Fonteneau JF, Bhardwaj N: Dendritic cells resurrect antigens from dead cells. Trends Immunol. 2001a, 22:141–148.

Larsson M, Fonteneau JF, Lee A, Bhardwaj N: Interations of viruses with dendritic cells. In: Dendritic Cells; 2001b.

Larsson M, Fonteneau J, Somersan S, Sanders C, Bickham K, Thomas E, Mahnke K, Bhardwaj N: Efficiency of cross presentation of vaccinia virus-derived antigens by human dendritic cells. Eur J Immunol 2001c, 31:3432–3442. The authors quantify the efficiency of human DC cross-presention of vaccinia virus-derived antigen by using an antigen-specific $CD8^+$ CTL clone as responder cells. They show that low quantities of necrotic or apoptotic infected cells are equally efficient sources of antigen for cross-presentation. Maturation of DCs, especially with CD40L, is required to achieve efficient cross-presentation that can approach the presentation efficiency of a high quantity of exogenously added peptides.

Lenz LL, Butz EA, Bevan MJ: Requirements for bone marrow-derived antigen-presenting cells in priming cytotoxic T cell responses to intracellular pathogens. J. Exp. Med. 2000, 192:1135–1142. Using a chimeric mice model, the authors show that BM-derived APCs are required to induce CTL responses against *Listeria monocytogenes* and vaccinia virus. Interestingly, CTL responses against LCMV can develop in the absence of BM-derived APCs, suggesting that CTL responses can be initiated by direct presentation of viral antigens by non-BM-derived cells.

Li M, Davey GM, Sutherland RM, Kurts C, Lew AM, Hirst C, Carbone FR, Heath WR: Cell-associated ovalbumin is cross-presented much more efficiently than soluble ovalbumin in vivo. J Immunol 2001, 166:6099–6103.

Lu Z, Yuan L, Zhou X, Sotomayor E, Levitsky HI, Pardoll DM: CD40-independent pathways of T cell help for priming of CD8(+) cytotoxic T lymphocytes. J. Exp. Med. 2000, 191:541–550. Using an in vitro system with DCs as APCs and $CD8^+$ and $CD4^+$ T cells from influenza hemagglutinin class I and II peptide-specific TCR transgenic mice, the authors show that both CD40-dependent and -independent CD4 help is required for efficient priming of $CD8^+$ T cells. This CD40-independent CD4 help comprises direct sensitization of DCs by $CD4^+$ T cells or $CD4^+$ T cell-derived lymphokines.

Machy P, Serre K, Leserman L: Class I-restricted presentation of exogenous antigen acquired by Fcgamma receptor-mediated endocytosis is regulated in dendritic cells. Eur. J. Immunol. 2000, 30:848–857.

Maecker HT, Ghanekar SA, Suni MA, He XS, Picker LJ, Maino VC: Factors affecting the efficiency of CD8+ T cell cross-priming with exogenous antigens. J Immunol 2001, 166:7268–7275.

Menges M, Rossner S, Voigtlander C, Schindler H, Kukutsch NA, Bogdan C, Erb K, Schuler G, Lutz MB: Repetitive injections of dendritic cells matured with tumor necrosis factor alpha induce antigen-specific protection of mice from autoimmunity. J Exp Med 2002, 195:15–21.

Norbury C, Princiotta M, Bacik I, Brutkiewicz R, Wood P, Elliott T, Bennink J, Yewdell J: Multiple antigen-specific processing pathways for activating naive CD8+ T cells in vivo. J Immunol 2001, 166:4355–4362.

Pfeifer JD, Wick MJ, Roberts RL, Findlay K, Normark SJ, Harding CV: Phagocytic processing of bacterial antigens for class I MHC presentation to T cells. Nature 1993, 361:359–362.

Ploegh HL: Viral strategies of immune evasion. Science 1998, 280:248–253.

Pooley JL, Heath WR, Shortman K: Intravenous soluble antigen is presented to CD4 T cells by CD8(-) dendritic cells, but cross-presented to CD8 T cells by CD8(+) dendritic cells. J. Immunol. 2001, 166:5327–5330. The authors show that the three mouse spleen subsets of DCs ($CD4^+/CD8^-$, $CD4^-/CD8^-$, and $CD4^-/CD8^+$) have the same efficiency in vivo for the uptake of soluble OVA. However, after injection of soluble OVA in vivo and purification of these DCs subsets, only the $CD4^-/CD8^+$ subset was able to process and present OVA epitopes to $CD8^+$-specific T cells, whereas all three DC subsets were able to present OVA epitopes to $CD4^+$-specific T cells with a higher efficiency of the $CD8^-$ DC subsets. The authors conclude that the $CD4^-/CD8^+$ DCs are specialized to cross-present antigens to $CD8^+$ T cells.

Ramachandra L, Song R, Harding CV: Phagosomes are fully competent antigen-processing organelles that mediate the formation of peptide:class II MHC complexes. J. Immunol. 1999, 162:3263–3272.

Restifo N: Building better vaccines: how apoptotic cell death can induce inflammation and activate innate and adaptive immunity. Curr Opin Immunol 2000, 12:597–603.

Riegler S, Herbart H, Einsele H, Brossart P, Jahn G, Sinzger C: Monocyte-derived dendritic cells are permissive to the complete replicative cycle of human cytomegalovirus. J Gen Virol 2000, 81:393–399.

Rodriguez A, Regnault A, Kleijmeer M, Ricciardi-Castagnoli P, Amigorena S: Selective transport of internalized antigens to the cytosol for MHC class I presentation in dendritic cells. Nature Cell Biol. 1999, 1:362–368. The authors used a mouse splenic-derived DC line to show that only DCs and not macrophages can cross-present OVA antigen to $CD8^+$ T cells from immune complexes. The pathway is TAP dependent and involves an active and size-dependent transport of specific antigens across endosomal membrane to the cytosol inside DCs.

Salio M, Cella M, Suter M, Lanzavecchia A: Inhibition of dendritic cell maturation by herpes simplex virus. Eur. J. Immunol. 1999, 29:3245–3253.

Sauter B, Albert ML, Francisco L, Larsson M, Somersan S, Bhardwaj N: Consequences of cell death. Exposure to necrotic tumor cells, but not primary tissue cells or apoptotic cells, induces the maturation of immunostimulatory dendritic cells. J. Exp. Med. 2000, 191:423–434.

Servet-Delprat C, Vidalain PO, Azocar O, Le Deist F, Fischer A, Rabourdin-Combe C: Consequences of Fas-mediated human dendritic cell apoptosis induced by measles virus. J Virol 2000a, 74:4387–4393.

Servet-Delprat C, Vidalain PO, Bausinger H, Manie S, Le Deist F, Azocar O, Hanau D, Fischer A, Rabourdin-Combe C: Measles virus induces abnormal differentiation of CD40 ligand-activated human dendritic cells. J. Immunol. 2000b, 164:1753–1760.

Sigal LJ, Crotty S, Andino R, Rock KL: Cytotoxic T-cell immunity to virus-infected non-haematopoietic cells requires presentation of exogenous antigen. Nature 1999, 398:77–80. Transgenic mice expressing the human poliovirus receptor only in nonhematopoietic cells were created to assess whether BM-derived APCs are essential for viral antigen presentation. After infection with polio virus, CTLs were elicited in mice reconstituted with wild-type BM but remained undetectable in mice reconstituted with BM from $TAP^{-}/^{-}$ mice. This study demonstrates that nonhematopoietic cells are unable to directly initiate a immune response against poliovirus and that TAP-dependent antigen cross-presentation by bone marrow pAPCs is sufficient for the priming of poliovirus specific CTLs.

Sigal LJ, Rock KL: Bone Marrow-derived Antigen-presenting cells are required for the generation of cytotoxic T lymphocyte responses to viruses and use transporter associated with antigen presentation (TAP)-dependent and -independent pathways of antigen presentation. J. Exp. Med. 2000, 192:1143–1150. Using transgenic mice, the authors show that cross-presentation by BM-derived APCs is required for the generation of CTL responses against LCMV and influenza antigens. A functional TAP is required to present the majority of studied epitopes. However, a TAP-independent pathway can be used by BM-derived APCs to generate certain epitopes when they are expressed in high quantities.

Singh-Jasuja H, Toes RE, Spee P, Munz C, Hilf N, Schoenberger SP, Ricciardi-Castagnoli P, Neefjes J, Rammensee HG, Arnold-Schild D, Schild H: Cross-presentation of glycoprotein 96-associated antigens on Major Histocompatibility Complex class I molecules requires receptor-mediated endocytosis. J. Exp. Med. 2000, 191:1965–1974.

Sotomayor EM, Borrello I, Rattis F-M, Abrams J, Staveley-O'Carroll K, Levitsky HI: Cross-presentation of tumor by bone marrow derived antigen presenting cells is the dominant mechanism in the induction of T cell tolerance during B-cell lymphoma progression. Blood 2001, 98:1070–1077.

Sotomayor EM, Borrello I, Tubb E, Rattis FM, Bien H, Lu Z, Fein S, Schoenberger S, Levitsky HI: Conversion of tumor-specific CD4+ T-cell tolerance to T-cell priming through in vivo ligation of CD40. Nat Med 1999, 5:780–787.

Subklewe M, Paludan C, Tsang ML, Steinman RM, Münz C: Dendritic cells cross-present latency gene products from Epstein-Barr Virus-transformed B cells and expand tumor-reactive CD8(+) killer T cells. J. Exp. Med. 2001, 193:405–412.

Tabi Z, Moutaftsi M, Borysiewicz LK: Human cytomegalovirus pp65- and immediate early 1 antigen-specific HLA class I-restricted cytotoxic T cell responses induced by cross-presentation of viral antigens. J Immunol 2001, 166:5695–5703.

Villadangos JA, Ploegh HL: Proteolysis in MHC class II antigen presentation: who's in charge? Immunity 2000, 12:233–239.

Yewdell J, Anton LC, Bacik I, Schubert U, Snyder HL, Bennink JR: Generating MHC class I ligands from viral gene products. Immunol Rev 1999a, 172:97–108.

Yewdell JW, Norbury CC, Bennink JR: Mechanisms of exogenous antigen presentation by MHC class I molecules in vitro and in vivo: implications for generating CD8+ T cell responses to infectious agents, tumors, transplants, and vaccines. Adv Immunol 1999b, 73:1–77.

Chapter 13

Dendritic Cells and HCMV Cross-Presentation

G. Arrode, C. Davrinche

Abstract. The outcome of a viral infection is the result of an endless fight between the organism whose task is to mount an antiviral response and the virus that adapts strategies to circumvent the host response. Human cytomegalovirus (HCMV), a latent herpesvirus, can be considered as a spearhead in exploiting co-existence with the host to develop numerous immuno-evasion mechanisms. The ability of the organism to initiate a primary immune response against viruses such as HCMV is highly dependent on the capacity of professional antigen-presenting cells (APCs), namely dendritic cells (DCs), to prime and activate specific effector T cells. Recent findings emerging from the murine cytomegalovirus (MCMV) animal model demonstrated that infection of murine DCs with MCMV impaired their capacity to prime an effective T cell response. Even though data on interference of HCMV with DC functions are still limited, immunosuppressive effects identical to those reported

G. Arrode, C. Davrinche (✉)
INSERM U 395, CHU Purpan, BP 3028, 31024 Toulouse Cédex, France

for MCMV can be suspected and we may then ask how a cytotoxic T lymphocyte (CTL) response is generated in these unfavourable conditions. In response to this question, cross-presentation of HCMV antigens by uninfected DCs to $CD8^+$ T cells could be considered a key process in initiating an immune response. In this chapter we discuss the mechanisms through which DCs could acquire HCMV antigens and how cross-presentation could be modulated throughout infection. Moreover, further knowledge of DC functions is key for the development of DC-based immunotherapy against HCMV.

1 Introduction

The outcome of a viral infection is the result of an endless fight between the organism whose task is to mount an antiviral response and the virus that adapts strategies to circumvent the host response. Human cytomegalovirus (HCMV), a latent herpesvirus, can be considered a spearhead in exploiting co-existence with the host to develop numerous immuno-evasion mechanisms. The ability of the organism to initiate a primary immune response against viruses such as HCMV is highly dependent on the capacity of professional antigen-presenting cells (APC), namely dendritic cells (DCs), to prime and activate specific effector T cells. Recent findings emerging from the murine cytomegalovirus (MCMV) animal model demonstrated that infection of murine DCs with MCMV impaired their capacity to prime an effective T cell response. Even though data on interference of HCMV with DC functions are still limited, immunosuppressive effects identical to those reported for MCMV can be suspected and we may then ask how a cytotoxic T lymphocyte (CTL) response is generated in these unfavourable conditions. In response to this question, cross-presentation of HCMV antigens by uninfected DCs to $CD8^+$ T cells could be considered a key process in initiating an immune response. In this chapter we discuss the mechanisms through which DCs could acquire HCMV antigens and how cross-presentation could be modulated throughout infection. Moreover, further knowledge of DC functions is key for the development of DC-based immunotherapy against HCMV.

2
HCMV Tropism and Pathogenesis

HCMV, also called human herpes virus 5 (HHV-5), is a β-herpesvirus whose genome is the largest (230–250 kbp) of the double-stranded DNA viruses. HCMV, like the other enveloped Herpesviridae, consists of an isocaedric nucleocapside surrounded by a tegument, also named matrix (for review see Mocarski 1996). HCMV infection is restricted to humans, and histopathological analyses have revealed a broad spectrum of target organs and cell types in which the virus is able to replicate, such as epithelial cells from various origins (lung, intestine, nephron, retina), endothelial cells, fibroblasts, macrophages, smooth muscle cells, astrocytes and neurons (for review see Sinzger and Jahn 1996). After a primary infection that is usually asymptomatic in healthy individuals, the nonproductive viral genome persists for the lifetime of the host and can be reactivated episodically.

HCMV infection is extremely prevalent and can lead to severe and often fatal disease in immunocompromised individuals, such as patients with chronic illness and organ transplant recipients. In the latter, HCMV infection can lead to rejection and in some cases to death. Commonly encountered HCMV-associated pathologies in immunodeficient patients include retinitis, hepatitis and pneumonitis.

Infection by HCMV is the most common congenital viral infection and carries a high risk of long-term morbidity and mortality. Intrauterine transmission to the fetus after primary infection in early and late pregnancy usually results in death or in severe neurological handicap, including sensorineural hearing loss with typical migrational anomalies, optic atrophy, disturbed myelination, cerebella hypoplasia and microcephaly (Mocarski 1996; Perez-Jiminez et al. 1998). Studies of MCMV tropism showed that all different cell types in the brain can be infected but with various levels of replication (van Den Pol et al. 1999). Unexpectedly, it has been shown that even in seropositive women, HCMV reinfection with a different strain can lead to intrauterine transmission and symptomatic congenital infection (Boppana et al. 2001). HCMV is one of the most harmful opportunistic infections faced by patients with AIDS and may contribute to neurological defects in immunodeficient adults after either primary infection or reactivation. Post-mortem examination of AIDS patients suggests that many neurons are latently infected by HCMV (Mocarski 1996; Cohen and Dix 1997).

Because under immunosuppression, reactivation from latency may cause disease, studies have been carried out in haematopoietic progenitor and peripheral blood to determine the cell type distribution of latent transcripts. In vitro infection experiments as well as studies in the naturally infected host provided evidence that HCMV remains latent in myeloid and DC progenitors (Hahn et al. 1998). After a primary infection, haematogenous dissemination of the virus throughout the body could involve endothelial cells, circulating monocytes and polymorphonuclear cells (Gerna et al. 1992; Grefte et al. 1993; Sinzger et al. 1995).

Thus, in immunocompetent people, HCMV persists in a latent state in reservoir cells until episodic reactivation occurs, whatever triggering mechanism is involved, and leads to viral shedding. During such a chronic infection, immunocompetent hosts are able to contain viral replication and spreading through activation and expansion of anti-HCMV T lymphocytes. In contrast, immunosuppressed hosts, whether or not of iatrogeneic origin, are no longer protected against the cytopathic effects of the virus and may then develop HCMV-associated disease (for review see Britt and Alford 1996).

Prevention of HCMV-associated complications in allogeneic bone marrow transplants after adoptive transfer of HCMV-specific T cells suggests a major role of T lymphocytes in protection against both primary infection and reactivation of the virus (Walter et al. 1995).

Furthermore, in vivo experiments carried out on mice showed that depletion of NK and T cells induced recurrence of latent murine MCMV, suggesting that latency is maintained in the host through a sustained monitoring by a cellular immune response (Polic et al. 1998, Steffens et al. 1998). Altogether, these findings demonstrate that cellular immunity plays a major role in controlling acute as well as latent and recurrent infection by HCMV.

3 Cellular Immune Response Against HCMV

3.1 IE1 (UL123) and pp65 (UL83) Are Immunodominant Antigens

Direct evidence that T cells of both $CD4^+$ and $CD8^+$ subsets were the main effectors involved in protection was provided by in vivo adoptive transfer of T cell both in mouse (Reddehase et al. 1985) and in bone marrow-transplanted humans (Riddell et al. 1992). Amongst the 200 potential open reading frames found in the HCMV genome only a few have been reported as antigenic proteins. Originally, studies aimed at identifying targets of T lymphocytes showed that the dominating $CD8^+$ T cell response was directed against the matrix protein pp65 (UL83), because these CTLs represented 70%–90% of the overall anti-HCMV response (Wills et al. 1996). Later on, more and more data were reported on the prevalence of pp65-specific CTL with the use of limiting dilution assays (Gyulai et al. 2000), and cytofluorimetric analyses based on either staining of intracellular cytokines or labelling with major histocompatibility complex (MHC) class I peptide tetramers (Kern et al. 1998; Gillespie et al. 2000). Surprisingly, the immediate-early regulatory protein IE1 that had been labelled as a target for CTL since 1988 (Borysiewicz et al. 1988) sank into oblivion until recent rehabilitation by several groups showing that magnitude of anti-IE1 precursor frequency could be similar to that of anti-pp65 (Gyulai et al. 2000, Kern et al. 1999).

We further learned from adoptive transfer experiments that $CD4^+$ T cells were required for long-term persistence of the transferred $CD8^+$ CTL in bone marrow transplants (Walter et al. 1995). These findings fit in with data showing that $CD4^+$ T cells directed against IE1 were present at a high frequency in blood donors (Davignon et al. 1995). Antiviral effects of TNF-α and IFN-γ secreted by anti-IE1 $CD4^+$ T cell clones as well as their direct cytotoxicity (Davignon et al. 1996, Vaz-Santiago et al. 2001), support the assumption that $CD4^+$ T cells may play a significant role in the control of spread of HCMV. Kinetics studies of $CD4^+$ T cells appearance after primary infection suggest that these cells play a primary protective role against HCMV (Rentenaar et al. 2000). Even though the $CD8^+$ T cell subset seems to be much in evidence, the role of $CD4^+$T cells is likely underestimated and needs to be extensively re-evaluated through determination of precursor frequency in healthy blood donors. In summary, protective T cell response against HCMV seems to be dominated by T cells directed against the regulatory protein IE1 and the structural protein pp65.

3.2
HCMV Immunosubversion

HCMV developed strategies for being ignored by the immune system throughout infection. Amongst these evasion processes, those inducing a down-regulation of MHC class I molecule transport to the cell surface involve viral proteins encoded by multiple independent loci within the unique short region (US). These immunosuppressive proteins are expressed from the immediate-early to the late phase of the virus cycle (for reviews see Ploegh 1998; Alcami and Koszinowski 2000; Loenen et al. 2001). Accordingly, we may ask how in these unfavourable conditions CTL directed against HCMV could be activated after primary infection and how frequency of these effectors could be maintained at such a high level in latently infected healthy donors. Indeed, an extensive study of anti-pp65 response in healthy seropositive donors showed that freshly isolated specific $CD8^+$ T cells were effector memory T cells as assessed by their phenotypes and their capacity to kill peptide-pulsed targets (Gillespie et al. 2000). These data suggested that in addition to latency, HCMV viraemia is maintained in immunocompetent donors because of sustained low-grade replication which consistently provides antigens for CTL activation. This is further supported by studies showing long-lasting anti-IE1 and anti-pp65 responses in numerous immunocompetent donors (Gyulai et al. 2000). The question remains whether the impact of viral tactics to down-regulate the host response is significant and it leads Reddehase to reconsider the term *immuno-evasion* for the more appropriate term *immunosubversion* (Reddehase 2000).

Surprisingly, the main targets of anti-HCMV T cell response are IE1 and pp65 proteins that are both available very early after infection. A recent study in mouse showed that MCMV-infected macrophages resist MHC class I down-modulating mechanisms (Hengel et al. 2000). Provided that this could occur in human macrophages, these findings give new clues in the understanding of elevated $CD8^+$ T cell responses directed against IE1 and pp65 in seropositive donors as the result of sustained viral reactivation (Reddehase 2000). Whereas this could be an explanation to support a secondary response in seropositive people, the question remains regarding the establishment of a primary response against the virus.

3.3 HCMV Infection of Dendritic Cells and Presentation of Incoming pp65

Because primary immune responses require the involvement of DCs as professional APCs there was some question about interaction of HCMV with DCs. It has been established that productive infection of DCs could be obtained in vitro provided that clinical HCMV strains have not been propagated in fibroblasts. Nevertheless, some clinical isolates did not replicate in DCs in vitro, pointing out important interstrain differences in the infection of these cells (Riegler et al. 2000). Because we do not yet have any direct proof of DCs infection in vivo, we may ask whether it is a prerequisite for the establishment of a primary immune response. In vitro as well as in vivo, permissive infection of mouse DCs with MCMV resulted in their functional paralysis through interference with their endocytic capacity, maturation process and ability to prime allogeneic T cells (Andrews et al. 2001). Even if no data have been published yet for HCMV, we expect that similar immunosuppressive mechanisms could be described for human DCs. In that case, infection of DCs would impair their capacity to sustain a primary T cell response except against antigens that are available before expression of viral immunosuppressors.

Interestingly, whereas the matrix protein pp65 is synthesised in the early phase of the viral cycle, it is delivered into the cytosol of infected cells in the first minutes after fusion of the viral envelope with the cell membrane (Topilko and Michelson 1994). This process makes incoming pp65 processable into peptides for MHC class I loading and presentation to $CD8^+$ T cells (McLaughlin-Taylor et al. 1994, Arrode et al. 2000, Gallot et al. 2001) and thus could provide a quick defence mechanism against the virus as described in the HIV infection (Buseyne et al. 2001). We may then ask whether this immediate pathway of antigen presentation could account for priming of CTL responses against pp65. Indeed, immediate entry of HCMV matrix proteins such as pp65 into the cytosol of DCs could provide an efficient way for MHC class I-associated peptide presentation irrespective of HCMV strain variability, provided that the virus adsorbs to and fuses with cell membrane. Analysis of pp65 expression level in DCs after infection at 5 m.o.i. with either laboratory strain AD169 or clinical strain TB40/E showed that it was either undetectable or strongly delayed compared with that observed in HCMV-infected fibroblasts (Arrode et al. 2000; Tabi et al. 2001). These data suggest that under these conditions, presentation of incoming pp65 unlikely accounted for activation of CTL by infected DCs. The reason that DCs are less efficient than non-professional APCs such as fibroblasts and astrocytoma cells (Arrode et al. 2000, Gallot et al. 2001) in in vitro presentation of incoming pp65 to $CD8^+$ T cells is still an open question. Nevertheless, it has been demonstrated that infection of immature DCs for 1 h with the laboratory strain AD169 at 50 m.o.i. allowed generation of anti-pp65

$CD8^+$ T cells from peripheral blood mononuclear cells (PBMCs) of HCMV-positive blood donors, providing evidence that DCs were able to process and present incoming pp65 to CTL (Cho et al. 2001). When comparing DCs versus skin fibroblasts as stimulators for anti-pp65 $CD8^+$ T cells, the authors observed identical efficiencies except that fibroblasts were infected with 10 m.o.i. for 6 h. Altogether these findings suggest that a very high multiplicity of infection is required for presentation of incoming proteins by DCs and raises the question of its physiological relevance. In conclusion, we could make the assumption that other mechanisms are involved in priming of CTL response directed against HCMV.

4
Cross-Presentation of HCMV Antigens to Overcome Viral Subversion

4.1
HCMV-Infected Dead Cells as an Antigen Source for DCs

DCs function as sentinels of the immune system and are characterised by their capacity to activate T cells through unusual pathways of antigen capture including phagocytosis of apoptotic and necrotic cells and transfer from live cells (for review see Banchereau et al. 2000, Harshyne et al. 2001). Because DCs can deliver exogenous antigens either in soluble or particulate form into the MHC class I pathway it is conceivable that anti-viral CTL priming might be achieved in this way in cases of infection by cytopathic viruses. Co-evolution of viruses and host cells resulted in the development of strategies to ensure their own survival. A first line of antiviral defence may involve various signal transduction mechanisms including programmed cell death of infected cells to avoid viral spreading. We showed that very early after infection with HCMV AD169 strain, MRC5 fibroblasts became sensitive to apoptosis mediated by ligands of the TNF family [TNF-α, Fas-L, TNF-related apoptosis-inducing ligand (Trail)] and then provided a suitable source of antigen for phagocytosis by immature DCs. We demonstrated that internalisation of purified apoptotic and secondary necrotic infected fibroblasts allowed cross-presentation of pp65 as assessed by activation of $CD8^+$ T cells (Arrode et al. 2000). The in vivo relevance of these findings was supported by data demonstrating that in Fas and TNF-R1 knockout mice, apoptosis of infected cells played a pivotal role in the clearance of murine CMV

(Fleck et al. 1998). Our data were confirmed by others except that the authors suggested that cross-presentation occurred through an unidentified but apoptosis-independent mechanism (Tabi et al. 2001). Later on, by using co-culture of immature DCs and infected fibroblasts, we confirmed our previous results with additional evidence that acquisition of pp65 from infected cells required cell-to-cell contact as observed by others (Tabi et al. 2001) and that cross-presentation was effective when DCs were added to recently infected fibroblasts (Arrode et al. 2002).

The fact that infected fibroblasts were sensitised to TNF-α apoptosis very early after infection, before viral replication, emphasised the importance of cross-presentation for efficient and rapid induction of CTL. Importantly, our data showing that pp65 was derived from the immediate delivery of the viral matrix into fibroblasts suggest that any cell able to interact with the virus and sensitive to apoptosis could serve as a source of pp65 antigen for DCs.

Mechanisms inhibiting TNF-R family-mediated apoptosis have been described involving the viral proteins UL37, an inhibitor of mitochondrial permeability (Goldmacher et al. 1999) and UL36, a suppressor of caspase-8 (Flice) activity (Skaletskaya et al. 2001). These processes may act in the immediate-early and early-late phases after HCMV infection, respectively, thus providing an escape mechanism allowing viral replication and spreading. This emphasises the importance of both the very immediate delivery of pp65 in infected cells and sensitisation to apoptosis before synthesis of these HCMV inhibitors. Together these observations argue in favour of cross-presentation of the incoming pp65 as a major mechanism for priming of CTL directed against pp65. We have not yet dissected the molecular mechanisms involved in pp65 acquiring by DCs, but preliminary data suggested that co-culture of early infected fibroblasts with immature DCs increased the number of apoptotic fibroblasts.

4.2
Temporal Regulation of HCMV Cross-Presentation by DCs

It has been demonstrated that Trail is inducible after infection of fibroblasts with HCMV AD169 and then allows sensitisation of infected cells to apoptosis (Sedger et al. 1999). This suggests that apoptosis may occur through an autocrine mechanism providing viral antigens for phagocytosis by DCs at local sites of infection.

After viral infections, a first line of defence involves innate immunity, mainly by NK cells that can exert anti-viral activity through either secretion of cytokines or direct cytotoxicity. Studies of mice infected with MCMV as well as the description of extreme susceptibility to HCMV infection in humans with a complete lack of NK cells support a key role for these cells (Bukowski et al. 1985; Biron et al. 1989). Accordingly, we can assume that killing of HCMV-infected cells by NK cells could provide starting material for cross-presentation by DCs. Some reports demonstrating that monocyte-derived DCs have potent cytotoxic activity in vitro on human tumour cells (Liu et al. 2001; Vanderheyde et al. 2001; Vidalain et al. 2001) suggested that cross-presentation of dead tumour cells by DCs could be an important innate effector mechanism in the elimination of malignant cells by $CD8^{+}$ T cells in vivo. We can assume that after 'danger' signals provided by bystander virus infection, DCs acquire the ability to kill infected cells under the action of cytokines such as Trail (Vidalain et al. 2000) and TNF-α or still-unknown mediators (Hubert et al. 2001), either through a bystander effect or via cell-to-cell contact. This field of research should be further investigated for a better understanding of the role of cytotoxic DCs at the interface between innate and adaptive immunity against viruses. Cytotoxicity as well as other effector functions are acquired by DCs through transmission of maturation signals involving dsRNA, CD40L, CpG motifs, LPS and type I interferon (Reis e Sousa 2001).After their recruitment, localisation and maturation at sites of pathogen invasion, migration to regional lymph nodes is an essential step in the immunobiology of DCs.

Obviously, all key processes involved in DC functions could be targeted by viruses to avoid an immune response (Klagge et al. 1999). We showed that late-HCMV infected MRC5 fibroblasts and their respective supernatants inhibited DC maturation through secretion of TGF-β and that cross-presentation of pp65 to $CD8^{+}$ T cells was impaired (Arrode et al. 2002). In contrast, maturation of DCs and cross-presentation were optimal in the presence of early-infected fibroblasts. These findings emphasise the significance of the timing of the recruitment of DCs at sites of infection. Interestingly, HCMV US28 gene is known to encode a viral homologue to CC chemokine receptors (Bodaghi et al. 1998) that could interfere with the chemokine-dependent recruitment of DCs at sites of infection (Sallusto and Lanzavecchia 2000). These data provide additional arguments favouring the idea that cross-presentation must occur early

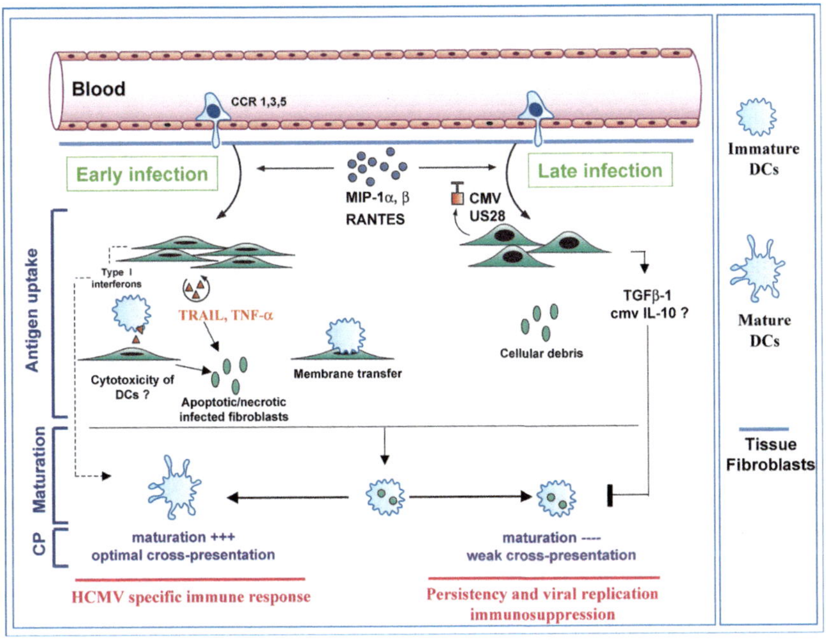

Fig. 1. Hypothetical model for temporal regulation of HCMV cross-presentation by dendritic cells (*DCs*)

after infection to overcome the viral subversive mechanisms and may explain why pp65 is the main target for CTL.

In summary, even though data from primary infected individuals are not available we can assume that cross-presentation of incoming pp65 is a major mechanism in priming $CD8^+$ T cells directed against HCMV (Fig. 1). Finally, we must take into account recent findings showing that the presence of $CD4^+$ T cells is a critical point in regulation of cross-presentation towards 'cross-priming' versus 'cross-tolerance' (Albert et al. 2001). Whether HCMV may exert a tolerising effect through, for instance, inhibition of CD40 signalling in DCs remains to be further explored.

5
HCMV Cross-Presentation for Immunotherapy

In light of data showing that recovery of HCMV-specific $CD8^+$ and $CD4^+$ T cells after allogeneic bone marrow transplantation was a critical point to control HCMV-related diseases (Walter et al. 1995), it appeared more and more evident that $CD8^+$ as well as $CD4^+$ T cell adoptive immunotherapy could be an effective way to reconstitute cellular immunity against HCMV. Questions then arose about the development of safety procedures to allow efficient production of effector T cells. Originally, anti-HCMV T cells were generated from donor PBMCs by using autologous HCMV-infected fibroblasts as stimulating APCs. These practices quickly appeared to be time consuming and did not exclude potential reinfection after cell transfer. To exclude live HCMV virions as well as live viral vectors, virus-free approaches based-on cross presentation of viral proteins by DCs have been developed.

In this context, the induction of proliferative and cytotoxic T cells directed against HCMV by using DCs pulsed with HCMV lysate or purified recombinant pp65 as a source of antigen has been reported (Maeker et al. 2001; Peggs et al. 2001). Convinced by the numerous findings demonstrating that cross-presentation could be an appropriate strategy to expand CTLs ex vivo we developed an approach to circumvent the use of infectious virus based on the simultaneous triggering of the anti-IE1 and anti-pp65 responses by means of a recombinant chimeric protein, IE1-pp65 (Vaz-Santiago et al. 2001). Using this recombinant protein we succeeded in stimulating and expanding both $CD4^+$ T cells directed against IE1 and $CD8^+$ T cells against pp65 from PBMCs of HCMV-seropositive blood donors. In our study, $CD4^+$ T cells from 75% of blood donors responded to IE1-pp65, as assessed by intracellular staining of IFN-γ. Furthermore, MHC class II-restricted $CD4^+$ T cell clones directed against IE1 were stimulated in the presence of IE1-pp65-pulsed targets, as determined through either production of IFN-γ or cytotoxic activity. Surprisingly, we demonstrated that anti-pp65 $CD8^+$ CTL were stimulated and expanded after stimulation of PBMCs from seropositive donors with IE1-pp65, regardless of HLA restriction. These results argued in favour of cross-presentation of IE1-pp65 by DCs or peripheral macrophages, even though this remains to be determined. Altogether our data showed that IE1 and pp65 were available within the chimera protein as in vitro targets for T cell responses.

Importantly, in the course of PBMC stimulation from various donors with peptides we observed that some of the PBMCs were not responsive to known dominant epitopes, suggesting that the use of an entire protein as opposed to peptides stands a better chance to expand CTLs. Nevertheless, we speculated that in some donors, some CTL clones were likely to be present at a very low frequency at the time of bleeding so that they were not detected under conditions used for ex vivo stimulation. This is in agreement with data suggesting that in the setting of latent infection HCMV may periodically reactivate and recruit $CD8^+$ CTL (Gillespie et al. 2000). A study on factors affecting the efficiency of $CD8^+$ T cell cross-presentation by isolated DCs with exogenous antigens reported that efficiency varied between individuals according to differences in the responding T cell repertoire, in the endocytic machinery and in HLA type, in time of blood drawing (Maeker et al. 2001). By quantitation of peptide-specific T cells with tetramer staining and cytokine flow cytometry under varied conditions of stimulation, the authors suggested that using complex sources of antigens such as HCMV lysates may lower chances of triggering all peptide specific T cells because of the presence in the T cell repertoire of few cells that can respond to low density antigens. Accordingly, approaches targeting IE1 and pp65 aimed to focus on triggering the immunodominant anti-CMV T cell repertoire of both the $CD8^+$ and $CD4^+$ subsets. This approach is supported by an increasing number of studies reporting characterisation of $CD4^+$ and $CD8^+$ T cell response against IE1 and pp65 (Gyulai et al. 2000, Bitmansour et al. 2001, Komanduri et al. 2001)

Whatever approach is considered, we expect that improving our knowledge of DC functions will provide new clues to improve efficiency of ex vivo stimulation and expansion procedures for adoptive immunotherapy with T cells, even using PBMCs from HCMV-seronegative donors.

References

Albert ML, Jegathesan M, Darnell RB. (2001) Dendritic cell maturation is required for the cross-tolerization of CD8+ T cells. Nat Immunol 2:1010–1017

Alcami A, Koszinowski UH. (2000) Viral mechanisms of immune evasion. Mol Med Today 6:365–372

Andrews DM, Andoniou CE, Granucci F, Ricciardi-Castagnoli P, Degli-Esposti MA. (2001) Infection of dendritic cells by murine cytomegalovirus induces functional paralysis. Nat Immunol 2:1077–1084

Arrode G, Boccaccio C, Abastado JP, Davrinche C. (2002) Cross-presentation of human cytomegalovirus pp65 (UL83) to CD8(+) T cells is regulated by virus-induced, soluble-mediator-dependent maturation of dendritic cells. J Virol 76:142–150

Arrode G, Boccaccio C, Lule J, Allart S, Moinard N, Abastado J, Alam A, Davrinche C. (2000) Incoming human cytomegalovirus pp65 (UL83) contained in apoptotic infected fibroblasts is cross-presented to CD8(+) T cells by dendritic cells. J Virol 74:10018–10024

Banchereau J, Briere F, Caux C, Davoust J, Lebecque S, Liu YJ, Pulendran B, Palucka K. (2000) Immunobiology of dendritic cells. Annu Rev Immunol 18:767–811

Biron C, Byron K, Sullivan J. (1989) Severe Herpesvirus infections in an adolescent without natural killer cells. N Engl J Med 320:1731–1735

Bitmansour AD, Waldrop SL, Pitcher CJ, Khatamzas E, Kern F, Maino VC, Picker LJ. (2001) Clonotypic structure of the human CD4(+) memory T cell response to cytomegalovirus. J Immunol 167:1151–1163

Bodaghi B, Jones TR, Zipeto D, Vita C, Sun L, Laurent L, Arenzana-Seisdedos F, Virelizier JL, Michelson S. (1998) Chemokine sequestration by viral chemoreceptors as a novel viral escape strategy: withdrawal of chemokines from the environment of cytomegalovirus-infected cells. J Exp Med 188:855–866

Boppana SB, Rivera LB, Fowler KB, Mach M, Britt WJ. (2001) Intrauterine transmission of cytomegalovirus to infants of women with preconceptional immunity. N Engl J Med 344:1366–1371

Borysiewicz LK, Graham S, Hickling JK, Mason PD, Sissons JG. (1988) Human cytomegalovirus-specific cytotoxic T cells: their precursor frequency and stage specificity. Eur J Immunol 18:269–275

Britt WJ, Alford CA. (1996) Cytomegalovirus. In:, Fields BN, Knipe DM (eds) *Fields Virology.* Lippincott-Raven, Philadelphia, PA, p. 2493–2523

Bukowski JF, Warner JF, Dennert G, Welsh RM. (1985) Adoptive transfer studies demonstrating the antiviral effect of natural killer cells in vivo. J Exp Med 161:40–52

Buseyne F, Le Gall S, Boccaccio C, Abastado JP, Lifson JD, Arthur LO, Riviere Y, Heard JM, Schwartz O. (2001) MHC-I-restricted presentation of HIV-1 virion antigens without viral replication. Nat Med 7:344–349

Cho HI, Han H, Kim CC, Kim TG. (2001) Generation of Cytotoxic T lymphocytes specific for human cytomegalovirus using dendritic cells in vitro. J Immunother 24:242–249

Cohen B, Dix R.(1997). AIDS and the nervous system. In *AIDS and the nervous system.* Berger J, Levy, RM, ed. Lippincott-Raven, Philadelphia, PA, p. 595–639

Davignon JL, Castanie P, Yorke JA, Gautier N, Clement D, Davrinche C. (1996) Anti-human cytomegalovirus activity of cytokines produced by CD4+ T-cell clones specifically activated by IE1 peptides in vitro. J Virol 70:2162–2169

Davignon JL, Clement D, Alriquet J, Michelson S, Davrinche C. (1995) Analysis of the proliferative T cell response to human cytomegalovirus major immediate-early protein (IE1): phenotype, frequency and variability. Scand J Immunol 41:247–255

Fleck M, Kern ER, Zhou T, Podlech J, Wintersberger W, Edwards CKd, Mountz JD. (1998) Apoptosis mediated by Fas but not tumor necrosis factor receptor 1 prevents chronic disease in mice infected with murine cytomegalovirus. J Clin Invest 102:1431–1443

Gallot G, Vivien R, Ibisch C, Lule J, Davrinche C, Gaschet J, Vie H. (2001) Purification of Ag-specific T lymphocytes after direct peripheral blood mononuclear cell stimulation followed by CD25 selection. I. Application to CD4(+) or CD8(+) cytomegalovirus phosphoprotein pp65 epitope determination. J Immunol 167:4196–4206

Gerna G, Zipeto D, Percivalle E, Parea M, Revello MG, Maccario R, Peri G, Milanesi G. (1992) Human cytomegalovirus infection of the major leukocyte subpopulations and evidence for initial viral replication in polymorphonuclear leukocytes from viremic patients. J Infect Dis 166:1236–1244

Gillespie GM, Wills MR, Appay V, O'Callaghan C, Murphy M, Smith N, Sissons P, Rowland-Jones S, Bell JI, Moss PA. (2000) Functional heterogeneity and high frequencies of cytomegalovirus- specific CD8(+) T lymphocytes in healthy seropositive donors. J Virol 74:8140–8150

Goldmacher VS, Bartle LM, Skaletskaya A, Dionne CA, Kedersha NL, Vater CA, Han J, Lutz RJ, Watanabe S, McFarland ED, Kieff ED, Mocarski ES, Chittenden T. (1999) A cytomegalovirus-encoded mitochondria-localized inhibitor of apoptosis structurally unrelated to Bcl-2. Proc Natl Acad Sci USA 96:12536–12541

Grefte A, van der Giessen M, van Son W, The TH. (1993) Circulating cytomegalovirus (CMV)-infected endothelial cells in patients with an active CMV infection. J Infect Dis 167:270–277

Gyulai Z, Endresz V, Burian K, Pincus S, Toldy J, Cox WI, Meri C, Plotkin S, Berencsi K. (2000) Cytotoxic T lymphocyte (CTL) responses to human cytomegalovirus pp65, IE1-Exon4, gB, pp150, and pp28 in healthy individuals: reevaluation of prevalence of IE1-specific CTLs. J Infect Dis 181:1537–1546

Hahn G, Jores R, Mocarski ES. (1998) Cytomegalovirus remains latent in a common precursor of dendritic and myeloid cells. Proc Natl Acad Sci USA 95:3937–3942

Harshyne LA, Watkins SC, Gambotto A, Barratt-Boyes SM. (2001) Dendritic cells acquire antigens from live cells for cross-presentation to CTL. J Immunol 166:3717–3723

Hengel H, Reusch U, Geginat G, Holtappels R, Ruppert T, Hellebrand E, Koszinowski UH. (2000) Macrophages escape inhibition of major histocompatibility complex class I-dependent antigen presentation by cytomegalovirus. J Virol 74:7861–7868

Hubert P, Giannini SL, Vanderplasschen A, Franzen-Detrooz E, Jacobs N, Boniver J, Delvenne P. (2001) Dendritic cells induce the death of human papillomavirus-transformed keratinocytes 1. FASEB J 15:2521–2523

Kern F, Surel IP, Brock C, Freistedt B, Radtke H, Scheffold A, Blasczyk R, Reinke P, Schneider-Mergener J, Radbruch A, Walden P, Volk HD. (1998) T-cell epitope mapping by flow cytometry. Nat Med 4:975–978

Kern F, Surel IP, Faulhaber N, Frommel C, Schneider-Mergener J, Schonemann C, Reinke P, Volk HD. (1999) Target structures of the CD8(+)-T-cell response to human cytomegalovirus: the 72-kilodalton major immediate-early protein revisited. J Virol 73:8179–8184

Klagge IM, Schneider-Schaulies S. (1999) Virus interactions with dendritic cells. J Gen Virol 80:823–833

Komanduri KV, Donahoe SM, Moretto WJ, Schmidt DK, Gillespie G, Ogg GS, Roederer M, Nixon DF, McCune JM. (2001) Direct measurement of CD4+ and CD8+ T-cell responses to CMV in HIV-1-infected subjects. Virology 279:459–470

Liu S, Yu Y, Zhang M, Wang W, Cao X. (2001) The involvement of TNF-alpha-related apoptosis-inducing ligand in the enhanced cytotoxicity of IFN-beta-stimulated human dendritic cells to tumor cells. J Immunol 166:5407–5415

Loenen WA, Bruggeman CA, Wiertz EJ. (2001) Immune evasion by human cytomegalovirus: lessons in immunology and cell biology. Semin Immunol 13:41–49

Maecker HT, Ghanekar SA, Suni MA, He XS, Picker LJ, Maino VC. (2001) Factors affecting the efficiency of CD8+ T cell cross-priming with exogenous antigens. J Immunol 166:7268–7275

McLaughlin-Taylor E, Pande H, Forman S, Tanamachi B, CR L, Zaia J, Greenberg P, Riddell S. (1994) Identification of the major late human cytomegalovirus matrix protein pp65 as a target antigen for CD8+ virus-specific cytotoxic T lymphocytes. J Med Virol 43:103–110

Mocarski Jr. ES. (1996) Cytomegaloviruses and their replication. In:, Fields BN, Knipe DM (eds) *Fields Virology*. Lippincott-Raven, Philadelphia, PA, p. 2447–2491

Peggs K, Verfuerth S, Mackinnon S. (2001) Induction of cytomegalovirus (CMV)-specific T-cell responses using dendritic cells pulsed with CMV antigen: a novel culture system free of live CMV virions. Blood 97:994–1000

Perez-Jiminez A, Colamaria V, Franco A, Grimau-Merino R, Darra R, Fontana E, Zullini E, Beltramello O, Dalla-Bernardina B. (1998) Epilepsy and disorders of cortical development in children with congenital cytomegalovirus infection. Rev Neurol 26:42–49

Ploegh HL. (1998) Viral strategies of immune evasion. Science 280:248–253

Polic B, Hengel H, Krmpotic A, Trgovcich J, Pavic I, Luccaronin P, Jonjic S, Koszinowski UH. (1998) Hierarchical and redundant lymphocyte subset control precludes cytomegalovirus replication during latent infection. J Exp Med 188:1047–1054

Reddehase MJ. (2000) The immunogenicity of human and murine cytomegaloviruses. Curr Opin Immunol 12:738

Reddehase MJ, Weiland F, Munch K, Jonjic S, Luske A, Koszinowski UH. (1985) Interstitial murine cytomegalovirus pneumonia after irradiation: characterization of cells that limit viral replication during established infection of the lungs. J Virol 55:264–273

Reis e Sousa C. (2001) Dendritic cells as sensors of infection. Immunity 14:495–498

Rentenaar RJ, Gamadia LE, van DerHoek N, van Diepen FN, Boom R, Weel JF, Wertheim-van Dillen PM, van Lier RA, ten Berge IJ. (2000) Development of virus-specific CD4(+) T cells during primary cytomegalovirus infection. J Clin Invest 105:541–548

Riddell SR, Watanabe KS, Goodrich JM, Li CR, Agha ME, Greenberg PD. (1992) Restoration of viral immunity in immunodeficient humans by the adoptive transfer of T cell clones. Science 257:238–241

Riegler S, Hebart H, Einsele H, Brossart P, Jahn G, Sinzger C. (2000) Monocyte-derived dendritic cells are permissive to the complete replicative cycle of human cytomegalovirus. J Gen Virol 81 Pt 2:393–399

Sallusto F, Lanzavecchia A. (2000) Understanding dendritic cell and T-lymphocyte traffic through the analysis of chemokine receptor expression. Immunol Rev 177:134–140

Sedger LM, Shows DM, Blanton RA, Peschon JJ, Goodwin RG, Cosman D, Wiley SR. (1999) IFN-gamma mediates a novel antiviral activity through dynamic modulation of TRAIL and TRAIL receptor expression. J Immunol 163:920–926

Sinzger C, Grefte A, Plachter B, Gouw AS, The TH, Jahn G. (1995) Fibroblasts, epithelial cells, endothelial cells and smooth muscle cells are major targets of human cytomegalovirus infection in lung and gastrointestinal tissues. J Gen Virol 76:741–750

Sinzger C, Jahn G. (1996) Human cytomegalovirus cell tropism and pathogenesis. Intervirology 39:302–319

Skaletskaya A, Bartle LM, Chittenden T, McCormick AL, Mocarski ES, Goldmacher VS. (2001) A cytomegalovirus-encoded inhibitor of apoptosis that suppresses caspase-8 activation. Proc Natl Acad Sci USA 98:7829–7834

Steffens HP, Kurz S, Holtappels R, Reddehase MJ. (1998) Preemptive CD8 T-cell immunotherapy of acute cytomegalovirus infection prevents lethal disease, limits the burden of latent viral genomes, and reduces the risk of virus recurrence. J Virol 72:1797–1804

Tabi Z, Moutaftsi M, Borysiewicz LK. (2001) Human Cytomegalovirus pp65- and immediate early 1 antigen-specific HLA Class I-restricted cytotoxic T cell responses induced by cross-presentation of viral antigens. J Immunol 166:5695–5703

Topilko A, Michelson S. (1994) Hyperimmediate entry of human cytomegalovirus virions and dense bodies into human fibroblasts. Res Virol 145:75–82

van Del Pol A, Mocarski E, Saederup J, Meier T. (1999) Cytomegalovirus cell tropism, replication, and gene transfer in brain. J Neurosci 19:10948–10965

Vanderheyde N, Aksoy E, Amraoui Z, Vandenabeele P, Goldman M, Willems F. (2001) Tumoricidal activity of monocyte-derived dendritic cells: evidence for a caspase-8-dependent, fas-associated death domain-independent mechanism. J Immunol 167:3565–3569

Vaz-Santiago J, Lule J, Rohrlich P, Jacquier C, Gibert N, Le Roy E, Betbeder D, Davignon JL, Davrinche C. (2001) Ex vivo stimulation and expansion of both CD4(+) and CD8(+) T cells from peripheral blood mononuclear cells of human cytomegalovirus-seropositive blood donors by using a soluble recombinant chimeric protein, IE1-pp65. J Virol 75:7840–7847

Vidalain PO, Azocar O, Lamouille B, Astier A, Rabourdin-Combe C, Servet-Delprat C. (2000) Measles virus induces functional TRAIL production by human dendritic cells. J Virol 74:556–559

Vidalain PO, Azocar O, Yagita H, Rabourdin-Combe C, Servet-Delprat C. (2001) Cytotoxic activity of human dendritic cells is differentially regulated by double-stranded rna and cd40 ligand. J Immunol 167:3765–3772

Walter EA, Greenberg PD, Gilbert MJ, Finch RJ, Watanabe KS, Thomas ED, Riddell SR. (1995) Reconstitution of cellular immunity against cytomegalovirus in recipients of allogeneic bone marrow by transfer of T-cell clones from the donor. N Engl J Med 333:1038–1044

Wills MR, Carmichael AJ, Mynard K, Jin X, Weekes MP, Plachter B, Sissons JG. (1996) The human cytotoxic T-lymphocyte (CTL) response to cytomegalovirus is dominated by structural protein pp65: frequency, specificity, and T-cell receptor usage of pp65-specific CTL. J Virol 70:7569–7579

Subject Index

Printing (Computer to Plate): Saladruck Berlin
Binding: Stürtz AG, Würzburg

Current Topics in Microbiology and Immunology

Volumes published since 1989 (and still available)

Vol. 233/I: **Tyler, Kenneth L.; Oldstone, Michael B. A. (Eds.):** Reoviruses I. 1998. 29 figs. XVIII, 223 pp. ISBN 3-540-63946-2

Vol. 233/II: **Tyler, Kenneth L.; Oldstone, Michael B. A. (Eds.):** Reoviruses II. 1998. 45 figs. XVI, 187 pp. ISBN 3-540-63947-0

Vol. 234: **Frankel, Arthur E. (Ed.):** Clinical Applications of Immunotoxins. 1999. 16 figs. IX, 122 pp. ISBN 3-540-64097-5

Vol. 235: **Klenk, Hans-Dieter (Ed.):** Marburg and Ebola Viruses. 1999. 34 figs. XI, 225 pp. ISBN 3-540-64729-5

Vol. 236: **Kraehenbuhl, Jean-Pierre; Neutra, Marian R. (Eds.):** Defense of Mucosal Surfaces: Pathogenesis, Immunity and Vaccines. 1999. 30 figs. IX, 296 pp. ISBN 3-540-64730-9

Vol. 237: **Claesson-Welsh, Lena (Ed.):** Vascular Growth Factors and Angiogenesis. 1999. 36 figs. X, 189 pp. ISBN 3-540-64731-7

Vol. 238: **Coffman, Robert L.; Romagnani, Sergio (Eds.):** Redirection of Th1 and Th2 Responses. 1999. 6 figs. IX, 148 pp. ISBN 3-540-65048-2

Vol. 239: **Vogt, Peter K.; Jackson, Andrew O. (Eds.):** Satellites and Defective Viral RNAs. 1999. 39 figs. XVI, 179 pp. ISBN 3-540-65049-0

Vol. 240: **Hammond, John; McGarvey, Peter; Yusibov, Vidadi (Eds.):** Plant Biotechnology. 1999. 12 figs. XII, 196 pp. ISBN 3-540-65104-7

Vol. 241: **Westblom, Tore U.; Czinn, Steven J.; Nedrud, John G. (Eds.):** Gastroduodenal Disease and Helicobacter pylori. 1999. 35 figs. XI, 313 pp. ISBN 3-540-65084-9

Vol. 242: **Hagedorn, Curt H.; Rice, Charles M. (Eds.):** The Hepatitis C Viruses. 2000. 47 figs. IX, 379 pp. ISBN 3-540-65358-9

Vol. 243: **Famulok, Michael; Winnacker, Ernst-L.; Wong, Chi-Huey (Eds.):** Combinatorial Chemistry in Biology. 1999. 48 figs. IX, 189 pp. ISBN 3-540-65704-5

Vol. 244: **Daëron, Marc; Vivier, Eric (Eds.):** Immunoreceptor Tyrosine-Based Inhibition Motifs. 1999. 20 figs. VIII, 179 pp. ISBN 3-540-65789-4

Vol. 245/I: **Justement, Louis B.; Siminovitch, Katherine A. (Eds.):** Signal Transduction and the Coordination of B Lymphocyte Development and Function I. 2000. 22 figs. XVI, 274 pp. ISBN 3-540-66002-X

Vol. 245/II: **Justement, Louis B.; Siminovitch, Katherine A. (Eds.):** Signal Transduction on the Coordination of B Lymphocyte Development and Function II. 2000. 13 figs. XV, 172 pp. ISBN 3-540-66003-8

Vol. 246: **Melchers, Fritz; Potter, Michael (Eds.):** Mechanisms of B Cell Neoplasia 1998. 1999. 111 figs. XXIX, 415 pp. ISBN 3-540-65759-2

Vol. 247: **Wagner, Hermann (Ed.):** Immunobiology of Bacterial CpG-DNA. 2000. 34 figs. IX, 246 pp. ISBN 3-540-66400-9

Vol. 248: **du Pasquier, Louis; Litman, Gary W. (Eds.):** Origin and Evolution of the Vertebrate Immune System. 2000. 81 figs. IX, 324 pp. ISBN 3-540-66414-9

Vol. 249: **Jones, Peter A.; Vogt, Peter K. (Eds.):** DNA Methylation and Cancer. 2000. 16 figs. IX, 169 pp. ISBN 3-540-66608-7

Vol. 250: **Aktories, Klaus; Wilkins, Tracy, D. (Eds.):** Clostridium difficile. 2000. 20 figs. IX, 143 pp. ISBN 3-540-67291-5

Vol. 251: **Melchers, Fritz (Ed.):** Lymphoid Organogenesis. 2000. 62 figs. XII, 215 pp. ISBN 3-540-67569-8

Vol. 252: **Potter, Michael; Melchers, Fritz (Eds.):** B1 Lymphocytes in B Cell Neoplasia. 2000. XIII, 326 pp. ISBN 3-540-67567-1

Vol. 253: **Gosztonyi, Georg (Ed.):** The Mechanisms of Neuronal Damage in Virus Infections of the Nervous System. 2001. approx. XVI, 270 pp. ISBN 3-540-67617-1

Vol. 254: **Privalsky, Martin L. (Ed.):** Transcriptional Corepressors. 2001. 25 figs. XIV, 190 pp. ISBN 3-540-67569-8

Vol. 255: **Hirai, Kanji (Ed.):** Marek's Disease. 2001. 22 figs. XII, 294 pp. ISBN 3-540-67798-4

Vol. 256: **Schmaljohn, Connie S.; Nichol, Stuart T. (Eds.):** Hantaviruses. 2001, 24 figs. XI, 196 pp. ISBN 3-540-41045-7

Vol. 257: **van der Goot, Gisou (Ed.):** Pore-Forming Toxins, 2001. 19 figs. IX, 166 pp. ISBN 3-540-41386-3

Vol. 258: **Takada, Kenzo (Ed.):** Epstein-Barr Virus and Human Cancer. 2001. 38 figs. IX, 233 pp. ISBN 3-540-41506-8

Vol. 259: **Hauber, Joachim, Vogt, Peter K. (Eds.):** Nuclear Export of Viral RNAs. 2001. 19 figs. IX, 131 pp. ISBN 3-540-41278-6

Vol. 260: **Burton, Didier R. (Ed.):** Antibodies in Viral Infection. 2001. 51 figs. IX, 309 pp. ISBN 3-540-41611-0

Vol. 261: **Trono, Didier (Ed.):** Lentiviral Vectors. 2002. 32 figs. X, 258 pp. ISBN 3-540-42190-4

Vol. 262: **Oldstone, Michael B.A. (Ed.):** Arenaviruses I. 2002, 30 figs. XVIII, 197 pp. ISBN 3-540-42244-7

Vol. 263: **Oldstone, Michael B. A. (Ed.):** Arenaviruses II. 2002, 49 figs. XVIII, 268 pp. ISBN 3-540-42705-8

Vol. 264/I: **Hacker, Jörg; Kaper, James B. (Eds.):** Pathogenicity Islands and the Evolution of Microbes. 2002. 34 figs. XVIII, 232 pp. ISBN 3-540-42681-7

Vol. 264/II: **Hacker, Jörg; Kaper, James B. (Eds.):** Pathogenicity Islands and the Evolution of Microbes. 2002. 24 figs. XVIII, 228 pp. ISBN 3-540-42682-5

Vol. 265: **Dietzschold, Bernhard; Richt, Jürgen A. (Eds.):** Protective and Pathological Immune Responses in the CNS. 2002. 21 figs. X, 278 pp. ISBN 3-540-42668-X

Vol. 266: **Cooper, Koproski (Eds.):** The Interface Between Innate and Acquired Immunity, 2002, 15 figs. XIV, 116 pp. ISBN 3-540-42894-1

Vol. 267: **Mackenzie, John S.; Barrett, Alan D. T.; Deubel, Vincent (Eds.):** Japanese Encephalitis and West Nile Viruses. 2002. 66 figs. X, 418 pp. ISBN 3-540-42783-X

Vol. 268: **Zwickl, Peter; Baumeister, Wolfgang (Eds.):** The Proteasome-Ubiquitin Protein Degradation Pathway. 2002, 17 figs. X, 213 pp. ISBN 3-540-43096-2

Vol. 269: **Koszinowski, Ulrich H.; Hengel, Hartmut (Eds.):** Viral Proteins Counteracting Host Defenses. 2002, 47 figs. XII, 325 pp. ISBN 3-540-43261-2

Vol. 270: **Beutler, Bruce; Wagner, Hermann (Eds.):** Toll-Like Receptor Family Members and Their Ligands. 2002, 31 figs. X, 192 pp. ISBN 3-540-43560-3

Vol. 271: **Koehler, Theresa M. (Ed.):** Anthrax. 2002, 14 figs. X, 169 pp. ISBN 3-540-43497-6

Vol. 272: **Doerfler, Walter; Böhm, Petra (Eds.):** Adenoviruses: Model and Vectors in Virus Host Interactions. Virion and Structure, Viral Replication, Host Cell Interactions. 2003, 63 figs., approx. 280 pp. ISBN 3-540-00154-9

Vol. 273: **Doerfler, Walter; Böhm, Petra (Eds.):** Adenoviruses: Model and Vectors in Virus Host Interactions. Immune System, Oncogenesis, Gene Therapy. 2003, 33 figs., approx. 280 pp. ISBN 3-540-06851-1

Vol. 274: **Workman, J. L. (Ed.):** Protein Complexes that Modify Chromatin. 2003, 38 figs., XII, 296 pp. ISBN 3-540-44208-1

Vol. 275: **Fan, Hung (Ed.):** Jaagsiekte Sheep Retrovirus and Lung Cancer. 2003, 63 figs., XII, 252 pp. ISBN 3-540-44096-3

GPSR Compliance

The European Union's (EU) General Product Safety Regulation (GPSR) is a set of rules that requires consumer products to be safe and our obligations to ensure this.

If you have any concerns about our products, you can contact us on ProductSafety@springernature.com

In case Publisher is established outside the EU, the EU authorized representative is:

Springer Nature Customer Service Center GmbH
Europaplatz 3
69115 Heidelberg, Germany

Batch number: 10755329

Printed by Printforce, the Netherlands